Levels of Organization in the Biological Sciences

The Evolution of Cognition, edited by Cecilia Heyes and Ludwig Huber, 2000

Origination of Organismal Form, edited by Gerd B. Müller and Stuart A. Newman, 2003

Environment, Development, and Evolution, edited by Brian K. Hall, Roy D. Pearson, and Gerd B. Müller, 2004

Evolution of Communication Systems, edited by D. Kimbrough Oller and Ulrike Griebel, 2004

Modularity: Understanding the Development and Evolution of Natural Complex Systems, edited by Werner Callebaut and Diego Rasskin-Gutman, 2005

Compositional Evolution: The Impact of Sex, Symbiosis, and Modularity on the Gradualist Framework of Evolution, by Richard A. Watson, 2006

Biological Emergences: Evolution by Natural Experiment, by Robert G. B. Reid, 2007

Modeling Biology: Structure, Behaviors, Evolution, edited by Manfred D. Laubichler and Gerd B. Müller, 2007

Evolution of Communicative Flexibility, edited by Kimbrough D. Oller and Ulrike Griebel, 2008

Functions in Biological and Artificial Worlds, edited by Ulrich Krohs and Peter Kroes, 2009

Cognitive Biology, edited by Luca Tommasi, Mary A. Peterson and Lynn Nadel, 2009

Innovation in Cultural Systems, edited by Michael J. O'Brien and Stephen J. Shennan, 2009

The Major Transitions in Evolution Revisited, edited by Brett Calcott and Kim Sterelny, 2011

Transformations of Lamarckism, edited by Snait B. Gissis and Eva Jablonka, 2011

Convergent Evolution: Limited Forms Most Beautiful, by George McGhee, 2011

From Groups to Individuals, edited by Frédéric Bouchard and Philippe Huneman, 2013

Developing Scaffolds in Evolution, Culture, and Cognition, edited by Linnda R. Caporael, James Griesemer, and William C. Wimsatt, 2013

Multicellularity: Origins and Evolution, edited by Karl J. Niklas and Stuart A. Newman, 2016

Vivarium: Experimental, Quantitative, and Theoretical Biology at Vienna's Biologische Versuchsanstalt, edited by Gerd B. Müller, 2017

Landscapes of Collectivity in the Life Sciences, edited by Snait B. Gissis, Ehud Lamm, and Ayelet Shavit, 2017

Rethinking Human Evolution, edited by Jeffrey H. Schwartz, 2018

Convergent Evolution in Stone-Tool Technology, edited by Michael J. O'Brien, Briggs Buchanan, and Metin I. Erin, 2018

Evolutionary Causation: Biological and Philosophical Reflections, edited by Tobias Uller and Kevin N. LaLand, 2019

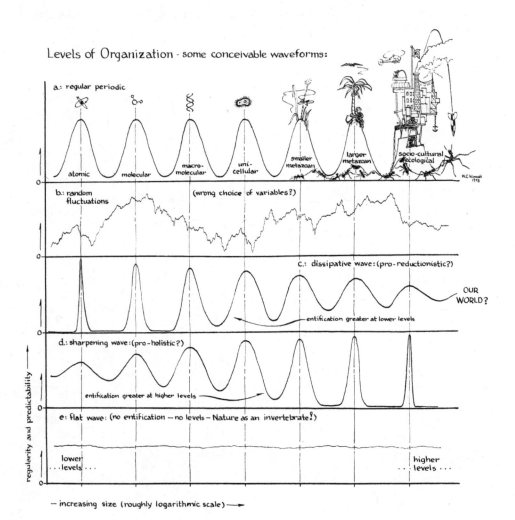

Levels of Organization - some conceivable waveforms:

a.: regular periodic

atomic   molecular   macro-molecular   uni-cellular   smaller metazoan   larger metazoan   socio-cultural ecological

W.C. Wimsatt 1978

b.: random fluctuations   (wrong choice of variables?)

c.: dissipative wave: (pro-reductionistic?)

OUR WORLD?

entification greater at lower levels

d.: sharpening wave: (pro-holistic?)

entification greater at higher levels

e.: flat wave: (no entification – no levels – Nature as an invertebrate!)

lower ...levels...   higher ...levels...

regularity and predictability ⟶

⟵ increasing size (roughly logarithmic scale) ⟶

Levels of Organization in the Biological Sciences

Edited by Daniel S. Brooks, James DiFrisco, and William C. Wimsatt

The MIT Press
Cambridge, Massachusetts
London, England

This book was set in Times New Roman by Westchester Publishing Services. Printed and bound in the United States of America.

Library of Congress Cataloging-in-Publication Data

Names: Brooks, Daniel S., editor. | DiFrisco, James, editor. | Wimsatt, William C., editor.
Title: Levels of organization in the biological sciences / edited by Daniel S. Brooks, James DiFrisco, and William C. Wimsatt.
Description: Cambridge, MA : The MIT Press, 2021. | Series: Vienna series in theoretical biology | Includes bibliographical references and index.
Identifiers: LCCN 2020035370 | ISBN 9780262045339 (paperback)
Subjects: LCSH: Biology--Philosophy. | Biological systems.
Classification: LCC QH331 .L5288 2021 | DDC 570.1--dc23
LC record available at https://lccn.loc.gov/2020035370

10 9 8 7 6 5 4 3 2 1

Contents

Series Foreword

Biology is a leading science in this century. As in all other sciences, progress in biology depends on the interrelations between empirical research, theory building, modeling, and societal context. But whereas molecular and experimental biology have evolved dramatically in recent years, generating a flood of highly detailed data, the integration of these results into useful theoretical frameworks has lagged behind. Driven largely by pragmatic and technical considerations, research in biology continues to be less guided by theory than seems indicated. By promoting the formulation and discussion of new theoretical concepts in the biosciences, this series intends to help fill important gaps in our understanding of some of the major open questions of biology, such as the origin and organization of organismal form, the relationship between development and evolution, and the biological bases of cognition and mind. Theoretical biology has important roots in the experimental tradition of early twentieth-century Vienna. Paul Weiss and Ludwig von Bertalanffy were among the first to use the term *theoretical biology* in its modern sense. In their understanding, the subject was not limited to mathematical formalization, as is often the case today, but extended to the conceptual foundations of biology. It is this commitment to a comprehensive and cross-disciplinary integration of theoretical concepts that the Vienna Series intends to emphasize. Today, theoretical biology has genetic, developmental, and evolutionary components, the central connective themes in modern biology, but it also includes relevant aspects of computational or systems biology and extends to the naturalistic philosophy of sciences. The Vienna Series grew out of theory-oriented workshops organized by the KLI, an international institute for the advanced study of natural complex systems. The KLI fosters research projects, workshops, book projects, and the journal *Biological Theory*, all devoted to aspects of theoretical biology, with an emphasis on—but not restriction to—integrating the developmental, evolutionary, and cognitive sciences. The series editors welcome suggestions for book projects in these domains.

Gerd B. Müller, Thomas Pradeu, and Katrin Schäfer

Preface

This book arose from the Thirty-Sixth Altenberg Workshop in Theoretical Biology, which took place at the Konrad Lorenz Institute for Evolution and Cognition Research in March 2018. The idea for a workshop on levels of organization originated from conversations among the three of us during a coffee break at the International Society for the History, Philosophy, and Social Studies of Biology (ISHPSSB) meeting in Montreal (2015). Dan and James then drafted a workshop proposal, Bill revised it, and it was eventually accepted by the KLI. Our guiding motivation for the workshop and book project has been to attend to a growing awareness that the levels concept remains critically underanalyzed, despite its major influences in the biological sciences and philosophy of science alike. We believe that the chapters collected in this volume address this lacuna and break new ground with a fresh reappraisal of some of the core issues endemic to levels. On several issues, surprising connections are drawn and robust consensus emerges.

The coherent vector of ideas that emerges from this book would not have been possible without multiple major collaborative efforts among contributors, which involved extensive discussion, revision, and sharpening of each other's work. We thank the participants for contributing to an intellectually productive project from which we, the editors, have drawn new ideas and inspiration. The finished product represents the crystallization of a discussion that we hope will continue to grow, deepen, and include even more participants interested in thinking through the nature and import of levels of organization.

We thank the Konrad Lorenz Institute for financial, logistical, and scientific support at all stages of planning and execution of this project. In particular, the assistance of Gerd B. Müller, Isabella Sarto-Jackson, and Eva Lackner made for a smooth-running workshop and well-organized subsequent book proposal. We also thank the editors at MIT Press for their patience and guidance in the preparation of the final book manuscript. We are grateful to Johannes Jaeger for his input and early encouragement in shaping our project. JD thanks the Research Foundation—Flanders (FWO) for financial support. Bill thanks Dan and James both for the original impetus and for their management and performance of the bulk of the editorial work. Our extended joint conversations as editors during this period, and the quality of the contributions has, we think, led to important synthetic insights into the significance of levels of organization as a theme of scientific philosophy.

Daniel S. Brooks
James DiFrisco
William C. Wimsatt

Participants of the Thirty-Sixth Altenberg Workshop in Theoretical Biology, "Hierarchy and Levels of Organization in the Biological Sciences," at the Konrad Lorenz Institute for Evolution and Cognition Research on March 11, 2018, Klosterneuburg, Austria. Top row, from left: Jim Woodward, Jon Umerez, Sara Green, Alan Love, Bob Batterman, Carl Gillett, Jan Baedke, and James DiFrisco. Front row, from left: Matthew Baxendale, Ilya Tёmkin, Angela Potochnik, Thomas Reydon, Bill Wimsatt, Jim Griesemer, Markus Eronen, and Dan Brooks.

Introduction
Levels of Organization: The Architecture of the Scientific Image

Daniel S. Brooks, James DiFrisco, and William C. Wimsatt

I.1   Introduction

What are levels of organization? According to the standard, informal version of the concept, the natural world comprises a number of vertically stratified layers of entities and processes, such as the molecular, cellular, tissue, organ, organism, population, and ecosystem levels. The contents of one level are said to "make up" the contents of the next higher level, a relationship that is typically understood in terms of composition or scaling relationships. This basic idea can be widely found throughout scientific and philosophical texts, including the introductory pages of most biology textbooks, as well as in reviews and research articles.

It is a common assumption that the hierarchical image of the world captures a variety of important methodological, epistemic, and ontological patterns in science and in nature. Methodologically, levels are sometimes thought to demarcate areas of scientific inquiry and to organize research practices around clusters of similar phenomena. Epistemically, levels of organization demarcate classificatory domains where different forms of scientific explanation are deployed (e.g., evolutionary versus molecular explanations). Ontologically, locating an entity at a level is sometimes thought to embed it in causal relations to entities at the same level, or noncausal dependence relations between entities at different levels. Unsurprisingly, these different dimensions of significance have attracted diverse usages of "levels" by scientists and philosophers of science alike. This book aims to embrace the diversity of usages as aspects of the big picture of levels of organization, in contrast to the narrower renderings one finds in specific debates involving the concept.

The entry point for the levels concept in mainstream philosophy of science was discussions of reductionism and explanation in the mid-twentieth century (Nagel, 1961, chap. 12; Oppenheim & Putnam, 1958). Through the intersection of multiple misguided assumptions, a simplistic "layer-cake" image of levels came to dominate philosophical attention thereafter as the default conception of levels of organization (Brooks, 2017). In this picture, higher-level and more complex entities were shown through reduction to be "nothing more than" aggregates of more basic entities, so that, as Roger Sperry impishly suggested, "eventually everything is held to be explainable in terms of essentially nothing" (Sperry, 1976, p. 167; Wimsatt, 1976). At the same time, the search for the ultimate "desert ontology" (Quine) was appended to a nonrealist claim that we could never access "things in themselves" (Kant) and to Russell's suspicion of "theoretical entities." These metaphysical and epistemological biases, supported by the formal construction of artificial languages, led philosophers of science to a skeptical retreat to a linguistically constituted replacement of phenomena, properties, and objects by data, predicates, and "posits." The real world seemed to be lost in this philosophical shell game. In parallel, however, biologists continued to rely on levels of organization as a robust structural feature of the real biological world. Instead of positing a rigid system of stepwise reductions, scientists were busy adapting the concept to diverse tasks as both a conceptual tool and an open-ended organizing principle (see, e.g., Lehrman, 1953; Novikoff, 1945; Rowe, 1961; Schneirla, 1951).

A particularly important problem in the history of the concept of levels was the explanation of the evolution of complexity. Herbert Simon's (1962/1996) influential argument in "The Architecture of Complexity" related this to evolutionary dynamics and relative stability. He argued that evolution would proceed far faster through the successive articulation of stable subassemblies than by the piecemeal accumulation of individual mutations and would thus produce an organismic architecture reflecting the former more stable process. Moreover, systems meeting his dynamical conditions for "near-decomposability" will show relative stability (after their rapid transient approaches to equilibrium) of the subsystems on the time scale appropriate for the whole system, with commensurate dynamics, thus meeting many of the criteria for a level of organization. This suggests that organization into levels is a key structural precondition for the evolution of complex systems under natural selection. In this line of thought, levels are not simply a theoretical vocabulary or abstract principle concerning the "ultimate" relations between scientific domains (Oppenheim & Putnam, 1958) but are a crucial empirical feature of evolutionary systems that is accessible and consequential to scientific inquiry (e.g., McShea, 2001).

Simon's approach to complexity and levels was taken further in Wimsatt's work (1976, 1994/2007), which excavated the broad significance of levels of organization for methodology, epistemology, and ontology of science. In Wimsatt's framework, the issues of reduction and scientific explanation—which had formerly dominated philosophical attention to levels—become partial aspects of a more multidimensional picture in which levels of organization are central to the ontological "architecture" of nature in the scientific image. We situate the present volume in this tradition (see Wimsatt, 2007, particularly chaps. 1–3), which simultaneously views levels as a rich descriptive resource for the study of complex systems, as a deep structural principle for understanding their dynamics, and as itself a complex problem.

I.2   The Levels Problem: A Curious Research Profile

When we turn to actively clarifying and developing the concept of levels of organization, it immediately becomes clear that it can be conceptualized in many different ways and recruited for many distinct tasks. This richness demands further analysis, as the diversity of conceptual possibilities and empirical applications brings a vast combinatorial landscape into relief.

As a consequence of its ubiquity and flexibility, the levels concept is responsible for a great deal of conceptual work throughout its domains of application. Scientists often apply the notion of levels to formulate claims about the phenomena they investigate and the explanations they construct. Additionally, scientists and philosophers of science routinely reference the stratified representation of the world in various long-standing debates, including debates about the nature of causation, reductionism and emergence, levels of selection, methodological individualism in the social sciences, and the ultimate Gordian knot of the mind–body problem. Regardless of the position one takes in these disputes, all parties tend to rely substantially on some understanding of levels to give their positions structure and content.

Yet, despite enthusiastic applications of the concept, consensus on a definitive account of levels has remained elusive. Indeed, despite numerous explicative efforts[1] (or perhaps *because* of them), the number of meanings and significances attributed to levels has now metastasized to such an extent that it is doubtful there will be a unique meaning or explication of the term. Instead, usage of "levels" continues to be governed mostly by its intuitive appeal, and its justification is often taken as self-evident. Wimsatt (1994/2007), observing this, writes, "The notion of a compositional level of organization is presupposed but left unanalyzed by virtually all extant analyses of inter-level reduction and emergence" (p. 203). It is not just reduction and emergence that presuppose and leave unanalyzed the levels concept: virtually all analyses and discussions involving the notion are guilty of passing the buck.

In areas of philosophy such as metaphysics and philosophy of mind, for instance, talk of levels of organization is bound up with discussions of (in-principle) reduction and physicalism. In these cases, "levels" is used as a shorthand reference to forms of systematic dependence, often vertically depicted, between, for example, objects, properties, events, and states. Most notorious here are dependence relations captured by supervenience or realization (e.g., Kim, 1999). In philosophy of science, on the other hand, "levels" often appears either as a primitive notion or as an auxiliary concept used to help articulate other ideas, such as mechanistic constitution between parts and wholes, reductive explanation, levels of selection, levels of biological individuality, upper-level dynamical autonomy, and downward causation (metaphysical ideas with social and biological origins; see Campbell, 1974). These ideas seem to derive expressive power from the notion of levels, yet the meaning and scope of "levels" remain underdetermined by these special problem contexts.

In the sciences, on the other hand, the concept of "levels" is often used interchangeably with the related idea of "hierarchy" (Allen & Starr, 1982; Bossort et al., 1977; Eldredge & Salthe, 1984; Pattee, 1973; Salthe, 2009; Weiss, 1971). Without properly conditionalizing what exactly "hierarchy" means, however, this association can be problematic (see especially the discussion of "nestedness" by Eronen, this volume; cf. Umerez, this volume). To a philosopher or mathematician, a hierarchy, without qualifier, is an abstract set-theoretic term that refers merely to the grouping of partially ordered elements into sets and subsets. Thus, "hierarchy" reiterates the challenges of dealing with levels, since the conditions and parameters involved in postulating a hierarchy (such as set membership and the relationships between these sets) themselves must first be properly defined in order to be meaningful. Furthermore, there are many distinct types of hierarchies (Craver, 2015; Salthe, 1985; Zafeiris & Vicsek, 2018), few of which are appropriate to discussing levels of organization. Dominance hierarchies, rankings, phylogenetic taxonomies, and authority structure are four common uses of hierarchical ordering that are plainly inapplicable to the more molar idea of "levels of organization."

Adding to these issues, philosophers who refer to "levels of organization" often claim to be importing the term as it is used in science. Again, this simply reproduces the original problem. With such a diversity of uses in science, it is difficult to know, without further explication, which aspect of "levels" one is importing from science.

"Levels of organization" thus exhibits a curious profile when viewed as a distinctive object of analysis: it is at once highly visible, actively figuring widely into scientific and philosophical claims and theories, and yet it resists basic consensus concerning its meaning(s) and significance(s) (Brooks, 2019). For this reason, another intention of this volume is to offer foundational considerations for conceptualizing a common context with which to understand the status of "levels" in philosophy and science (see especially Brooks, this volume).

I.3 The Challenges of Levels

At least four principal challenges face any systematic analysis of "levels." We refer to these interrelated challenges under the headings of *skepticism, ambiguity, entanglement,* and *ubiquity*.

> *Skepticism. Can a positive account of levels of organization be offered that deals with the critical worries that skeptics bring against the notion?*

Owing perhaps to the theoretical lacuna surrounding the levels concept, recent work in philosophy of science has turned critical, giving rise to a nascent "levels skepticism" (Brooks, 2017). This collection of critical stances generally claims that the concept of levels is a misleading or even vacuous notion for understanding how scientists represent the natural world (see especially Eronen, 2015; Guttman, 1976; Potochnik & McGill, 2012; Potochnik, this volume). One line of argument alleges that levels talk is applied in the literature without sufficient clarity: "levels" appears to simply mean too many things at once (Eronen, 2013, 2015; Rueger & McGivern, 2010). Another line of argument asserts exactly the opposite—namely, that the concept imparts a view of nature and science that is too rigid to be fruitful, or even that it is radically false (Guttman, 1976; Potochnik, 2017, chap. 6; Potochnik & McGill, 2012). Finally, another line of criticism holds that the importance attributed to the notion of "levels" has been exaggerated by philosophers and that its importance to actual science is unsupported (Thalos, 2013; also implicit in Waters, 2008).

Levels skepticism seems to depend on adherence to a strong epistemic precautionary principle, according to which the potential errors or harmful idealizations that threaten users of the concept outweigh its apparent epistemic benefits (e.g., organizing domains of knowledge, capturing reliable causal generalizations). In contrast to "levels skepticism," a "levels enthusiasm" can be envisaged that has its precautions running the other direction: the potential epistemic gains that the concept promises outweigh the oversimplifications and confusions it may engender. Obviously, whether one stance is more appropriate, and what exactly the harms and benefits are, is influenced by the version(s) of the levels concept one is using. Among philosophers of science, levels skepticism has often been justified by the specific implausibility of Oppenheim and Putnam's "layer-cake" levels and its attendant reductionist doctrines (see Brooks, 2017), though not always (see Potochnik, this volume).

Ambiguity. How can we account for the conceptual variation of "levels" when seeking to understand its general character and significance across diverse contexts of usage?

Sometimes, "levels" expresses dependencies such as compositional part–whole relations between objects. At other times, it expresses forms of independence of dynamics at different scales or descriptive resolutions. Sometimes the term is applied to the whole of nature or science, and sometimes it is restricted to particular collections of objects with no generalizability beyond. And so on. None of these meanings are necessary or sufficient for defining "levels," although they occur in different instances of usage and may be freely combined into distinct expressed meanings of "levels of organization" across these instances. The ambiguity of "levels" can be used to support skepticism (see above), but it can also be interpreted in a more positive light. Brooks (2019) has argued that this challenge of "levels" should be recast as one concerning *conceptual variation*. That is, we should *not* ask whether there is a core, essential meaning of the levels concept but should instead view "levels" as an important case study for seeing how semantic variation in scientific concepts figures into scientific reasoning. A related approach takes the ambiguities surrounding levels as important *characteristics* that reflect the real complexity of hierarchical organization in nature—in other words, "the complexity of complexity" (Griesemer, this volume)—rather than as rival defining properties (Wimsatt, 1976, 2007). "Levels" in this approach can be construed as "cluster kinds" rather than kinds defined by essential properties (DiFrisco, this volume), as has been done for other biological concepts (Ereshefsky, 2017).

Entanglement. How do we extricate "levels" from its entanglement within various debates that either co-opt or assume the meaning of "levels"?

"Levels" as treated in science and philosophy is almost invariably entangled within one or another sprawling debates (see above). Accordingly, most characterizations of the term are couched within the terminology of a particular debate (see especially Brooks, 2017, §5). This is unsurprising given that the concept can be highly sensitive to the circumstances of its usage (Brooks, this volume). However, in cases of entanglement where the term becomes *interdefined* with terms of the embedding debate, it can obscure work that the levels concept per se is doing. For instance, Oppenheim and Putnam (1958) defined their idea of microreduction in terms of levels of organization ("reductive levels"), which over time caused the levels concept to become associated with assumptions about the unity of science (see Brooks, 2017). One such assumption was their central idea that levels neatly correspond to scientific disciplines, now recognized as a dubious claim.[2] Their "layer-cake" account of levels slowly became reified as the default conception of levels in philosophy of science. "Levels" as such came to embody layer-cake ideas and corollaries, such as the expectation that reductionism (or antireductionism) treats scientific disciplines in a simplistic level-ordered, stepwise fashion (see, e.g., Waters, 2008).

Another co-option of "levels" that deserves special scrutiny occurred in the causal exclusion debate in the late 1990s and early 2000s (Block, 2003; Kim, 1998). The canonical form of the argument, expressed by Jaegwon Kim (1998), pictures the "mental" and "physical" as competing for causal efficacy. Although Kim's argument does not always refer to "levels," Block (2003) extends the exclusion argument to explicitly encompass levels of organization. The extended version concerns the causal efficacy of all nonfundamental natural phenomena, perched at "macro-levels," which threaten to "drain away" to an all-encompassing "micro-level" where all causal powers of the world are grounded. Through this debate, "levels" became strongly associated with metaphysical doctrines of supervenience and the causal closure of physics.

These associations are problematic for a number of reasons, however. Consider first supervenience. Given that supervenience is a relation between properties only, it is uninformative about how those properties are distributed among compositional levels of objects and is even compatible with there being no levels at all. Moreover, relations between properties alone cannot capture the categorially rich descriptions of complex systems we find in science—which can include entities, processes, interactions, dispositions, masses, and more (see Gillett, 2013).

In a similar vein, the "causal closure of physics"—the idea that every physical effect has a sufficient physical cause—is at odds with the fact that scientific models of interlevel causation picture levels of organization as *nonclosed* (Green, this volume; Woodward, this volume; Batterman, this volume; although see Gillett, this volume). Maintaining the closure doctrine therefore requires a heavily idealized and in-principle rendering of "the physical" that only tenuously resembles the contemporary science of physics (Wilson, 2018). As a consequence, philosophers are misled into wondering why there are sciences besides "physics" and levels other than "the physical" at all. All of this, while in real science, it could be argued (as Dresden did in 1974) that even physics (hydrodynamics, shock waves, and turbulence) can't be reduced to physics (statistical mechanics)! At the base of these problems, we suspect, is the fateful methodological decision to philosophize about levels of organization in nature without a serious appreciation of the workings and the deliverances of science (Chemero & Silberstein, 2008; Ladyman & Ross, 2007).

Thus, supervenience and physical causal closure don't tell us anything about compositional levels of organization and have no significant effect on the scientific issues related to the study of levels. This highlights an additional challenge related to the entanglement of levels in existing philosophical debates. Those debates may generate multiple, incompatible expectations that cannot all be met at once and that may hinder progress on fruitful explication of the levels concept. But the expectations deriving from certain debates (e.g., stepwise correspondences between sciences and their objects, or supervenience and causal closure of physics) may also be incommensurate with the scientific image of levels. The entanglement of "levels" with questionable assumptions puts additional pressure on the task of explicating the notion on its own terms and in serious engagement with science. But this task is not without further challenges.

> **Ubiquity**. *How can we account for the fact that "levels" is ubiquitous across multiple scientific fields but is not explicated as a technical term within any one particular scientific field?*

The relative lack of development and resulting skepticism surrounding "levels" stand in contrast to the widespread and enthusiastic reference to the term and its satellite ideas in the scientific literature. However, this presents its own problems for keying into the levels concept as an object of study. Philosophers often claim to be appropriating the term as it is used in science, but scientists also frequently take "levels" to be a primitive term. Furthermore, while reference to "levels" is ubiquitous in the biological sciences especially, the concept is not proprietary to any one subfield of biology. Instead, "levels" concerns the *relationships* between subfields, as well as between their objects of study. Interestingly, when the notion of levels is used to frame relationships between subfields of biology or between broader categories of science, it can be used to express theses about unity, reduction, and integration (Mitchell, 2003; Nagel, 1961; Oppenheim & Putnam, 1958) but also contrary theses about disunity and antireductionism (Grene, 1987; Montalenti, 1976).

This gives the levels concept a "cosmopolitan" profile and also blocks one potential strategy that can be used to deal with scientific concepts that exhibit semantic variation or ambiguity. This is to split the concept into different versions according to a subdivision of fields that invoke it. For example, instead of speaking of "genes" as such, one can speak of "classical genes" or "molecular genes." In attempting the same splitting strategy with "levels," however, one finds that particular fields don't have a rich and worked-out notion of levels that can simply be picked up by theorists. Some areas of the life sciences may lean more heavily on "levels" than others (e.g., evolution, development, and physiology versus biochemistry and purely compositional molecular biology[3]), but an understanding of levels of organization per se requires placing oneself at the boundaries between these fields and examining their interrelationships. In evolutionary biology, for example, the notion of levels of selection is defined against an unarticulated background notion of levels of organization (Griesemer, 2005; see DiFrisco, this volume), giving rise to the familiar "buck-passing" on the topic of levels.

The ubiquity of "levels" presents multiple challenges for theoretical work on the topic. First, the fact that "levels" crosses subfields without having a ready-made articulation in the form of general levels of organization indicates that more work is required than just describing local disciplinary patterns of usage and explanatory styles that invoke the term. Here, the task of philosophers and theorists is *constructive* and not merely descriptive. Second, work on levels should account for the ability of scientists to freely change resolution between general levels of organization (e.g., as found in textbook presentations) and more local disciplinary renderings of levels. One possibility is that general levels of organization should be viewed as *robust* points of convergence across multiple discipline-specific decompositions of the biological world (Wimsatt, 1972, 2007). Third, work on levels should be able to explain what allows levels of organization to simultaneously support theses about unity and disunity, as well as reduction and antireduction alike. On the one hand, compositional relationships put entities at different levels into an integrative relationship, but on the other hand, levels demarcate quasi-independent channels of causal interaction and are sites of scale-specific dynamics and causal generalizations. It is hard to avoid the conclusion that there may be deep and as-yet unearthed principles underlying this dual character of levels: namely, that the "same" material exhibits qualitatively different properties and behaviors at different spatial and temporal scales. A common result of several chapters in the present volume is to focus critical attention on this core theme as an open problem area for investigations of levels of organization, which we explore in the next section.

I.4   Looking Forward: A Unified Front on Downward Causation, Dynamical Autonomy, and Compositionality

Although this book exhibits a diversity of ideas and approaches that is typical of topical collections of essays (see "chapter summaries" below), many of the chapters also collectively articulate into a strikingly coherent picture. A large part of this unified front concerns the connection between compositional and causal aspects of levels of organization. A specific area of rough consensus from different contributors that we would like to highlight first concerns *downward causation*. Downward causation—the idea that causes from a higher level influence effects at a lower level—has received sustained attention from philosophers due to its connection to mental causation, especially in the context of the causal exclusion discussion (see above; see Baumgartner, 2010; Campbell, 1974; Craver & Bechtel, 2007; Eronen, 2013; Kim, 1992). A major source of perplexity is the fact that downward causation is often present, implicitly or explicitly, in the causal explanations used in many sciences (especially, but not exclusively, the life and social sciences), and yet fairly standard philosophical reconstructions of downward causation make it seem incoherent.

One such reconstruction runs as follows: levels comprise parts and wholes, and thus downward causation would seem to involve causation from a whole to its parts. But our concept of causation requires that causes and effects are distinct: causation is irreflexive and doesn't relate items that are compositionally or logically connected. This is why it seems wrong to say that a person's walking across the room caused their component cells to move across the room or (to use the example in Lewis, 2000, p. 78) that a person's saying "hello" caused their saying "hello" loudly (see Woodward, this volume; Gillett, this volume). Assuming that levels of organization are intimately connected to composition, then, there should be no causal relationships between higher levels (wholes) to lower levels (parts). Instead, higher and lower levels should stand in some variety of noncausal dependence relationships such as composition or constitution, with causation operating only within a level or across different level-bound individual parts and wholes.

How do we square this negative result with the fact that successful scientific explanations routinely rely on downward causation? Consider the following putative examples of downward causation. Certain butterflies undergo generational changes in wing coloration patterns in response to changing environmental temperature and photoperiod (Nijhout, 1994). The flight trajectory of an individual in a flock of starlings is influenced by changes in the local average velocity vector and position of the group (Hildenbrandt et al., 2010; see also Mitchell, 2009; Zafeiris & Vicsek, 2018). The gating of ion channel proteins in action potential propagation is influenced by cellular-level potential (Hodgkin & Huxley, 1952; Woodward, this volume). In biological development, tissue shapes and interactions can exert control over cell fates and gene expression activity (Soto et al., 2008). Relatedly, in carcinogenesis, biomechanical features of the tissue microenvironment have been shown to influence cell proliferation (Bissell & Hines, 2011; see Green, this volume). Are these examples somehow mistaken or only temporary placeholders to be eventually supplanted by physical explanations that stick to a single lower level?

Fortunately, a simpler interpretation is available. The main philosophical problem arises when we recognize that levels relate parts and wholes, as well as infer from this that interlevel causation must *also* relate parts and wholes. Careful attention to the scientific explanations that involve downward causation reveals that this is generally not what is at stake. Instead, in the above examples, causation is described as a relationship between *variables* or *features* that belong to wholes and parts rather than between the wholes and parts themselves (see Woodward, this volume). In dynamical contexts, downward causes are often modeled as *boundary conditions* that constrain a lower-level behavior (Ellis &

Kopel, 2019; Noble, 2012) and can also be treated as causal variables. Causation between variables at different levels seems to not violate the irreflexivity requirement: there is no objectionable logical or compositional connection between photoperiod and wing color, for example, or between tissue shape and cell fate.

However, matters get more complicated when we recognize that certain variables too, like individual parts and wholes, can tolerate some forms of logical or "compositional" operations indicating rearrangements or juxtapositions without changing in value. The latter class comprises the *aggregative* variables such as mass, energy, momentum, and net charge, which figure into the conservation laws of physics (Wimsatt, 2007, chap. 12). The mass of an object is not independent of the mass of its parts, for example. Due to their aggregative relationship, they cannot be manipulated independently (Woodward, this volume), and the whole cannot transfer mass (or energy) to one of its parts against fixed background conditions and in a closed system.

From these distinctions, we begin to see that two very different notions of causation can be invoked in discussions of causation across levels. One concerns the transfer of conserved physical quantities, and another concerns counterfactual relationships that may or may not involve such quantities. Regarding the first notion, the regularities embedded in physical conservation laws may be extended to the result that conserved quantities are not transferred across compositional levels—although this will depend on how we individuate compositional levels (Gillett, this volume).[4] This form of "levels closure" is not to be confused with the thesis that the lowest microphysical level is "causally closed" without qualification. By contrast, counterfactual relationships like those described by the interventionist approach to causation are *not* closed within a level (Woodward, 2003, this volume; Green, this volume), as long as the counterfactual does not describe a transfer of conserved quantities between aggregative variables. In such an "aggregative" case, as Woodward's chapter argues, the values of the variables cannot be fixed independently of one another. The downward causal relations embedded in well-established scientific explanations rarely belong to this objectionable type.[5]

Although this does not resolve all of the problems surrounding downward causation, we believe it is an important result of the present volume that can aid in orienting future work on levels while also aiding in framing concrete research programs (see Green, this volume).

Another forward-looking result concerns the relationship between composition and notions of same-level causation and dynamical autonomy. In Batterman's chapter, levels are envisaged roughly as spatiotemporal scales at which a system displays dominant behaviors or robust properties that can be faithfully modeled independently of its properties at other scales—for example, the representative spatial volumes at which a steel beam begins to display continuum behavior despite its being composed of discrete particles at a lower level or scale. This multiscale modeling perspective on levels is not conceptually based on part–whole relations, although they are certainly connected. Similarly, in Woodward's chapter, levels are pictured as sets of causal variables that are characterized by "conditional independence" from other variables at other levels. He notes that, although composition and scale-based considerations are frequently related to considerations of independence, these two notions of level only imperfectly track one another.

These insights raise new conceptual and empirical questions. First, if levels centrally involve autonomy of behaviors or independence of variables, how should we identify which entities at which compositional levels are the appropriate units to define behaviors over or measure variables on? In some sense, even if dynamical or causal levels are not defined compositionally, compositional levels of objects may be necessary for determining "higher" and "lower" and even for recognizing the fact that they are not always tracked by levels of conditional independence or dynamical autonomy. In turn, the dynamics or causal interactions an entity participates in may determine its placement at a compositional level (DiFrisco, this volume). Here again, still deeper connections between the two aspects remain open to be explored. Alternatively, it might be argued that causal or dynamical significance of levels of organization is better captured in terms of spatial and temporal scales, without reference to compositional criteria at all (Baedke & McManus, 2018; DiFrisco, 2017; Eronen, 2015).

Assuming we do keep causal and compositional aspects of levels together, however, a further question is whether we can explain when and why they dissociate. DiFrisco's chapter attributes biological examples of this dissociation to the decoupling of structure and function in systems evolving under selection. From a more general perspective, Wimsatt's chapter illuminates how causal robustness of higher-level properties is a systemic outcome of the interrelations between the paradigmatic features of levels of organization. Specifically, he explores how processes organize by energy level of interactions—a foundational principle in Simon's (1962/1996) thinking about levels—and this organization correlates with size scales and relaxation times, giving rise to "near-decomposable" systems with scale- or level-specific dynamics. Differences between these scale- or level-specific processes yield different stability properties at different time scales. As a result, a high-level feature can be stable under perturbations to low-level features, including perturbations that change the structural identity or even the compositional level of a part. In evolutionary systems, for example, stabilizing selection can maintain a homologous trait while molecular processes such as genetic drift, gene co-option, duplication, and subfunctionalization alter the genes or networks that are involved in producing it over evolutionary time (Haag & True, 2018).

Although the present volume offers new resources for understanding and clarifying these issues, our hope is that it will also serve to adumbrate a frontier of open themes for future work on levels of organization. As the connection between compositional and causal aspects highlights, questions about the nature of levels of organization involve both conceptual and empirical aspects, together with more relatively fixed and more flexible elements, interacting in a reciprocal, iterative fashion. This interplay is reflected in several chapters that coalesce into another unified front that is worth highlighting, on the nature of levels of organization as a progressive scientific concept.

Since there are reasons to expect that no ultimate, unitary definition of levels is waiting to be discovered, we believe it is warranted to evaluate "levels" not by means of an idle definition but rather in terms of its roles in scientific reasoning as a dynamic, open-ended idea capable of performing multiple, overlapping functions in distinct empirical settings. Progress on the nature of levels, instead, reveals a compelling case study of *conceptual advance*.

One key initial observation here is that "levels" does not comprise a fixed representation of the workings of nature but rather captures many different empirical structures, relations, and reasoning patterns found in the biological sciences. Several chapters in this book (e.g., Brooks, Love, Green) converge for this reason on identifying clusters of overlapping scientific practices to capture what is meant by referring to levels in scientific usage. These approaches exploit the expectation that putative levels of organization (such as cells, tissues, and multicellular organisms) comprise "local maxima of regularity and predictability" (Wimsatt, 1976, 1994/2007) within the assemblage of physical matter into stable configurations. Importantly, scientists rely upon but also discover these configurations in their experimental manipulations and not only in their representations (see Love, this volume). However, given that these investigations mostly concern *local* practices, it remains to be seen whether or to what extent such configurations generalize or cumulate into an account of levels that accommodates the full list of features that the concept appears to elicit (see especially Wimsatt, this volume). For this, we turn to the more general promise that the concept of "levels" confers to continued scientific efforts as a whole.

To understand what we mean to capture by the "promise" of the levels concept, consider what Brigandt (2020) describes as the "*forward-looking nature*" of scientific concepts:

All too often concepts are merely seen as the outcome of science—a term being coined once a new biological entity has been discovered, or a mature definition being established once the relevant scientific knowledge is in. But concepts also continuously undergo transformation, and they function by guiding ongoing scientific practice. A biology concept can motivate future scientific efforts, and it can also provide a scaffold to direct the generation of new knowledge and the organization of complex knowledge.

Applied to levels, this perspective captures how the levels concept is not exactly a finished product of scientific efforts but actively figures into these efforts as a dynamic term whose reference changes both between usage instances and in response to the emergence of new problems (for example, the problem of identifying and explaining the "major transitions in individuality" in evolution).[6] Thus, while the conceptual activity of proposing and rejecting definitions may play a role in analyzing the nature of levels, it should be seen neither as the default nor the definitive source of insight into the nature of levels, nor the usefulness of the concept.[7] Instead, continuing attention to levels in biology will be valuable if it adds insight or helpfully orients research on a biological problem by connecting it to the relatively more fixed principles embedded in the concept (e.g., dynamic autonomy, compositionality, causal segregation, or interlevel causation). As such problems are typically *open* (Brooks, 2019; Brooks & Eronen, 2018), the concept applies not only to current or expected scientific knowledge but also to unresolved challenges and gaps therein, which remain targets for scientific investigation. This has the consequence that the nature of levels of organization is itself partly an open empirical question.

The contributions of this volume document a multitude of such "forward-looking" functions of the levels concept, which will surely persist beyond definitional debates concerning what "levels" means. For example, the term appears in scientific efforts as, variously, a scaffold or target for scientific discovery (Griesemer), a key term in constructing or explicating compositional explanations (Gillett), constraints (Umerez), and natural kinds (Reydon), and is a point of departure for scientific experimentation (Love) and a means of identifying and contrasting causal relations (Woodward), as well as an empirical trend in evolution (Tëmkin), among many other things.

I.5   Chapter Summaries

William Wimsatt starts the volume with a discussion of the background motivations that spurred his original work on levels (Wimsatt, 1976; see also Wimsatt, 1994/2007)—particularly the disconnect between philosophical currents of eliminativism, linguistic philosophy, and "desert ontology," as well as the "ontology of the tropical rainforest" one finds in the sciences that study upper-level phenomena. He then proceeds to develop a systematic account of nineteen features of levels, building from dynamics of compositional systems at different spatiotemporal and energy scales to considerations of robustness and the evolution of levels. This analysis yields a new ("Waring blender") criterion on emergence articulating the divergent analyses of Wimsatt and Batterman, "level leakage" and manipulation of levels, and the relations between levels, perspectives, causal thickets (Griesemer, this volume), manipulation (Woodward, this volume; Love, this volume), and multiscale modeling (Batterman, this volume; Woodward, this volume).

Dan Brooks turns to characterizing the levels concept as a proper object of analysis. Brooks begins by noting that the basic character of the levels concept (e.g., how and where it is used and what its role in biological thought is) remains widely underspecified, creating tension with the concept's ubiquitous use in biological science. To relieve this tension, Brooks identifies two profiles attached to the levels concept in biology, each of which accounts for different patterns of usage exhibited by the concept in different types of scientific texts. One profile posits "levels" as a kind of conceptual tool used in research contexts for various practical purposes, while the other profile posits "levels" as an overarching scientific doctrine that serves more general purposes in science akin to cell theory in biology or the germ theory of disease in pathology. Brooks reconciles these two usage profiles by identifying the "erotetic" capacities of the levels concept in aiding to organize, structure, or articulate biological problems.

In a turn from the enthusiastic treatments of the first two chapters, Angela Potochnik spearheads a critical perspective, arguing that reference to levels may actually mislead or obscure the scientific efforts to which the concept should be contributing. For this, Potochnik develops a dilemma facing those who wish to address endemic challenges facing levels (particularly its ambiguity). This dilemma concerns two strategies prominent in recent literature toward clarifying the concept: specifically, one may either seek to offer a more precise notion of "levels," which in Potochnik's view fails to generate a well-defined ordering in almost all circumstances, or one may embrace many local definitions as heuristically useful, which leaves hanging what work, if any, the concept itself actually confers to scientific efforts. That we have reached such a fork in the road concerning levels, Potochnik concludes, is indicative of the concept's *inability* to deliver on work it promises as an organizing concept.

Markus I. Eronen's contribution offers another critical but constructive perspective on levels, concerning the widely held assumption that the notion relies on the hierarchical idea of *nestedness*. Nestedness refers to the idea that the parts comprising a system are taken to be contained within a superordinate, housing whole. In this way, multicellular organisms, say, are composed of organs, then tissues, then cells. Eronen details several problems with this assumption, arguing especially that the *branching* structure of composition violates the expectation that a series of levels be neat and orderly. Instead, not only are biological systems heterogeneously composed of different things at each descending level (thus branching between different subseries of compositional elements), but these subunits in fact do not comfortably "line up" together to form a unitary level structure at all. Eronen then traces several consequences of this for different notions of downward causation, drawing attention especially to the underdetermined status of "levels" informing these different notions.

James Griesemer's contribution offers a sophisticated analysis of the role of levels in the development and evolution of complex systems. Beginning with a reflection on Herbert Simon's watchmaker parable, Griesemer finds the thought experiment wanting, since although it offers an accessible means for conceptualizing the evolution of complex systems via modular hierarchies, the depiction of the two watchmakers is missing the *scaffolding* elements that guide the production of complex order. Griesemer then attends to developing a framework that centralizes scaffolding, which emphasizes an "epistemic view" of Wimsatt's ontology "from the top down." In the face of nature's complexity, Griesemer argues, all natural phenomena begin as causal thickets before they can be winnowed down to emerging and increasingly robust regularities, seen as Wimsatt's perspectives and, finally, levels. Key to Griesemer's insights is a process-oriented, development point of view, which emphasizes the interactivity that is core to the types of complexity that are embodied by changing systems in ontogeny. Finally, he concludes with the thought that, like major transitions in evolution, we may also find similarly important transitions in biological and cognitive development.

James DiFrisco investigates the interplay between compositional and causal-dynamical aspects of levels of organization with a view toward understanding the role of levels in biological development and evolution. The chapter begins by showing how the dissociability of compositional and causal criteria creates ambiguities in the placement of things at levels and then argues that this is an unresolved problem for both structural and functional accounts of levels. To integrate compositional and causal-dynamical aspects, he develops a framework in which general levels of organization (a la Wimsatt) can be restricted to levels decomposed under a specific causal process. In this framework, levels include just the components and processes that contribute to a focal process, such as development, morphogenesis, adaptation, or evolution. The central conceptual strategy in this "process-relative" approach to levels is to rely on a notion of constitution between processes to individuate causally relevant components and determine their placement at levels. He then explores a proposed decomposition of the biological process of developmental evolution and shows how a central feature of these levels is variational independence from other levels. The resulting "levels of developmental evolution" are then contrasted with two defective decompositions of evolution into levels: levels of selection and replicator–interactor hierarchies.

Alan Love's chapter investigates how levels of organization fit into experimental practices in developmental biology. Levels are not only meant to represent biological phenomena, Love argues, but also figure into experimental manipulations scientists conduct as part of their investigation of these phenomena. Drawing on the examples of mixed cell aggregation and tissue engineering, Love reasons that the stable configuration states that comprise targets of intervention in these examples motivate a modest realism about levels. This realism is grounded in the success of these manipulation practices, especially their independence from one another, which corroborates an underlying robustness attributable to the notions of levels at work in these practices. This approach thus offers a middle-range basis for the reality of levels, one that avoids problems attached to global, "layer-cake" accounts of levels, as well as difficulties associated with accounts that are too localized to extrapolate to different instances where the term is used.

Jan Baedke points out that most conceptions of levels or hierarchical organization in the philosophy of science literature, particularly accounts given by the new mechanism, posit a hierarchical layout in biological systems in terms of only composition or constitution between levels. This view, Baedke argues, is too static to account for the dynamic changes that systems undergo over time, as it (1) assumes that levels are always given and (2) does not account for how levels come into existence, or change in kind, during development and evolution. To meet these new challenges, Baedke develops the idea of dynamic hierarchies, which exhibit changes in both the compositional and temporal organization ordered by levels over time. This approach to thinking about biological hierarchies better accommodates questions concerning changes in levels, including their creation, particularly in the context of evolutionary developmental biology.

James Woodward's chapter develops an account of levels and downward causation within an interventionist framework. Central to this account is the notion of *conditional independence*. A variable X is causally independent of Y conditional on Z when X is causally relevant (in the standard interventionist sense) to Z, Y is causally relevant to Z, but conditional on the values of Z, and changes in the value of X make no further difference to Y. For example, conditional on the temperature of a gas (Z) at equilibrium, variations in the individual kinetic energies of the gas's particles (X_n) make no difference to the gas's pressure (Y). Typically, this holds when the Z variables are coarse-grainings of or have lower dimensions than the X variables. Woodward then formulates a conception of "levels" from this notion: levels are natural groupings of variables that possess conditional independence from other groupings of variables across a range of explananda, where variables at an "upper" level may be related to those at a lower level on the basis of many-to-one mappings. When this sort of independence holds, causal explanations formulated in terms of upper-level variables don't miss any causally relevant detail from lower levels, in addition to being more computationally tractable. From here, Woodward notes that most philosophical objections to downward causation assume that downward causation relates parts and wholes, which are not suitably distinct to stand in causal relationships. However, when formulated as a relation between upper- and lower-level *variables*, downward causation is unobjectionable. For these variables to be suitably distinct, he argues, it is enough that they are "independently fixable"—they can take on values via interventions independently of one another.

Sara Green engages an important issue with levels, concerning how theoretical assumptions about the most relevant level of analysis influence choices of experimental design. Although interlevel and downward causation are often thought to be questions that trouble only philosophers, Green argues that the directional influence of variables between higher and lower scales has profound practical implications. Focusing on biomechanical features of the tumor microenvironment in cancer research as a case study, Green argues against a unidirectional, "bottom-up" approach to understanding carcinogenesis (a view privileged by the so-called oncogene paradigm or SMT—somatic mutation theory). Instead, Green contends that a place for downward influences can also be robustly identified in the role of solid-state tissue properties in tumor development. Such influences are present, for example, in the role of tissue stiffness (a higher-scale feature) in disturbing lower-scale features (like cell adhesion within tissue architecture), which influences the progression of breast cancer.

Robert Batterman attends to multiscaling modeling in active and inactive materials. For this, he draws upon sophisticated case studies in physical and material sciences to explore how issues of scale (and, more tentatively, level) can be applied to biological cases. For instance, whereas materials such as steel beams are treated as inactive (i.e., composite systems existing at quasi-equilibrium), active materials self-regulate their stresses and movements, as Batterman shows in the case of mitotic spindles in the cytoskeleton of eukaryotic cells, which aid in the segregation of daughter cells in mitosis. Central to Batterman's analysis is the idea of a representative volume element (RVE), which comprises a statistically representative, scale-relative sample of the material being investigated. Focusing on RVEs perched at different salient (micro, meso, and macro) scales often reveals that the dominant behavior of a material or system is autonomous from behaviors at scales deviating from the RVE.

Carl Gillett's contribution delivers an analysis of the compositional levels we find scientists discussing in the plural "compositional explanations" they offer in physiology, cell biology, and molecular biology. Focusing on muscle contraction as a case study, Gillett develops an "integrative account" that frames such levels as the integrated ontological commitments of many models in compositional arrays, roughly, the productively closed but compositionally ordered layers of individuals, activities, and properties we find posited in our plural but integrated, compositional models/explanations. Gillett argues this pluralist but integrative account captures the features of levels in this scientific context. Additionally, Gillett answers several general criticisms levied against levels by philosophers, including the charge that scale can replace levels and that the levels concept is too vague to be of any use. He concludes by motivating the need to relate his integrative account of compositional levels to the notions of level in other areas of science, such as evolutionary biology and particularly the levels of selection debate.

Thomas Reydon examines the close relationship between the notions of levels and *natural kinds*, noting that levels are characterized by kinds and that kinds are located at levels of organization. On the basis of this tight connection, he shows how the same kind of response to skepticism about kinds also applies to skepticism about levels. Focusing specifically on functional kinds like "genes," Reydon argues that they can be shown to be nonarbitrary if the aspect of functioning targeted by our investigative aims is grounded in some real feature of the world. Interestingly, this grounding relation can involve the dependence of a lower level on a higher level rather than the other way around. This is what we find with genes, he argues, which are metaphysically dependent on genomes. Genes are individuated by their functional roles within genomes, as difference-making causes of phenotypic differences, and accordingly they can be viewed as comprising a (nonarbitrary) functional kind and a functional level of organization.

Jon Umerez takes a theoretical biology approach to the questions posed of hierarchy and levels in this volume, arguing that *organization* comprises a more basic category from which insight into these notions can be gleaned. For this, Umerez attends to the idea of a "constraint-mediated control hierarchy," which is developed from the works of Howard Pattee. Particular attention is given to Pattee's ideas of constraint and control and how they can be used to identify different kinds of hierarchical relations among levels. This approach meets two challenges of levels: first, to construct an alternative to universal notions of levels, which are widely seen as problematic, but also, second, to develop an aim-related framework in which to cast levels, which address levels-skeptical problems associated with other accounts of levels in the literature. Umerez also offers some historical contextualization for the contemporary discussion on levels and hierarchy, noting some commonalities with earlier discussions in the 1960s and 1970s between researchers such as Pattee, Marjorie Grene, and Mario Bunge.

Ilya Tëmkin's chapter explores how levels extend beyond biological individuals and into the sociocultural realm. Tëmkin surveys a number of information-carrying apparatuses that are introduced when cultural transmission is integrated with evolution at higher levels of organization (such as residential bands or even geopolitical entities), which differs categorically from biological means of inheritance. This evolutionary hierarchy is traced through the human individual (organism) into a unique sociocultural/cognitive series of organizational units, which is distinct but ultimately complementary to the biological evolutionary series (community, deme, population) of higher levels. This perspective, importantly, connects human behavior and cultural activities with the biological notions of replication and self-maintenance, albeit with attendant differences. Tëmkin concludes by describing how this revised hierarchical framework will be useful for generating further insight into hominid evolution.

Notes

1. A fair number of other conferences and edited volumes have been devoted, in part or in toto, to explicating levels and hierarchical organization in biological (and physical) systems (Eldredge et al., 2016; Greenberg & Tobach, 1984, 1987; Hood et al., 1995; Koestler & Smythies, 1969; Pattee, 1973; Pumain, 2006; Redfield, 1942; Sellars et al., 1949; Whyte et al., 1969).

2. Although superficially similar ideas, such as the idea that clusters of phenomena associated with atomic nuclei, with chemical bonding, and with many classical molar mechanical interactions, are driven by forces with disjoint or largely disjoint energy levels from each other would have many defenders (e.g., Batterman, 2001, this volume; Griesemer, this volume; Simon, 1962/1996; Wimsatt, 1976, 1994/2007, this volume; Woodward, this volume).

3. Thus, if one allowed structure here, rather than just composition (of whatever scope), we have primary, secondary, and tertiary folding structure in hemoglobin subunits, plus quaternary structure (the hemoglobin molecule is a tetramer with two alpha and two beta chains) plus the "quintenary" structure formed by sickle-cell hemoglobin (a hemoglobin variant with a valine substituted for a glycine at position 6 in the beta chains) stacking with alternating orthogonal pairs of tetramers into a long crystalline chain under low oxygen conditions in the cell and all on their appropriate time scales. This leads to the characteristic sickle-cell distortions leading to erythrocyte logjams and ruptures and further oxygen depletion in peripheral capillaries, as well as runaway disease crisis and anoxia in humans under oxygen stress. There are clearly multiple causal layers here and arrangement of parts (which makes higher-level properties emergent) (sensu Wimsatt, 2007, chap. 12, this volume) and transmission of causes all the way from the sequence substitution to the molar sickle-cell crisis affecting morbidity and mortality of the affected individual. But should we regard the primary through quintenary configurational structures as "levels" in the same sense as the scale-relative compositional hierarchies including valine, hemoglobin, erythrocyte, and peripheral capillaries? (See also Mitchell & Gronenborn 2017, for more on investigations of the structure of hemoglobin.)

4. The conservation laws governing aggregative variables hold for systems that are *closed* to matter and/or energy exchange with their environment. An interesting consequence is that, for open systems like organisms, if compositional levels are to be energetically closed, then tracking them through time will mean including parts of the environment in a given organismic level. Part of the difficulty of conceptualizing such cases is that standard philosophical notions of composition are synchronic, whereas the transmission of energy takes time (as required by special relativity). Thanks to Jim Woodward for insightful discussion on this point.

5. A different objection can be raised at this point that variables belonging to wholes and parts are not distinct enough to stand in causal relations because of *supervenience* (e.g., Baumgartner, 2010; Kim, 1992)—even if the variables are not aggregative. By invoking supervenience against scientific examples of downward causation, however, we run into the same epistemological problems of "in-principle" posturing against successful scientific explanations (see "Entanglement" above). While it is true that computational or practical limitations factor into the use of high-level variables in scientific explanations, not all cases of higher-level causation and downward causation can be attributed to epistemic finitude. In some cases, high-level variables are simply not well defined at lower levels (the "supervenience base") (Dresden, 1974). When they are well defined, high-level features often have

different individuation conditions than their lower-level correlates and so cannot always be tracked in terms of lower-level features (DiFrisco, 2018). Thus, eye spots and other characteristic markings on a butterfly wing are selected for species recognition or predator avoidance. *They signify something specific to a potential songbird predator other than what produces them or what they are composed of*—for example, the large eyes of a songbird-predating owl when suddenly deployed, as when the butterfly opens its wings, or the nauseating taste and milkweed-based poison of the monarch, so effective in averting potential predators that it is mimicked (e.g., by the viceroy). These signals are causally effective at the molar level because of the perceptual search image of the potential mate or predator, with relational satisfaction conditions that are thus massively multiply realizable.

6. Although less attended to, there are equally challenging issues in development, reproduction, and host–biome relationships.

7. See also Brigandt and Love (2012) for a complementary discussion on the concept of "evolutionary novelty."

References

Allen, T. F. H., & Starr, T. B. (1982). *Hierarchy: Perspectives for ecological complexity.* Chicago, IL: University of Chicago Press.

Baedke, J., & McManus, S. (2018). From seconds to eons: Time scales, hierarchies, and processes in evo-devo. *Studies in History and Philosophy of Science Part C: Studies in History and Philosophy of Biological and Biomedical Sciences, 72,* 38–48.

Batterman, R. W. (2001). *The devil in the details: Asymptotic reasoning in explanation, reduction, and emergence.* Oxford, UK: Oxford University Press.

Baumgartner, M. (2010). Interventionism and epiphenomenalism. *Canadian Journal of Philosophy, 40*(3), 359–383.

Bissell, M. J., & Hines, W. C. (2011). Why don't we get more cancer? A proposed role of the microenvironment in restraining cancer progression. *Nature Medicine, 17*(3), 320–329.

Block, N. (2003). Do causal powers drain away? *Philosophy and Phenomenological Research, 67*(1),133–150.

Bossort, A. K., Jasieniuk, M. A., & Johnson, E. A. (1977). Levels of organization. *BioScience, 27*(2), 82.

Brigandt, I. (2020). How are biology concepts used and transformed? In K. Kampourakis & T. Uller (Eds.), *Philosophy of science for biologists* (pp. 79–101). Cambridge, UK: Cambridge University Press.

Brigandt, I., & Love, A. C. (2012). Conceptualizing evolutionary novelty: Moving beyond definitional debates. *Journal of Experimental Zoology B: Molecular and Developmental Evolution, 318*(6), 417–427.

Brooks, D. S. (2017). In defense of levels: Layer-cakes and guilt by association. *Biological Theory, 12*(3), 142–156.

Brooks, D. S. (2019). A new look at levels of organization in biology. *Erkenntnis.*

Brooks, D. S., & Eronen, M. I. (2018). The significance of "levels of organization" for scientific research: A heuristic approach. *Studies in History and Philosophy of Biological and Biomedical Sciences, 68,* 34–41.

Campbell, D. T. (1974). Downward causation in hierarchically organised biological systems. In F. J. Ayala & T. Dobzhansky (Eds.), *Studies in the philosophy of biology: Reduction and related problems* (pp. 179–186). London/Basingstoke, UK: Macmillan.

Chemero, A., & Silberstein, M. (2008). After the philosophy of mind: Replacing scholasticism with science. *Philosophy of Science, 75,* 1–27.

Craver, C. F. (2015). Levels. In T. Metzinger & J. M. Windt (Eds.), *Open MIND* (pp. 1–26). Frankfurt am Main, Germany: MIND Group.

Craver, C. F., & Bechtel, W. (2007). Top-down causation without top-down causes. *Biology & Philosophy, 22*(4), 547–563.

DiFrisco, J. (2017). Time scales and levels of organization. *Erkenntnis, 82*(4), 795–818.

DiFrisco, J. (2018). Token physicalism and functional individuation. *European Journal for Philosophy of Science, 8*(3), 309–329.

Dresden, M. (1974). Reflections on "fundamentality and complexity." In C. P. Enz & J. Mehra (Eds.), *Physical reality and mathematical description* (pp. 133–166). Dordrecht, Netherlands: Riedl.

Eldredge, N., Pievani, T., Serrelli, E., & Tëmkin, I. (Eds.), 2016. *Evolutionary theory: A hierarchical perspective.* Chicago, IL: University of Chicago Press.

Eldredge, N., & Salthe, S. (1984). Hierarchy and evolution. *Oxford Surveys in Evolutionary Biology, 1,* 184–208.

Ellis, G. F. R., & Kopel, J. (2019). The dynamical emergence of biology from physics: Branching causation via biomolecules. *Frontiers in Physiology, 9,* 1966.

Ereshefsky, M. (2017). Species. In E. N. Zalta (Ed.), *The Stanford encyclopedia of philosophy*. Retrieved from https://plato.stanford.edu/archives/fall2017/entries/species/

Eronen, M. I. (2013). No levels, no problems: Downward causation in neuroscience. *Philosophy of Science, 80*(5), 1042–1052.

Eronen, M. I. (2015). Levels of organization: A deflationary account. *Biology and Philosophy, 30*(1), 39–58.

Eronen, M. I., & Brooks, D. S. (2018). Levels of organization in biology. In E. N. Zalta (Ed.), *The Stanford encyclopedia of philosophy*. Retrieved from https://plato.stanford.edu/archives/spr2018/entries/levels-org-biology/

Gillett, C. (2013). Constitution, and multiple constitution, in the sciences: Using the neuron to construct a starting framework. *Minds and Machines, 23*(3), 309–337.

Greenberg, G., & Tobach. E. (Eds.). (1984). *Behavioral evolution and integrative levels*. Hillsdale, NJ: Lawrence Erlbaum Associates.

Greenberg, G., & Tobach, E. (Eds.). (1987). *Cognition, language, and consciousness: Integrative levels*. Hillsdale, NJ: Lawrence Erlbaum Associates.

Grene, M. (1987). Hierarchies in biology. *American Scientist, 75*(5), 504–510.

Griesemer, J. R. (2005). The informational gene and the substantial body: On the generalization of evolutionary theory by abstraction. *Poznan Studies in the Philosophy of the Sciences and the Humanities, 86*(1), 59–116.

Guttman, B. S. (1976). Is "levels of organization" a useful concept? *BioScience 26*(2): 112–113.

Haag, E. S., & True, J. R. (2018). Developmental system drift. In L. Nuño de la Rosa & G. B. Müller (Eds.), *Evolutionary developmental biology*. (pp. 1–12) Cham, Switzerland: Springer.

Hildenbrandt, H., Carere, C., & Hemelrijk, C. K. (2010). Self-organized aerial displays of thousands of starlings: A model. *Behavioral Ecology, 21*, 1349–1359.

Hodgkin, A. L., & Huxley, A. F. (1952). A quantitative description of membrane current and its application to conduction and excitation in nerve. *Journal of Physiology, 117*, 500–544.

Hood, K., Greenberg, G., & Tobach, E. (1995). *Behavioral development: Concepts of approach/withdrawal and integrative levels*. New York, NY: Routledge.

Kim, J. (1992). Downward causation. In A. Beckermann, H. Flohrn, & J. Kim (Eds.), *Emergence or reduction? Prospects for nonreductive physicalism* (pp. 119–138). Berlin: De Gruyter.

Kim, J. (1999). Making sense of emergence. *Philosophical Studies, 95*(1–2), 3–36.

Koestler, A., & Smythies, J. R. (1969). *Beyond reductionism: New perspectives in the life sciences*. London, UK: Hutchinson.

Ladyman, J., & Ross, D. (2007). *Every thing must go: Metaphysics naturalised*. Oxford, UK: Oxford University Press.

Lehrman, D. S. (1953). A critique of Konrad Lorenz's theory of instinctive behavior. *Quarterly Review of Biology, 28*(4), 337–363.

Lewis, D. (2000). Causation as influence. In J. Collins, N. Hall, & L. Paul (Eds.), *Causation and counterfactuals* (pp. 75–106). Cambridge, MA: MIT Press.

McShea, D. W. (2001). The hierarchical structure of organisms: A scale and documentation of a trend in the maximum. *Paleobiology, 27*, 405–423.

Mitchell, S. D. (2003). *Biological complexity and integrative pluralism*. Cambridge, UK: Cambridge University Press.

Mitchell, S. D. (2009). *Unsimple truths: Science, complexity, and policy*. Chicago, IL: University of Chicago Press.

Mitchell, S. D., & Gronenborn, A. M. (2017). After fifty years, why are protein X-ray crystallographers still in business? *British Journal for the Philosophy of Science, 68*(3), 703–723.

Montalenti, G. (1976). From Aristotle to Democritus via Darwin. In F. Ayala and T. Dobzhansky (Eds.), *Studies in the philosophy of biology: Reduction and related problems* (pp. 3–20). London, UK: MacMillan.

Nagel, E. (1961). *The structure of science*. San Diego, CA: Harcourt Brace.

Nijhout, H. F. (1994). *Insect hormones*. Princeton, NJ: Princeton University Press.

Noble, D. (2012). A theory of biological relativity: No privileged level of causation. *Interface Focus, 2*, 55–64.

Novikoff, A. B. (1945). The concept of integrative levels and biology. *Science, 101*(2618), 209–215.

Oppenheim, P., & Putnam, H. (1958). Unity of science as a working hypothesis. In H. Feigl, M. Scriven, & G. Maxwell (Eds.), *Concepts, theories, and the mind-body problem* (pp. 3–36). Minneapolis: University of Minnesota Press.

Pattee, H. H. (Ed.). (1973). *Hierarchy theory: The challenge of complex systems*. New York, NY: G. Braziller.

Potochnik, A. (2017). *Idealization and the aims of science*. Chicago, IL: Chicago University Press.

Potochnik, A., & McGill, B. (2012). The limitations of hierarchical organization. *Philosophy of Science, 79*, 120–140.

Pumain, D. (Ed.) (2006). *Hierarchy in natural and social sciences*. Dordrecht, Netherlands: Springer.

Redfield, R. (Ed). (1942) *Levels of integration in biological and social systems*. Lancaster: The Jaques Cattell.

Rowe, J. S. (1961). The level-of-organization concept and ecology. *Ecology, 42*(2), 420–427.

Rueger, A., & McGivern, P. (2010). Hierarchies and levels of reality. *Synthese, 176*, 379–397.

Salthe, S. (1985). *Evolving hierarchical systems: Their structure and representation*. New York, NY: Columbia University Press.

Salthe, S. N. (2009). A hierarchical framework for levels of reality: Understanding through representation. *Axiomathes, 19*(1), 87–99.

Schneirla, T. C. (1951). The "levels" concept in the study of social organization in animals. In J. Rohrer & M. Sherif (Eds.), *Social psychology at the crossroads* (pp. 83–120). Harper.

Sellars, R. W., McGill, V. J., & Farber, M. (Eds.). (1949). *Philosophy for the future*. New York, NY: Macmillan.

Simon, H. (1996). The architecture of complexity: Hierarchic systems. In *The sciences of the artificial* (3rd ed., pp. 183–216). Cambridge, MA: MIT Press. (Original work published 1962)

Soto, A. M., Sonnenschein, C., & Miquel, P. A. (2008). On physicalism and downward causation in developmental and cancer biology. *Acta Biotheoretica, 56*(4), 257–274.

Sperry, R. W. (1976). Mental phenomena as causal determinants in brain function. In G. Globus, G. Maxwell, & I. Savodnik (Eds.), *Consciousness and the brain* (pp. 163–177). New York, NY: Plenum.

Thalos, M. (2013). *Without hierarchy: The scale freedom of the universe*. Oxford, UK: Oxford University Press.

Waters, C. K. (2008). Beyond theoretical reduction and layer-cake antireductionism: How DNA retooled genetics and transformed biological practice. In M. Ruse (Ed.), *The Oxford handbook of philosophy of biology* (pp. 238–261). Oxford, UK: Oxford University Press.

Weiss, P. A. (1971). *Hierarchically organized systems in theory and practice*. New York, NY: Hafner

Whyte, L.L., Wilson, A. G., & Wilson, D. M. (1969). *Hierarchical structures*. Dordrecht, Netherlands: Elsevier Publishing Company.

Wilson, M. (2018). *Physics avoidance: And other essays in conceptual strategy*. Oxford, UK: Oxford University Press.

Wimsatt, W. (1972). Complexity and organization. In K. F. Schaffner and R. S. Cohen (Eds.), *PSA 1972*, (pp. 67–86). Dordrecht, Netherlands: D. Reidel.

Wimsatt, W. C. (1976). Reductionism, levels of organization, and the mind-body problem. In G. Globus, G. Maxwell, & I. Savodnik (Eds.), *Consciousness and the brain* (pp. 205–267). New York, NY: Plenum.

Wimsatt, W. C. (2007). The ontology of complex systems: Levels of organization, perspectives, and causal thickets. In *Re-engineering philosophy for limited beings: Piecewise approximations to reality* (pp. 193–240). Cambridge, MA: Harvard University Press. (Original work published 1994)

Wimsatt, W. C. (2007). *Re-engineering philosophy for limited beings: Piecewise approximations to reality*. Cambridge, MA: Harvard University Press.

Woodward, J. F. (2003). *Making things happen: A theory of causal explanation*. Oxford, UK: Oxford University Press.

Zafeiris, A., & Vicsek, T. (2018). *Why we live in hierarchies? A quantitative treatise*. Dordrecht, Netherlands: Springer.

1 Levels, Robustness, Emergence, and Heterogeneous Dynamics: Finding Partial Organization in Causal Thickets

William C. Wimsatt

Overview

In this chapter, I review the philosophical and scientific contexts the led me to argue for the importance of levels of organization (and other real causal structures in nature—perspectives and causal thickets) and their implications for other exploratory, manipulative, and explanatory practices. I then propose a systematic account of nineteen different characteristics of levels, divided into three groups relating to dynamics, robustness, and evolutionary elaboration of levels. From this emerge new discussions of emergence, level leakage and manipulation, and relations between levels and perspectives, and with multiscale modeling. I will briefly explore other approaches to levels to indicate how they differ. This analysis articulates particularly closely with the chapters of Batterman, DiFrisco, Griesemer, Love, and Woodward, as well as resonates with the complementary accounts of Brooks and Eronen, and Green, emphasizing the heuristic role of invoking levels in scientific activities and Baedke's discussions of temporal change in levels through development and evolution.

1.1 History and Orientation

The discussion of levels of organization that I initiated in Wimsatt (1976a) was originally intended to characterize realist claims about the organization of nature, following those made by some physicists, biologists, and social scientists, who persisted in taking upper levels of organization—their objects, regularities, processes, and phenomena—seriously in a context dominated by eliminative reductionist philosophers and "nothing-but" reductionist scientists, allied only in their suspicion of macroscopic theory and objects. For the latter scientists, talk about ontology was always a search for the bottom-most stable objects and their interactions out of which everything else was built (although they seemed unable to agree on what these were—for neurophysiologists, neurons; for geneticists, genes; for chemists, atoms; etc.). The former eliminativist philosophers were suspicious of any posited upper-level things, motivated by examples of earlier falsified theories whose posited entities turned out not to exist. Both groups assumed that a future complete and correct apocalyptic science would express all in terms of the ultimate lowest-level entities of an all-encompassing physics, one that would successfully "cash in" all of the "in principle" claims made to date. This foundationalist attitude seemed inappropriate to most of the science I knew (Wimsatt, 1976b, 2007).

In addition, philosophers, secure in the belief that we could access nature only through linguistic concepts, persisted in talking about levels of theory, predicates, and law statements as if vocabulary could substitute for natural objects, relations, and phenomena entirely. This seemed truly strange! Philosophers may talk about vocabulary, but scientists did not. I was convinced that there was a way through the veil of language using robustness. Although any single means of access might be instrument, concept, and (for cultural objects) even language dependent, *the convergence of multiple at least partially independent lines of access or detection seemed capable of penetrating gaps in the linguistic armor of our concepts with real objects simply being the consilient coincidence of multiple properties within boundaries that are stable over the appropriate time scale* (Wimsatt, 1974, 1980). Robustness, used as a criterion for what is real, also proved to be a way to reverse the rush to eliminative reductionism, to recognize higher-level objects interacting with an autonomous dynamics, and to embrace multiple realizability and a form of emergence consistent with reductionism.[1]

But levels don't capture everything. Biology[2] is also characterized by multiple complementary systematic partial perspectives on systems not ordered by level, like anatomy, physiology, and genetics (Wimsatt, 1974). These and the background and sometimes foreground anarchy of causal thickets (Wimsatt, 1976a, 1994, 2007; Griesemer, this volume) together affect our practices of explanation and other related activities in all of the compositional sciences. (Figure 2 of 1976a and 1994 depicts a kind of phylogenetic ontology exploring the articulations of these different kinds of causal structures.) These have intertwined ontological, epistemological, and heuristic import richly clarified by Griesemer's penetrating analysis in this volume.

It is hard now to remember the schizophrenic context in which "responsible reductionists" embraced a desert ontology from which everything else was *in principle* derivable, while at the same time employing their theoretical and experimental procedures on molar processes and objects. This sort of ontological "nothing-but-ism," combined so happily with a pragmatic molar realism, seemed to me to be bizarre. It clearly also concerned Herbert Simon in 1962 in his classic paper, "The Architecture of Complexity." This awareness led in 1972[3] to my paper, "Complexity and Organization" which began with Simon's puzzle of how an *in principle* reductionist should conduct themselves as a pragmatic holist.

I had strong realist sympathies (only exacerbated by my work as an engineer, fascination with technology, and the creative use of approximations), extending to a menagerie of complex forces, processes, and objects, including chromosomal mechanics, hormonal regulatory systems, groups, social organizations, institutions, and much more as units,[4] as well as biological and cultural selection and social cohesion and the like, not as "theoretical constructs" but as real molar forces. I argued for what I later called the "ontology of the tropical rainforest" against the adequacy of a "desert ontology." Quine's ontological desert, a philosophical chimera first explored by Russell, was populated by entities whose explanatory power was after all only said to be usable *in principle*[5] (a dead giveaway that they couldn't deliver anything *in practice*), and I wanted to understand the effectiveness of the tools we actually used. I wanted room for the importance of heuristic and approximate methods, and scientific processes of discovery (stimulated by Simon's work on problem solving with computers[6]) rather than the reconstruction and justification of the static formal icons of finished theory characteristic of positivism.

1.2   Levels

I also felt that composition alone did not exhaust the notion of a level, so I sought other handles to capture their riches. I found multiple others, and one was particularly forceful to me. This was Plato's observation about the desirability of "carving Nature at its Joints" (Plato, *Phaedrus* 265e). And levels seemed to me to be at least major vertebrae in the body of nature. But how to capture this intuition in modern terms and to mine and relate it to the resources of other characterizations? (Unlike Potochnik, 2017, I saw these characterizations as complementary, rather than competing, sources of confusion.) Inspired by statistical mechanics (where I had learned about the concept of a phase space[7]) and by the notion of a fitness topography from population genetics, I speculated that perhaps one could imagine levels as local maxima of regularity and predictability in a phase space of all of the alternative modes of organization of matter. (This ambitious characterization in search of grounding also seemed for its scope to be dangerously metaphorical and metaphysical.) Attempts to characterize it further led to my imprecise but suggestive diagram of levels and a diagrammatic attempt at a phylogenetic ontology of alternative states of matter in figures 1 and 2 of Wimsatt (1976a). Turning from this characterization to the many heterogeneous and productive ways of characterizing levels led to the relatively disordered form of my analysis in both 1976a and 1994, as a list of properties of levels interspersed with articulating comments. I did not expect or find an analysis in terms of necessary and sufficient conditions. Here I will attempt a more systematic account to relate them.

Many relate to the ordering relationships that Bob Batterman has addressed in his discussions of multiscale modeling. Batterman's work and that of Sara Green (Green & Batterman, 2017) and Mark Wilson (2016) provide a grounding for much of my discussion of levels of organization in physical relationships that are important and of wide applicability to systems studied in different disciplines.

I place the characterizations of levels into three groups. The first emerges from considering the energetics of interactions among elements of a system and yields a perspective I attribute to Herbert Simon. These provide suggestive connections between Batterman's account and mine. The second emerges from noting that objects are paradigmatically robust, with levels as products of clustered interactions among robust objects yielding a systematic domain in which the entities and processes at levels provide relatively economical explanations of a wide variety of phenomena. This is what makes levels primary "joints" in the body of nature. This also connects with Batterman's approach (see Batterman, 2017), through his concern with dynamical autonomy at the molar level. The third group of characterizations arises from an ampliative and speculative exploration of how levels, as characterized according to these first two sets of indicators, could over evolutionary time grow in extent (in kinds of objects at that level) and complexity, in their interactions with other entities at the same level, and increasingly, at other levels, through "level leakage" and selection-driven accretion. (See also Baedke's important characterizations of the evolution and development of levels on phylogenetic and ontogenetic time scales.)

1.3   Levels and Strength of Interaction

So what do we use to identify and to argue for the existence of levels? Composition, size, dynamics, robustness, a kind of completeness, and a kind of decomposability all figure into it. First, the energetic characterization:

1. Composition: clearly levels are compositional,[8] but this is not sufficient. Why are some collections of objects levels and other not? Why cannot we pick any contiguous set of objects and declare it a level?[9]

2. Processes at the same level commonly have comparable relaxation times.

3. Processes at the same level commonly have roughly the same size scale.

4. Processes organize by energy level of interactions, which tend to cluster by size scale.

5. From 1 to 4, we get the possibility of a hierarchical near-decomposability account,[10] yielding levels, and

6. Level-specific dynamics, each with its own range of roughly comparable relaxation times,[11] and from that,

7. A simple form of multiple realizability (of the stable higher-level weak interactions over the variable lower-level strong interactions) once they go to equilibrium.

Energy level, and the physical clustering of interactions by energy level, is apparently Herbert Simon's primary criterion (Simon, 1962), and he argues that energy level commonly corresponds roughly to relaxation times (time to equilibrium) and to size scale. Composition is presupposed, so for him, items 1 to 4 go together, and indeed, textual comments imply that he thinks that compositional ordering would be determined by relative energy levels. Since near-decomposability relates to relative strengths of interaction, when a system is nearly decomposable (often through multiple levels of strength of interaction), one also gets decomposition of a system into hierarchically characterized parts, each with a level-specific dynamics, so the first seven characteristics go together. This could even fit the mechanism-levels account in some cases, although Griesemer's account (in this volume) shows that Simon needs scaffolding as well. Bechtel and Richardson claim a strong debt to Simon in the introduction to their classic and important work (Bechtel and Richardson, 1993, 2010) in their accounts of mechanism (see also Wimsatt, 2018, on the "Chicago Mechanists"). So Simon's is an extremely powerful and rich multifaceted account of level leveraged ultimately in terms of energy levels. If I had to backslide to a more conservative position, I would favor something like this, but even it is inadequate to the richness we find in the dynamics and ontologies of biological and sociotechnological cultural systems, and its power depends too much on *in principle* promissory notes when applied to such systems. Systems also may simply fail to be nearly decomposable because their interaction strengths overlap too much and do not cluster, so no levels, and we are most probably faced with a causal thicket.[12] When the dynamical separation fails, levels are (correctly) shown to be nonuniversal. If the power of general physical principles is to be extended further, I would expect insight from the kinds of multiscale modeling emerging from condensed-matter physics (see Batterman, 2017, this volume).

1.4   Levels and Robustness

I turn now to criteria for levels relating to inferential and dynamical robustness (Wimsatt, 1981). Inferential robustness is the detectability, measurability, or derivability of an object or property by multiple independent and individually fallible means. Although individually fallible, the fact that their modes of failure are independent of each other means that their collective probability of failure declines exponentially with increasing numbers of fallible links. They are also keying into different properties of the detected object, thus getting a richer characterization of the causal network involving its properties and their detectors. The kind of cross-checking provides both knowledge about the object (it is the kind of entity that can be detected by these diverse procedures) and also about the procedures. (They commonly don't produce exactly the same results, but these differences, as reflections of the same entity, tell us more about how the procedures interact with the entity.) This yields a lovely bootstrapping of knowledge both about the object under study and about the means for studying it that circumvents the classical argument from illusion for the unreality of sensory information.

Robustness is primarily applied to objects—a criterion from the trenches, one used by scientists as a working criterion for trustworthiness, but also serves for the reality of a process or property. It is a worthy successor to the philosopher's fruitless search for certainty. (See also Eronen, 2015, for an extended defense of the connection between robustness and reality.) This is clearly related to (but not identical with) the idea of physical robustness or stability or invariance of a physical property across differences in state description (see Batterman, 2003;[13] Wagner, 2005; Wimsatt, 1981; Woodward, 2003). An object will be inferentially robust because it is central to our concept of an object that it has many properties and thus that there are multiple ways of detecting it. This is richly imbedded in scientific practice. But the multiple ways of detecting something are also multiple ways of interacting with it. Most of these will involve other objects at the same level because we would also locate at that level other things with commensurate dynamics. And this has more implications that I flesh out below.

8. The idea of cutting nature at its joints suggests that cutting it at other places will be less successful in finding ordered patterns and objects to be used to construct theory. Thus, the metaphor of different theories as having sieves of different sizes (and properties) that yielded objects and their causal relations with different degrees of predictive power and simplicity—thus levels as local maxima of regularity and predictability in a phase space of all alternative modes of organization of matter.

9. Thus, levels on this account ought to be where we can construct theories that are both powerful and simple—where we get this biggest bang for the smallest buck.[14]

10. A robust object is an indicator of a level through the richness of its connections with other objects.

11. Since levels are composed of robust objects that are rich in relations with one another, levels are themselves a peculiar sort of robust object.

12. Levels suggest a kind of order in the causal thicket: a scope of same-level causality and a kind of local explanatory and descriptive completeness—everything detectible at that size scale ought to be characterizable at that scale.

13. Given that we have multiple realizability (from 7, but also from the stability of upper-level objects and relations and compositionality with smaller parts), yielding a more robust multiple realizability, a level-specific dynamics ought to be quasi-autonomous.[15]

1.5   The Evolution of Levels

In "The Architecture of Complexity,"[16] Herbert Simon suggested a mode of evolution of complex systems through the "aggregation of stable subassemblies." He used this to argue against a view that selection processes did not have time to generate the complex systems that we see. With this mode of composition, he suggested, we could get exponential increases in complexity in linear time.[17] He also imagined that we would in effect get a new level of organization with each aggregation operation. We could imagine this process in the engineering of complex technologies, and it would fit well the Bechtel or Craver accounts of levels. It also suggests some cases of biological evolution, such as the evolution of eukaryotes (if we ignore the massive coevolution of their components), of metazoa (as "aggregates of cells"—ignoring development), or evolution of sociality (as "aggregates of people"). But these claims fail where there is a significant developmental or a scaffolding process mediating growth in complexity.[18] This includes individual ontogeny, where complexification involves morphological change and differentiation. (Thus, even the evolution of metazoa as "aggregates of cells" fails as an example, and characterizing sociality in this way leads too easily to a simplistic and incorrect methodological individualism.) But cases not simply characterizable as "aggregation of stable subassemblies" also include processes like those discussed below, which could act to facilitate the articulation of stable subsystems to form Simon's next level up.

In 1994, I went further in a speculative mode to look to the evolution of levels. These approaches, which I propose to call "Levinsian," all presuppose levels as characterized in terms of the above list of properties:

14. Richard Levins (1968) suggests that organisms evolve under selection to minimize the uncertainty in their environments. Thus, for example, an organism should evolve sufficient reserve food and water storage to reliably reach the next resource site, and changes in the size and distribution of resources should drive changes in the capabilities of organisms.

15. Because of their regularities, levels are attractors for such systems that key into and use their patterns, often at multiple levels simultaneously, so the populations of entities at a given level should evolve over time.

16. Differential selection processes should tend to expand the scope of dynamical autonomy—increasing the range of multiple realizability and of robustness—still further in cases where a macro-level property contributes positively to fitness (Wagner, 2005; Wimsatt, 1994, p. 253).

17. Thus, levels should themselves evolve over time, through accretion, elaboration, and coevolution of their component parts. This has broader consequences:

18. As the objects at a level increase in size and complexity, their increased number of degrees of freedom should lead more commonly to "level leakage," in which they interact with objects or processes at other levels.

19. For this reason, higher levels should become increasingly less well defined, and what patterning occurs might increasingly often fit perspectives or causal thickets.[19] Levels might thus be less relevant, well defined, or a thinner (more limited) cut on the phenomena for some scientists and sciences than others—for example, ecology, psychology, or social sciences, especially the intersecting varieties of cultural evolution—and suggests the need for more relevant perspectives for a fuller account.

This evolution of levels (or of the causal thickets from which they emerge) could involve the generation of new forms of regular patterns, what I call perspectives, and in some cases new higher levels. This has happened in biological evolution (for the emergence of multicellularity and at the ecosystem level), especially at the levels of human social and technological evolution with greatly expanded powers and modes of construction, manipulation, and communication (Arthur, 2009; Wimsatt, 2013b, 2019; Wimsatt & Griesemer, 2007). This requires enough complexity—sufficient degrees of freedom—to allow not only new order-finding interactions but also new levels. Nuclear physics (beginning with the discovery of X-rays) was a level detected initially by level leakage.[20] Or in new technology, consider the systematic conjoint exploitation of semiconductor devices and photolithography to produce a hierarchically expanding range of complex devices beginning with integrated circuits of different types (perhaps a perspective, within which there are levels?).[21] One could imagine in the future the emergence of a new constructional technology driven by the systematic use of CRISPR technology (CRISPR DNA sequences from procaryotic genomes allow manipulation of genomes of all kinds with heretofore unprecedented precision). These also promise the potential for more frequent intersection with causal thickets.

1.6   One Type of Emergence, or Two?

Use of robustness as a criterion for the reality of objects (Wimsatt, 2007, chap. 4) gives us not only stones, organisms, and cathedrals. These objects are related, dynamically and compositionally, to others. With assembly, some properties may show no particular qualitative changes (as a cathedral is composed of arranged quarried and shaped stones, whose total mass is nonetheless preserved), but new properties may emerge, as when stones are arranged into an arch, permitting a loadbearing opening in a wall, and changing the stability of their arrangement and how it interacts with other parts of the cathedral and with the people who use it.

One form of emergence arises when a system property depends upon the organization of the parts (Wimsatt, 2007, chap. 12). Given the open-ended multiplicity of forms of organization, characterizing it seems an impossible task, so I turned the task around and asked what was required for a system property to be *nothing more than* a sum or aggregate of parts properties, as, for example, for mass, energy, and charge. I found four conditions for aggregativity that could be met (or violated) independently, thus giving a total of $2^4 = 16$ possible combinations.[22] Aggregativity results if all conditions are met.[23] But fifteen different ways in which the combination of conditions may fail to be met yield kinds of organization-dependent emergent properties. Diverse combinations illustrate how different properties such as critical mass (for U-235 or other fissionable materials), genetic epistasis, and being an oscillator (for a series-connected resistor, capacitor, inductance, and voltage source) were emergent properties. In each case, realization of the property is dependent upon the arrangement of the parts and their modes of interaction (relative to each other and to other parts, either as possible substitutes or as external to the system but affecting its behavior). Thus, whether a chunk of U-235 is a critical mass depends upon its arrangement (in a sphere or in a long linearly extended wire), its compression (so a shaped charge surrounding it is used for detonation in some nuclear bombs), and the presence of neutron absorbers in its environment (so neutron-absorbing control rods regulate reaction rates in some power reactors). Many different examples are found in Wimsatt (2007, pp. 278–279, table 12.1).

But Bob Batterman has convinced me that I must deal with a second concept of emergence that arises not when a system property is highly dependent upon the organization of its parts but just the reverse—when it is robust, invariant, or insensitive to the arrangement of parts (but where it seems nonetheless misleading to describe it as just an aggregate of parts' properties). The difference is quite striking—are these approaches polar opposites? The concept I have elaborated as non-aggregativity is particularly appropriate to machines and mechanisms where the system property arises through the structured interaction of a number of differentiated parts. But there is obviously more in the emergence tent, so I return to robustness.

The four criteria for aggregativity lead heuristically to what I call the "Waring Blender" criterion for emergence: take the system, and disrupt it in a Waring Blender. The emergent properties are the ones that disappear.[24] In effect, just such a criterion is used by those in biology who prepare their specimens with an ultracentrifuge. You can tell the scientific specialty by the rotation speed! For biological specimens, lower speeds (lower Gs) yield whole cells, and higher speeds produce objects at lower levels, cellular fragments that allow studying cellular ultrastructure. I return to this criterion later.

But emergence is also striking for generic properties and multiple realizability. Systematicity of behavior that remains invariant or nearly so at the macro-level under widespread changes at the micro-level is what I have elsewhere called "dynamical autonomy" (Wimsatt, 1976a, 2007). Robustness can arise for multiple reasons and, in biology, is a frequent product of selection (e.g., Wagner, 2005). In fact, widespread robustness of phenotypic properties across changes in genetic constitution in sexual species is a necessary condition for heritability—and thus is a requirement for evolution even to be possible (Wimsatt, 2007, pp. 218–219).

In a more familiar classical case, we have molecular gases governed at a microscale by collisions and exchange of momentum, but also, on a macroscale, they are the continuous incompressible fluids of hydro- and aerodynamics manifesting properties like pressure, viscosity, and laminar versus turbulent flow that seem qualitatively different from anything found at the microscale. And so also with the multiple different emergent properties of solids and their crystals, alloys, and mixtures that arise from structures at meso-levels between the micro- and the macro-level.

I accept Batterman's claim that this is what is often meant by scientists who talk about emergence. Their notion applies to macroscopically homogeneous bulk matter, whose macro-properties remain robust, stable, or invariant under almost all changes at more micro-levels. But at one (or more) intermediate mesoscales, such bulk matter *has* structure reflected in the contents of a "representative volume element" or RVE on a given intermediate size scale far above that of the micro-level and far below that of the macro-level.

For a metal, this might be characterized in terms of local misalignments in the crystal structure, yielding fractures or empty volumes and also inclusions of other elements. The RVE is a statistical concept reflecting the occurrence and distribution of other structural elements on an intermediate size scale, with properties that can be related both upward and downward to explain the upper-level emergent properties. These intermediate properties are commonly invariant over changes at the microstructural level and can be characterized in terms of one or more correlation lengths for the distribution of different kinds of matter at their level. See Batterman (2002, 2017, this volume) for more exposition of these ideas.

So the concept of emergence as non-aggregativity depends upon a highly specific arrangement of the parts of the system to generate the emergent property or behavior, while Batterman's concept applies to emergent properties or behaviors that are highly robust and multiply realizable over micro-level rearrangements of its parts. But despite their enormous difference, these two notions have a deeper connection: both disappear with application of the transformation of the "Waring Blender" criterion, although in different kinds of ways. Indeed, does not the emergence of different kinds of order—of whole cells at the higher level and of cellular ultrastructure at the lower level—just in effect generate different RVEs through the application of different energy levels to disrupt biological materials? And would not the application of higher energies (in an imagined ultra-ultracentrifuge) yield macromolecules, small molecules, atoms, and ions? Are the higher energy accelerators of particle physics simply an extension of the Waring Blender criterion? So Batterman's emergence involves a statistically characterized failure of homogeneity assumptions at intermediate levels and thus seems to be a special kind of case of emergence as non-aggregativity in my sense.[25]

It is special in that in order to develop appropriate theory to characterize it, the kinds of deviation from homogeneity on all scales (the continuum) are best described in terms of correlation lengths rather than the four dimensions of aggregativity. But both appear to be covered by the Waring Blender criterion, which interestingly, has a deep connection with the energy required for disrupting the order. Perhaps neither the four dimensions of aggregativity nor the formulation in terms of correlation lengths are exclusive fundamental characterizations of emergence but formulations adapted to different specific ways of characterizing order and raises the question whether there might be other formulations particularly suited to other kinds of order, such as fractals or turbulence.[26]

1.7   Levels as Explanatory Attractors: The Level Relativity of Explanations

I start with a quote from Wimsatt (2007, pp. 214–216) that describes Brownian motion (hereafter BM) in terms of the characteristics of relevant entities at lower and upper bounding levels. It is intriguing because no ontology is reified at the same (BM) level, but explanations are done in terms of entities and processes at lower and upper levels. Levels in this case are attractors of explanations, because of the higher robustness of entities and processes there. Perspectives also show a similar feature and for similar reasons.

There is a general level-centered orientation of explanations that can be explained in terms of the greater stability and robustness of entities at levels of organization, and probably more globally, in terms of the consequent robustness of levels themselves. This is a general and important meta-principle for the organization of explanations that is usually taken for granted and seldom commented on. It facilitates explanatory clarity, but occasionally misfires.[27] The robustness of levels tends to make them stable reference points that are relatively invariant across different perspectives and therefore natural points at which to anchor explanations of other things. *Explanations of the behavior of between-level entities tend to be referred upwards or downwards in level, or both—rather than being pursued in terms of other between-level things. Even the fine tuning of the exact "altitude" of the between level entity—its size and thus the distance it is above the lower and the distance it is below the upper levels—is motivated by concerns originating at one or the other of the levels.* The robustness of levels makes the level-relativity of explanations a special case of the phenomenon referred to in the preceding section-the explanation of that which is not robust in terms of that which is robust. I will consider the case of Brownian motion as a between-level phenomenon, which, by its very nature requires very special relations to the level below and the level above. (For a more technical exposition of some of the details, see Jeans, 1940 [and Batterman 2017].)…

[A] revealing indicator that Brownian motion particles are between levels is that they are given no intrinsic characterizations—as is indicated by the fact that things as diverse as dust motes and bacteria can all be Brownian motion particles. *Between-level entities tend to be defined functionally [as we must for their mass and surface area to be detectable] rather than in terms of their intrinsic properties—it is almost as if they have no intrinsic properties to use in such a definition.* If so, this suggests the paradoxical conclusion that we may recognize the intrinsic properties of things, at least in part, due to characteristic interactions they have with other same-level things, since only levels have the intensity of different kinds of interactions among entities to fix unique sets of intrinsic properties as being causally relevant. Multiple realizability in between-level contexts washes out the causal salience of most specific [lower-level] intrinsic properties.

This discussion of Brownian motion provides a starting place for assessing the strengths and limitations of using levels as a heuristic approach to problems where it does not fit exactly. This must start with a characterization of situations where an ideal characterization of levels as complete and isolated from other levels does not apply, for the failure of this isolation is the realistic general case.

1.8   Level Leakage

We must consider how *levels can leak interactions or effects* to other levels,[28] in ways that can facilitate our understanding of their relationship (Wimsatt, 2007, chap. 10).

Many gateway phenomena or experiments, like Brownian motion, have the form of finding level leakage and exploring ways to exploit it to manipulate lower-level variables, which can give far-reaching control of processes and entities at the other level. This is endemic in the biological sciences and in technology as well. Indeed, the richness of inter-actions in both of these areas makes it seem likely that level leakage is far more common in such cases and pushes them in the direction of causal thickets.[29]

In the Hewlett-Packard HP-41C programmable calculator, released in 1979, an error in early software allowed manipulation and movement of the supposedly fixed boundaries (the "curtain") between registers that were supposed to be visible and under user control and those that were not. Through this error, ingenious hackers were able to reach through the operating level to the machine language level of the system. They rapidly developed a lan-guage using "synthetic instructions" that they called "synthetic programming," which allowed new and powerful ways to manipulate the calculator by doing new things in newly accessible registers giving new control functions. These sped up many operations and allowed entirely new kinds of others, such as the manipulation of individual pixels in a display that normally permitted only whole alphanumeric characters. But this came with a danger of new and dangerous kinds of system crashes. The HP-41 users club produced a custom PPC ROM and a 500-page manual that allowed use of, and systematically documented, these capabilities, as well as other applications created using it.[30] Indeed, this story seems generalizable and important to the emergence of many new technologies and scientific fields.

1.9   Other Accounts of Levels

Perhaps one should argue, as Potochnik does, that we should just recognize that there are multiple distinct concepts of level that we fail to distinguish at our peril. To some extent, this is true. Thus, "levels of abstraction" is an entirely separate concept relating to the detail of a description. And David Marr's influential notion of levels are more like my perspectives (McClamrock, 1991. Levels of functional organization are plausibly assimilated to mechanistic levels, as characterized by Bechtel and Craver, which makes their account more broadly important. For them, the parts of a system entering in causal role explanations of system behavior are lower level to the system behaviors that they explain. A graph-theoretic account of functional organization in my dissertation (Wimsatt, 1971) and published much later (Wimsatt, 2002) would add some formal structure to that account, although unlike Cummins's account of function, I consider (Wimsatt, 2013a) the role of selection to be crucial. (A powerful critique of psychology's attempts to do without selection is Chiramuuta, 2018.)

But how are parts determined? Physical forces should tend to aggregate natural objects into sizes determined by their strengths and that of aggregating forces (like gravity, electrostatic attraction, and surface tension). For engineered objects, which invariably require substantial scaffolding in their production, it seems more that we see the *limits* in their sizes as determined by physical forces. Mechanisms may use heterogeneous materials and, for metals, special alloys, with properties appropriate to their roles. In the case of machinery (with transmitted forces and velocities determined by contact), shape and anchor points for motion are of crucial import. But mechanisms would include also electrical, electronic, and chemical interactions, in an open-ended manner, and I cannot see a systematic but general way to analyze how the diverse physical forces play a role in addressing their behavior.

A major difference between engineered objects and biological ones is that for engineered objects, the parts are constructed separately, with all of the scaffolding necessary for their individual creation, and then appropriately assembled, requiring significant additional scaffolding. Biological development requires coordinated scaffolding and generation operations that are for the most part self-directed as long as appropriate resources are provided. Griesemer's (this volume) account of scaffolding in managing causal thickets is of central importance here. Batterman's discussion of the role of multiscale modeling in "inactive matter" and "active matter" is also a very promising and complementary beginning. When targeting a complex system, some aspects of that organization will fit levels well, some perspectives, and some undisciplined causal thickets. Even when there are diverse differentiated parts involved in a complex mechanism, some aspects of its behavior should yield to the kind of analysis by Needleman and Dogic (2017) for mitotic spindle formation and cytoskeletal fiber networks and described by Batterman (this volume) as an example of multiscale modeling in active materials. Newman (2013) has systematically championed the role of generic forces in formation of biological structures. Evolution is nothing if not opportunistic, so we would expect selection to use order produced by generic forces wherever possible for increased robustness. However, the heterogeneity of biological structures and processes acting on different time scales suggests that multiple different RVEs ("representative volume elements") may be required to deal even with different aspects of a complex process at the same scale. In this way, we slice perspectives out of causal thickets in order to rearticulate them in search of an explanation.

1.10   Perspectives

The idea of a perspective is introduced in Wimsatt (1974), further discussed in Wimsatt (1976a), and probably most fully in Wimsatt (1994; see 2007, pp. 227–242). A new perspective emerges when a kind or small number of kinds of descriptive terms or causal interactions allow a systematic treatment of a class of behaviors of the system in relative isolation from other aspects of the system. Thus, anatomy, physiology, genetics, and network analysis define perspectives on the system and its behavior that are capable of explaining a class of behaviors of the system. Then the tools for accessing these interactions within a perspective will constitute a methodology, and the resulting behaviors define a kind of worldview or perspective on the system or, less metaphorically, a descriptive causal and explanatory niche for or cross section on the system. Rational decision theory, but also a more realistic heuristic and biases-laden behavioral decision theory, would represent (different) theories about how an individual's *beliefs, desires, attitudes, and personality characteristics* mediate how their inputs lead to their choice among possible actions. Perspectives are by definition incomplete and usually mutually complementary in the analysis of more complex behavior of the multiperspectivally characterized system.[31] We must not assume that the articulated sum of perspectives on a system will exhaust its description or causal interactions. There may remain an unsystematized skein of causal interactions constituting a causal thicket.

A perspective may become applicable also through a physical or conceptual manipulation of a system. In 1973, Richard Levins compared the complexity of a recently killed organism with a living one. He argued that the dead organism is much more complex than the living one, because in the latter, several hundred enzymes speed up key reactions by orders of magnitude, and these rates escalating above background mean that they dominate and control the rest. It looks as if this defines a perspective on the system (perhaps primary metabolism?), and it also looks like the addition of enzymes to a system not containing them (a thought experiment only—such a system without enzymes would not be living) would at least massively increase the near-decomposability of the system, suggesting that it might define a level on Simon's account.

Could a level be a perspective that also meets certain compositional relations? This might sometimes be revealing but is likely more confusing, since I take it that the notion of a perspective is less clear and less well defined than that of a level or the things involved in characterizing it above. A perspective that is not a level would presumably require at least the following: (1) a set of causally specific interactions that affect target entities selectively that are (2) rate dominant for the class of things that they affect and (3) acting fairly broadly across the parts of the system as characterized from that perspective, (4) which include entities at more than one level. I earlier (Wimsatt, 1974) characterized perspectives as (5) corresponding to the reach of a technology or set of technologies and claimed that (6) each perspective had a class of problems that could be solved within that perspective, (7) as well as some that required going beyond it, thus recognizing its incompleteness (see Wimsatt, 2007, chap. 10). Thus, the emergence of a new communication system or information channel (as in the neural, hormonal, or immune systems) and frequently in technological evolution, with new waves generated by speech, written language, printing, telegraphy, telephone, TV, and internet communications, would each presumably constitute new perspectives (or parts of one). (For technological evolution, see Arthur, 2009; Wimsatt, 2013b; Wimsatt, in Love & Wimsatt, 2019.)

Other researchers (particularly Dan Brooks, 2017; Brooks and Eronen, 2018) wish to resist the ontological claims of levels and regard them as heuristic tools in describing and organizing phenomena. But there is no conflict here. On my view, there is a kind of entanglement of epistemology and ontology once we can speak of processes of detection or interaction (Wimsatt, 1976a, part III). Before genes were ever physically detected, classical genetics provided a rich set of tools for inferences about heredity and led some writers to adopt an operationalist stance toward genes (see Wimsatt, 2007, chap. 6). Heuristic use of entities that may also have an ontological status might be indicated for entities with different degrees of robustness or that have ambiguous or context-specific boundaries. (Is the boundary layer of air stabilized around an organism part of the organism or part of its environment? Is the organism the right evolutionary unit for biological species or is it the mating pair?) Given that objects can be regarded as sets of processes with relatively slower relaxation times while still recognizing their ultimately ephemeral character, time scale changes may also indicate transformations in how objects are regarded. They can be regarded as particularly effective "false models" in organizing explanations, but don't forget that "false models" can hide dangerous simplifications and lead to problematic characterizations if we forget about their clay feet. I think that this is correct, but even a multiplicity of heuristic uses need not rule out realist claims as well.

1.11 Prologue as Epilogue

People will debate the role of levels of organization in the ontology of biological systems and their role as heuristics in investigating them. For all of these disputes we engage in, I am perhaps most intrigued by the following conjecture—a kind of empirical transcendental argument that reaches beyond biology through all of the sciences: if we did not have regularities in the universe, and indeed, regularities like what I have called levels of organization and perspectives, not only we but also all organized forms of life could not exist. We could not have evolved, nor could we persist. Selection needs developmental processes producing heritable order to work on, and relatively well-behaved small variations in it, to generate a kind of robust adaptive topography that is not so full of holes as to make a progressive accumulation of organized dynamical structures impossible. We are not inevitable (and surely not immortal) products of nature, but at least until our sun becomes a red giant or we succeed in practicing a comparable form of self-immolation, levels of organization seem unavoidable.

Acknowledgments

In constructing this chapter, I have benefited particularly from reading the extensive historical and critical writings of Daniel Brooks, James DiFrisco, and Markus Eronen. Coediting with Dan and James has been a pleasure and a privilege. Jim Griesemer and I have practiced a form of coevolution now spanning four decades that has produced in his chapter an elaborative inspiration and extension that substantially strengthens the account offered here and does much more. Alan Love's particularly acute reading led to several important organizational changes, and his "manipulationist" account interfaces nicely with the account offered here. But no one has had greater influence through multiple drafts and discussions than Bob Batterman in my attempts to articulate my account with his. And I am delighted if my actions have played a role in his emerging status as closet philosopher of biology. My year at the Pittsburgh Center for Philosophy of Science allowed me to attend Bob's superb graduate seminar on multiscale modeling, so I owe the center a debt of thanks for this as well. The final version of this chapter was edited by James DiFrisco.

Notes

1. The need for an adequate account of emergence consistent with reductionism would not have occurred to me at first, but my assignment at the conference for which I wrote my first levels paper (Wimsatt, 1976a) was to comment on the views of Roger Sperry. My invitees regarded this as a problematic challenge: Sperry easily skirted back and forth between reductionist claims and asserting emergence. They thought I would just have to tell Sperry that he was being inconsistent: every philosopher knew that to be reducible was not to be emergent. But a working principle of mine has been to take scientists seriously, and surely so for Sperry, whose surgical experiments I already knew and admired (well before the work on commisurotomy that got him fame and the Nobel Prize in 1981). And indeed, the more I looked at his position, the more I thought he was right and I could justify it, so my account of levels was developed from my new position as an "emergent reductionist." More recently, Bob Batterman has noted that there are *two* distinct (and equally necessary) strands in my discussions of emergence and defended the other. I discuss their relations below.

2. And of course, the human sciences and technologies (such as advertising and the manipulation of human behavior in elections and decision making practiced by Cambridge Analytica) reflect the role of our diverse but causally relevant subjectivities (see also Ervin Goffman's work, the *Dramaturgical Image of Self in Everyday Life* (1956), and von Uexkull's notion of Umwelt). It is not a surprise that perspectives should be crucial in the human sciences. What made my work (Wimsatt, 1974, 2007, chap. 10) unusual was that they could be found also in "objective" biology (and in ecologies' species niches). Our deep problem here is in our reductionistic biases in favor of monadic properties and massively underestimating the role of complex relational properties.

3. The "Complexity and Organization" paper was extracted from my dissertation (Wimsatt, 1971) and given at PSA-1972 but not published until 1974. The levels material was originally devised for a conference on Consciousness and the Brain in April 1973 (the date of the illustrations) but not published until 1976.

4. The thought that realism toward objects could go as far as social (and later toward biological) groups was almost certainly bolstered by Campbell (1958), which introduced me to his notion of triangulation (i.e., robustness as developed in Wimsatt, 1981) and applied it to social groups. But I had also been introduced to and fascinated by Lewontin's models and accounts of group selection as early as 1965 in a book, *Ideas in Modern Biology*, edited by John Moore, given to me by my father. See Wimsatt (2019) for my latest account.

5. As I noted in Wimsatt (1976b), any mathematician would see this use of "in principle" as unfounded unless it could be mapped onto the natural numbers, which it could not.

6. I first encountered Simon in 1964, along with early papers by Don Campbell and John Holland in Frank Rosenblatt's remarkably broad and deep course on Brain Models and Mechanisms. Rosenblatt was the grand-

father of the connectionism or "deep learning" of the 1980s for his work (especially in the period 1958–1962) on "Perceptrons." And his idea that one should seek "genetic" properties (i.e., properties realized or realizable in virtually all randomly constructed neural networks) anticipated Kauffman's (1969) work on "generic properties" of randomly constructed gene control networks, "simulated annealing," and essentially all work on emergent multiple realizability since.

7. In graduate school, I began to think of generalized property spaces as conceptual tools in biology and the social sciences by analogy with the idea of a phase space. This was also furthered by discussions of state spaces with trajectories and search trees in chess. Particularly stimulating were Lewontin's remarks on the dimensionality of genetic state spaces, in his classic work (Lewontin, 1973, chap. 6) and Simon's classic (1973) article on ill-structured problems.

8. Batterman (this volume) disagrees that levels are necessarily compositional. His development of many of the same points using "correlation lengths" has the advantage of articulating well with developing physical theory. Woodward's discussion (this volume) of conditional independence and his argument in favor of lower- versus upper-level variables rather than parts in articulating interlevel causation are also crucial to explicating the role of compositionality. Batterman will be analyzing the relations between their views in a current book manuscript.

9. Thus, if compositionality were the only criterion, wouldn't it be more elegant, for example, to go up a level simply by doubling the number of basic objects and then chunking them? This is so obviously false as to provide a *reductio*. Nature intervenes with its own sized chunks!

10. The interacting elements here might be specified either in terms of variables or in terms of parts. The former would coordinate with Batterman's interpretation and is most plausibly Simon's (see also Woodward's important discussion, this volume). The interpretation of levels as relating to a system and to its material parts is most congenial to the mechanists' interpretation of levels.

11. This picture is suggestive of Alfred Lotka's (1924) writings on stability, in which if one focused on a set of processes with roughly commensurate relaxation times in a complex dynamical system, other processes going to equilibrium significantly faster could be treated as transients and ignored, and others with significantly slower relaxation times would define "moving equilibria" in the variables under study. Simon was influenced by Lotka, who was also read by von Neumann, Wiener, von Bertalanffy, Weiss, and others in the intersection of mathematics, physics, and biology. Lotka's formulation is essentially the same as Simon's "near-decomposability" analysis.

12. This is one path to causal thickets, but there are others. Two or more perspectives showing too much interactional complexity would degenerate into a causal thicket because it would yield too much ambiguity as to how to locate or characterize the interacting objects and would produce conceptual confusion, as well as conflicts about which methodologies are appropriate to analyzing the phenomena. These methodological conflicts could also be a sufficient cause of a causal thicket (perhaps nature has too much scar tissue!), though, by making the very nature of the operative causes unclear. Is this a mechanistically explicable characterization of how a mechanistic account could fail (Wimsatt, 1976a)?

13. Batterman distinguishes stability of two types: stability under changes in initial conditions and stability under changes in the parameters of the system that he calls structural stability. He regards the latter as most centrally connected with robustness. See Batterman (2002, pp. 58–61). Since "multiple means of detection" is intended to refer most centrally to different causal or inferential processes in detecting or measuring, this is a different approach. Still not yet analyzed is the relation between the two approaches.

14. Thus, theoretical vocabularies come in levels because that's where the phenomena are, not the reverse, as was suggested by the linguistic turn in philosophy.

15. See also the discussion of "dynamical autonomy" for levels working through this argument in greater detail in Wimsatt (2007, pp. 220–221) and earlier in Wimsatt (1976a). It is an independent (although weaker) argument for the autonomy of upper-level dynamics that complements Batterman's in 2017.

16. This article has almost everything. It is the most important essay on the analysis and application of levels and hierarchical organization and their dynamics in the second half of the twentieth century. To say merely that it is a classic is an underestimate.

17. I speculate that even when development and scaffolding are taken into account, the rate of growth in complexity is greater than linear and nonuniform. It could be qualitatively stepwise logistic, beginning exponentially and slowing down when a limiting factor appears (as has happened at least twice, for cell size and for organism size), until a mode of scaffolding allows chunking of the top-level systems as units, which then may undergo higher-level articulation processes. I think that at the scale of the differentiated group ("super-organism" or "society," which may itself have multiple strata), at least one (or more) additional levels have become possible (markets? nation-states?) through the development of our rapid electronic broadband communication.

18. As Jim Griesemer argues in a searching analysis of Simon's "watchmaker" argument concerning evolution through stable subassemblies (Griesemer, this volume), Simon ignores the crucial role of scaffolding in the aggregation process, and this is part of what goes wrong in his failure to treat development adequately in the evolution of complexity.

19. There is a countervailing force here, however, when factors such as increased means of systematic variation production (e.g., Wimsatt, 2013b) and communication speeds can generate conditions for new higher-level selection processes acting upon newly chunked units of selection. This is what has happened with the emergence of cultural selection processes and correlative cultural units. See Wimsatt in Love and Wimsatt (2019).

20. Hacking (1983) notes that a scientific experiment may lead to construction of a new entity, phenomenon, or system that did not exist before. Technological creations like chip architectures or communication networks fit Hacking's characterization unproblematically like a glove.

21. Indeed, the construction of integrated circuits, with multiple layering of different semiconductor elements to realize their functions, is the construction of an interactionally complex object (Wimsatt, 1974) whose layers are like the material embodiments of multiple partial perspectives except that none of the layers are causally active by themselves until they are integrated and produce complete circuits!

22. In fact, the classification is even richer, since these four conditions can be treated as degree properties, allowing aggregativity as an approximation.

23. Aggregativity must seem like a very uninteresting property (or meta-property), but the aggregativity of mass, energy, and charge is embodied in the great conservation laws of physics discovered in the nineteenth century and has significant implications for theories using them.

24. This criterion has an unfortunate origin that cuts to the core of aggregativity and emergence. Jack Cowan (personal conversation) tells how the president of the University of Chicago (John Wilson, a psychologist) called him in for a talk. Cowan was then chair of the department of theoretical biology. Wilson, who didn't like small departments, said, "You and biophysics both use a lot of math and physics to study organisms—you must be doing the same thing." Cowan said, "Let me explain the difference. Take a rat and drop it into a Waring Blender. The biophysicist is interested in those properties that are invariant under that transformation." Wilson didn't appreciate the joke, or the need for the distinction, and merged the two departments, to the detriment of less well-funded theoretical biology.

25. One still needs to consider whether the energy necessary to create a given kind of order is the same as the energy necessary to disrupt it. In general, it is not.

26. Fractals could plausibly yield to an account in terms of a distribution of correlation lengths.

27. See the discussion of "perceptual focus" in the last two sections of Wimsatt (1980), where I discuss the biasing effect of the tendency to refer group phenomena down to the individual level of description or to describe groups as "collections of individuals," as if they had no organizational properties of their own, in the units of selection controversy.

28. I owe this fortuitous phrase to Stuart Glennen. Other key gateway experiments in which upper-level changes could effect specific kinds of lower-level changes would include Millikan's "oil drop" experiment to discover the charge of the electron and Boveri's "dispermic fertilization" experiments to show the individuality of the chromosomes in heredity.

29. I morphed the term "gateway experiment" from Mark Bedau's "gateway technology" (Bedau, 2019) where he applies the term to patent technologies that open a whole new adaptive niche to applications of that basic idea in often quite different contexts and morphologies. His key example is the development of inkjet technology.

30. My favorite (and my own programming contribution designed for the ROM) allowed the plotting of families of curves and up to nine functions at a time with different symbols, vastly extending the power of the standard plotter, which only allowed a single curve for a single function.

31. Since my 1974 paper, the first to argue for the importance of perspective-like things, a number of others have adopted some form of perspectivalism. These would include Ronald Giere, Sandra Mitchell, Helen Longino, Kenneth Waters, and Alan Love.

References

Arthur, B. (2009). *The nature of technology*. New York, NY: Free Press.

Batterman, R. (2002). *The devil in the details: Asymptotic reasoning in explanation, reduction, and emergence*. Oxford, UK: Oxford Studies in Philosophy of Science.

Batterman, R. (2017). Autonomy of theories: An explanatory problem. *Nous, 54*(4), 858–873.

Bechtel, W., & Richardson, R. C. (1993). *Discovering complexity: Decomposition and localization as strategies in scientific research*. Princeton, NJ: Princeton University Press.

Bechtel, W., & Richardson, R. C. (2010). *Discovering complexity: Decomposition and localization as strategies in scientific research* (2nd ed.). Cambridge, MA: MIT Press.

Bedau, M. (2019), Patented technology as a model system for cultural evolution. In A. C. Love & W. C. Wimsatt (Eds.), *Beyond the meme: The roles of development and population structure in cultural evolution* (Minnesota Studies in Philosophy of Science, Vol. 22) (pp. 237–260). Minneapolis: University of Minnesota Press.

Brooks, D. S. (2017). In defense of levels: Layer cakes and guilt by association. *Biological Theory, 12*(3), 142–156.

Brooks, D. S., & Eronen, M. I. (2018). The significance of levels of organization for scientific research: A heuristic approach. *Studies in History and Philosophy of Science Part C: Studies in History and Philosophy of Biological and Biomedical Sciences, 68*, 34–41.

Campbell, D. T. (1958). Common fate, similarity, and other indices of the status of aggregates of persons as social entities. *Behavioral Science, 3*, 14–25.

Chirimuuta, M. (2018). Marr, Mayr, and MR: What functionalism should now be about. *Philosophical Psychology, 31*(3), 403–418.

Eronen, M. (2015). Robustness and reality. *Synthese, 192*(12), 3961–3977.

Goffman, E. (1956). *The presentation of self in everyday life*. New York, NY: Doubleday.

Green, S., & Batterman, R. (2017). Biology meets physics: Reductionism and multi-scale modeling of morphogenesis. *Studies in History and Philosophy of Science Part C: Studies in History and Philosophy of Biological and Biomedical Sciences, 61*, 20–34.

Hacking, I. (1983). *Representing and intervening*. Cambridge, UK: Cambridge University Press.

Kauffman, S. A. (1969). Metabolic stability and epigenesis in randomly constructed genetic networks. *Journal for Theoretical Biology, 22*, 437–467.

Levins, R. (1968). *Evolution in changing environments*. Princeton, NJ: Princeton University Press.

Levins, R. (1973). In *Hierarchy theory: The challenge of complex systems* (H. Pattee, Ed.). New York, NY: Braziller.

Lewontin, R. (1973). *The genetical basis of evolutionary change*. New York, NY: Columbia University Press.

Lotka, A. J. (1924). *Elements of physical biology*. Baltimore, MD.: Williams and Wilkins Co. (Later reprinted 1957 by Dover as *Elements of mathematical biology*).

Love, A. C., & Wimsatt, W. C. (Eds.). (2019). *Beyond the meme: The roles of development and population structure in cultural evolution* (Minnesota Studies in Philosophy of Science, Vol. 22). Minneapolis: University of Minnesota Press.

McClamrock, R. M. (1991). Marr's three levels: A re-evaluation. *Minds and Machines, 1*, 185–196.

Moore, J., (1965). *Ideas in Modern Biology*. Garden City, NY: Natural History Press.

Needleman, D., & Dogic, Z. (2017). Active matter at the interface between materials science and cell biology. *Nature Reviews Materials, 2*(17048),1–14.

Newman, S. (2013). Excitable media in media res: How physics scaffolds metazoan development and evolution. In L. Caporael, J. R. Griesemer, & W. Wimsatt (Eds.), *Scaffolding in evolution, culture, and cognition* (pp. 109–123). Cambridge, MA: MIT Press.

Potochnik, A. (2017). *Idealization and the aims of science*. Cambridge, MA: Harvard University Press.

Simon, H. A. (1962). The architecture of complexity. *Proceedings of the American Philosophical Society, 106*(6), 467–482.

Simon, H. A. (1973). The structure of ill-structured problems. *Artificial Intelligence, 4*(3–4), 181–201.

Wagner, A. (2005). *Robustness and evolvability in living systems*. Princeton, NJ: Princeton University Press.

Wilson, M. (2016). *Physics avoidance and other essays in conceptual strategy*. Oxford, UK: Oxford University Press.

Wimsatt, W. (1971). *The conceptual foundations of functional analysis*. Unpublished doctoral dissertation, University of Pittsburgh.

Wimsatt, W. (1974). Complexity and organization. In K. F. Schaffner & R. S. Cohen (Eds.), *PSA-1972* (pp. 67–86). Dordrecht, Netherlands: Reidel.

Wimsatt, W. (1976a). Reductionism, levels of organization, and the mind-body problem. In G. G. Globus, G. Maxwell, & I. Savodnik (Eds.), *Consciousness and the brain* (pp. 199–267). New York, NY: Plenum.

Wimsatt, W. (1976b). Reductive explanation: A functional account. In A. C. Michalos, C. A. Hooker, G. Pearce, & R. S. Cohen (Eds.), *PSA-1974* (pp. 671–710). Dordrecht, Netherlands: Reidel.

Wimsatt, W. (1980). Reductionistic research strategies and their biases in the units of selection controversy. In T. Nickles (Ed.), *Scientific discovery: Vol. 2. Case studies* (pp. 213–259). Dordrecht, Netherlands: Reidel.

Wimsatt, W. (1981). Robustness, reliability, and overdetermination. In M. Brewer & B. Collins (Eds.), *Scientific inquiry and the social sciences* (pp. 124–163). San Francisco, CA: Jossey-Bass.

Wimsatt, W. (1994). The ontology of complex systems: Levels, perspectives, and causal thickets. *Canadian Journal of Philosophy, 20*, 207–274.

Wimsatt, W. (2002). Functional organization, functional inference, and functional analogy. In A. Ariew, R. Cummins, & R. Perlman (Eds.), *Function* (pp. 174–221). New York, NY: Oxford University Press.

Wimsatt, W. (2007). *Re-engineering philosophy for limited beings: Piecewise approximations to reality*. Cambridge, MA: Harvard University Press.

Wimsatt, W. (2013a). Evolution and the stability of functional architectures. CNRS conference on Function and Teleology. In P. Huneman (Ed.), *Functions: Selection and mechanisms* (pp. 19–41). Dordrecht: Springer.

Wimsatt, W. (2013b). Scaffolding and entrenchment. In L. Caporael, J. R. Griesemer, & W. Wimsatt (Eds.), *Scaffolding in evolution, culture, and cognition* (pp. 77–105). Cambridge, MA: MIT Press.

Wimsatt, W. (2018). Foreword. In S. Glennan & P. Illari (Eds.), *Routledge handbook of mechanisms and mechanical philosophy* (pp. xiv–xvi). New York, NY: Routledge.

Wimsatt, W. (2019). Articulating Babel: A conceptual geography for cultural evolution. In A. C. Love & W. Wimsatt (Eds.), *Beyond the meme: Development and population structure in cultural evolution* (Minnesota Studies in Philosophy of Science, Vol. 22, pp. 1–41). Minneapolis: University of Minnesota Press.

Wimsatt, W., & Griesemer, J. (2007). Reproducing entrenchments to scaffold culture: The central role of development in cultural evolution. In R. Sansome & R. Brandon (Eds.), *Integrating evolution and development: From theory to practice* (pp. 228–323). Cambridge, MA: MIT Press.

Woodward, J. (2003). *Making things happen: A theory of causal explanation.* Oxford, UK: Oxford University Press.

2 Levels of Organization as Tool and Doctrine in Biology

Daniel S. Brooks

Overview

'Levels of organization' is a familiar but widely uncharacterized concept that traffics in major commitments and auxiliary assumptions throughout the biological sciences.[1] In this chapter, I characterize the concept of levels, arguing that it encompasses dual profiles in scientific usage: one profile exhibits application as a conceptual tool used to collect and order descriptive content of a system or phenomenon of interest while another profile points to the concept's usage as a foundational doctrine of biology itself. Both reveal an erotetic core to the concept of levels, where the notion figures into conceptualizing and interacting with scientific problems that scientists face as investigative objects or as challenges to be taken up in future research. As such, 'levels' finds currency for both novice and experienced users alike as a pedagogical device and major organizing concept for biology. With this account in place, the contributions of the levels concept can be better contextualized to their appropriate circumstances of usage.

Levels and other modes of organization cannot be taken for granted, but demand characterization and analysis.
—William C. Wimsatt (1994/2007a, pp. 203–204)

2.1 Introduction

The concept of 'levels of organization' (hereafter 'levels') is an evocative notion that needs little introduction here (for that, see Brooks, DiFrisco, and Wimsatt, this volume). The stratified image of the world it presents impresses uniquely on scientists and philosophers alike in our thinking about the biological world: chains of part–whole relations stretching from the smallest to the largest natural phenomena; nature divided into classes of natural units that populate the strata of a layered world. Its prominence is apparent in the staggering ubiquity of the concept across biological texts of all stripes and specializations. The term is widely found alike in textbooks, monographs, research articles, reviews, and scientific op-eds. And still, the nature and significance of 'levels of organization' remain largely open questions. Although this is beginning to change, laments concerning the term's lack of clarity and calls for further attention to the concept's precise contributions to biological thought have always been, and remain, endemic of the discourse surrounding 'levels' (Eronen & Brooks, 2018). Referring to 'levels,' it appears, has ironically become a significant source of obscurity, rather than clarity, for the issues the term is nominally supposed to treat (Eronen, 2015; Guttman, 1976; Waters, 2008; see also Potochnik, this volume).

I seek in this chapter to alleviate some of this tension surrounding the concept of levels. In this endeavor, I will be especially concerned with the *character* of the concept: How is it used in the biological sciences? Where is the concept applied? What are its roles in biological thought? One of the principal issues with the levels concept is that these general questions have no immediate, familiar answers to contextualize the contributions that flow from the concept, which are numerous and rather disorganized. Consequently, the levels concept appears to be doing too many different things to be a viable, actively productive notion for scientific work (Potochnik, this volume). I seek to remedy this by offering an epistemic characterization of the conceptual profile of 'levels,' with the hope of contextualizing the concept's rampant influences into different aspects of that characterization.

To this end, I begin by offering a dual account of the character of 'levels of organization.'[2] One prong of this account concerns the usage of 'levels' as a conceptual tool. The approach I advocate directly incorporates the local circumstances of the concept's usage into our understanding of the term and its significance in biological usage. Specifically, I contend that 'levels' acts in such contexts to sculpt and focus scientific investigations by offering a means of articulating the structure and layout of natural phenomena and the way we investigate these phenomena.

I then articulate a second, hitherto underdeveloped side of the levels concept, arguing that the notion also possesses a more general character as a major organizing concept in science. Specifically, I contend that 'levels' possesses a profile and history of usage that warrants its acknowledgment as a *foundational doctrine* of the biological sciences. Roughly, this means that 'levels' designates a cluster of ideas, issues, and theses that underpin major commitments (both explicit and implicit) and observations in biological thought. I will appeal along the way to other examples of established doctrines in biology to inform and motivate this doctrinal character.

Following this characterization, I consider the three questions posed above regarding the character of 'levels.' The first two of these questions, how and where the concept is used, will receive the most direct treatment. The third question, the roles of 'levels' in biological thought, will require a more thorough and distinct analysis, which I can only partially provide here. Nonetheless, an important component to answering this question is given by the analysis I offer—namely, that a main locus of the concept's enduring impact in biology can be identified in its interaction with scientific problems (this is noted also in Brooks & Eronen, 2018, p. 36). In this way, an *erotetic core* is revealed of the notion, where the concept figures into the reasoning involved in posing scientific problems, the means to investigate them, and in providing their solutions.

2.2 The Fragmentary Character of the Levels Concept

In scientific usage, 'levels' is sometimes applied as a device to sculpt or focus scientific work. In this 'tool' form, 'levels of organization' is typically invoked as part of a scientific problem-solving task to elicit a descriptive map of a system or phenomenon of interest. Already at this juncture, 'levels' can exhibit behavior as a concept of epistemic or onto-logical significance, depending in part on the extent to which the system being investigated is known or unknown to a research community. In cases where the details of the system(s) being engaged are *unknown*, then the levels concept aids in furnishing a tentative descrip-tion until more details about the system's specific workings and composition are gathered. This occurs as the familiar structures that generally compose a living system (e.g., mole-cules, cells, tissues) are transferred to the target system or phenomenon of interest, and their description in turn becomes tasks for scientific investigation. Correspondingly, such sce-narios point to an epistemic role for the levels concept in that attending to these tasks com-prises investigative problems associated with a phenomenon of interest. When seen as part of an overall agenda that scientists wish to engage (such as constructing an explanation for the phenomenon), the problems composing the research area in question tend in such sce-narios to be relatively ill-structured or ill-defined. In more mature problem scenarios where the structure and organization of a target system or phenomenon are more well established, or *known*, the levels concept often takes on a more ontologically significant role by acting as a kind of reservoir of descriptions of the system. That is, 'levels' gains new significance in summarizing the layout of the system (through increasingly robust descriptions), with which scientists can more readily navigate a system or pose questions of the same (i.e., formulate or uncover more sophisticated problems to be solved).

Of course, none of this fully determines what 'levels' will express in a given instance; each experimental system will comprise its own descriptions, guided by distinct investiga-tive questions and interests, even concerning general structures like molecules, cells, and tissues. Instead, 'levels' in scientific research often expresses knowledge heavily contex-tualized within a particular system of interest, with its attendant methods, techniques, theo-ries, and scientific questions guiding the expressed content of the concept. This, in turn, means that 'levels' is highly sensitive to the circumstances of its usage (Brooks, 2017). This pertains to its expressed content but also to connotations the concept may acquire over vari-ous instances and repeated usage. Indeed, the circumstances of the concept's usage are not always contained within a given instance or even lineage of uses in a certain research context; it may also be part of an overarching debate, figuring centrally as a key concept. However, many of the debates that rely on the concept often coopt its meaning by interdefining 'levels' with terms or ideas native to these discussions (see Brooks, 2017, for further discussion). This *conceptual* embeddedness, or entanglement, is one of the main problems facing a pro-ductive analysis of the levels concept (Brooks, DiFrisco, and Wimsatt, this volume).

Against this trend of embedding the concept in coopting frameworks and debates, it is clear that disentangling the concept of levels requires that we incorporate directly the circumstances proximate to the concept's usage into our understanding of the notion of levels. In a series of papers, I have worked to explicate such an account under an umbrella framework I will call the *fragmentary account of levels* (Brooks, 2017, 2019; Brooks & Eronen, 2018). The core idea here is that 'levels' constitutes a '*fragmentary concept*,' meaning that the expressed meaning of the term in a given instance is composed of semi-independent and mutually explicative subunits of semantic content, which I call *content fragments* (CFs; see Brooks, *2020*). CFs are distinct core attributes that together contribute to the concept's expressed meaning in a given instance. Importantly, the content of each

CF can vary between instances of the concept's usage, depending on the user's context. The central CFs comprising different instances of 'levels' are summarized in table 2.1.

One consequence of this 'fragmentary' treatment is that there may be series of organizational levels outside of the purview of the canonical series portrayed in biology (atoms, molecules, cells, tissues, organs, organisms, communities, populations, ecosystems, biosphere). Instead, some levels strata will be idiosyncratic or, perhaps better, *native* to a particular type of structure *among other structures of the same type*. This emphasizes that claims involving levels will generalize to different *scopes* or breadths of applicability concerning different structural types. For example, as Alan Love (2012) points out, the four-level (primary, secondary, tertiary, quaternary) structure of proteins generalizes only partially across their overall domain (i.e., macromolecules). Specifically, the four-level layout extends only to proteins and nucleic acids but not to other major macromolecules such as lipids and sugars (Love, 2012, p. 117). This, Love further notes, does not detract from the *robustness* of such levels; rather, this is established according to the evidential weight accrued by reliably and independently generalizing these levels to their appropriate structures (see especially Love, this volume). Another "idiosyncratic" series of levels are those that compose bone structure (Weiner & Wagner, 1998). Reznikov et al. (2018) recently used electron tomography to describe in detail a surprising *twelve* distinct organizational levels that comprise the structure of bone, all the way from the tissue level down to the "atomic level" (see especially their figure 8, p. 8).

Perhaps unsurprisingly, this means that 'levels' in its tool form is never meant to exhaustively express the full descriptive extent of a given phenomenon or system of interest. Instead, *pivotal* levels are usually selected in a research community for investigative attention. The selection process by which certain levels (and not others) are endorsed as salient usually focuses on the promise these exhibit for generating insight or producing innovative research (see also Brooks, 2019). Moreover, and again, these endorsements are often enforced by the established effectiveness of methods and techniques used to investigate the phenomenon in question. To make this clearer, consider an example I have used before: levels of organization in the fly brain involved and investigated in motion detection research. In a classic review article, Martin Egelhaaf and Alexander Borst (1993) identify four pivotal levels of organization: these include the whole fly; several organs (specifically

Table 2.1
Central content fragments comprising the expressed content of 'levels' in a given instance

Content fragment	Supplied content	Example
Extension(s)	Thing(s) to which 'levels' refers	*Heavily context dependent.* Typical textbook series includes atoms-molecules-cells-tissues-organs-organisms-communities-ecosystems
Characterizing criteria	Means by which putative levels are characterized and identified	For example, *scale* (temporal or spatial), *composition*, or *explanatory relevance*
Scope	Breadth over which 'levels' is implemented	*Global* (whole world), *meso* (wider but restricted generalizability), *local* (specific system or its components)
Mode	Manner in which 'levels' is expressed	"Levels" invoked *ontologically* or *epistemically*

the eyes and muscle complex in the back); directionally selective, motion-sensitive lobula plate tangential cells (LPTCs); and synaptic connections between LPTCs. These are depicted in the top part of figure 2.1.

What should stick out here is that the levels involved in motion detection research are far from descriptively exhaustive of the fly's full neurophysiology and neuroanatomy: entire regions of the fly visual system (hence possible level demarcations) are *back-grounded* from the focus of the review. Consider two components of the visual system missing from Egelhaaf and Borst's depiction of pivotal levels in motion detection, which demonstrate the intentional incompleteness of the depiction (**bold** represents backgrounded components; *italics* represents the relation of level):

- LPTCs *compose* the **lobula plate** (interlevel).
- Retinal cells *terminate* into **large monopolar cells** (intralevel).

Both bolded components comprise important structures in the fly visual system: the lobula plate is a major part of the fly visual system, comprising the main computational center of

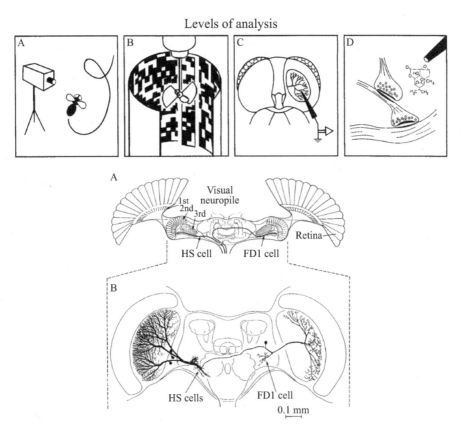

Figure 2.1
Implicit levels within level depictions. The top part of this image depicts pivotal levels selected by the authors to highlight as particularly salient or promising to then current research into insect motion detection. The bottom part of this image depicts another figure from the same paper, heavily implying intermediate levels that are not explicitly selected for their investigative attention in the review (specifically the lamina, medulla, and lobula; i.e., first, second, and third neuropils composing the visual system). Both images adapted from Egelhaaf and Borst (1993).

motion information (Hausen, 1984). In fact, its importance as a "relay station" between sensory input and motor output has earned it the label "the cockpit of the fly" (Pierantoni, 1976, p. 116), which Egelhaaf and Borst (1993) also include in the title of their review. Thus, it is noteworthy that Egelhaaf and Borst focus on LPTCs (the cellular level) to review new research on "the cockpit of the fly": instead of focusing on the behavior of the neuropil itself (arguably an organ-level entity), they choose instead to focus on the behavior of the cells that compose this neuropil. Simultaneously, though, it is obvious that they are aware of the lobula plate's importance and even that it comprises its own organizational level worthy of analysis. This is implicitly depicted in a later figure in their review, which shows the place of LPTCs within the lobula plate among the other neuropils (lamina, medulla, and lobula) that comprise the fly's visual system. I include this as the lower half of figure 2.1.

Large monopolar cells, on the other hand, putatively part of the cellular level to which LPTCs also belong, are not mentioned at all in Egelhaaf and Borst's (1993) review. This owes to the fact that they are not typical targets for experimental intervention, at least the kinds summarized in their review. Instead, these cells are important for tracing the pathways by which visual information is transferred from the eye to other optic centers, particularly the connection between the lamina and medulla. For this reason, such cells (as are many others, like bushy T cells) have received a good deal of attention in other research areas concerned with uncovering and documenting the anatomy and physiology of the fly visual system (e.g., Douglass & Strausfeld, 1996). This highlights the observation that other research communities investigating the same organism, and even same system, will for different purposes choose different structures and processes to populate a common levels framework.

Viewing the expressed content of the levels concept in terms of this fragmentary approach highlights a number of advantages as an interpretative framework. For one thing, it resolves a glaring problem facing analysis of the levels concept, namely, the *ambiguity* surrounding the concept (Brooks, 2019, *2020*). The charge of ambiguity is a serious criticism levied against the levels concept by so-called levels skeptics and appears centrally in critical arguments against the usefulness of the concept (e.g., Potochnik, this volume). For instance, in a short but excoriating commentary, Burton S. Guttman (1976) colorfully states that "if ['levels of organization'] is stated in any but the sloppiest and most general terms, it is a useless and even misleading concept" (p. 112). In other words, the levels concept appears, following Guttman, to traffic only in vague generalities but not in specifics. Using the approach I advocate, however, shows that this is not at all the case and that in fact determinate content is often easily reconstructible from different instances of usage. The key for reconstructing this determinate content, I have argued, lies in showing how the proximate factors of the concept's usage imbue 'levels' with the observational and inferential content gathered in pursuit of the problem-solving scientific tasks and activities into which users would have the concept contribute.

In fact, my approach turns the charge of ambiguity on its head, transforming the problem of accounting for the concept's putative ambiguity into a constructive *challenge* to pursue concerning *how conceptual variation figures into scientific reasoning* (Brooks, 2019). Thus, another major upshot of my approach is that it emphasizes the concept's operational potential for expressing many different ideas under the guise of one concept. That is, it reveals 'levels of organization' to be an exceptionally flexible concept customizable to a variety of empirical or scientific settings, interests, and perspectives. The price of this flexibility is an eminently tolerable initial ambiguity, which is easily reconciled with determinate content once the circumstances of the concept's usage have been incorporated into a reconstruction of its expression.

2.3   The Doctrinal Character of the Levels Concept

Another facet to the character of 'levels' encompasses a more general, overarching profile in its scientific usage. This profile, which is complementary to but ultimately distinct from the notion's use as a conceptual tool in scientific practice, indicates a more programmatic, principled character that cites and applies 'levels of organization' as a major organizing principle in biological science. That is, it reveals a *doctrinal* character to the levels concept, one that warrants its acknowledgment as a foundational element to the study of biology. This doctrinal status of 'levels,' I will show, accounts for patterns of usage outside of its application as a conceptual tool, which nonetheless constitute key uses of the term in certain types of scientific contexts. In a nutshell, 'levels of organization' is applied as a broad thesis by the scientific community regarding how the world is basically structured, while also hinting at how scientific investigation should, or at least *can*, proceed (see also Love, this volume).

When speaking of levels as a scientific doctrine, I intentionally mean to elevate the idea to something comparable to other broad, foundational ideas in science like cell theory in biology, the neuron doctrine in neuroscience, and the germ theory of disease in pathology. Considering the substantial reputations these ideas exhibit for their respective fields (and for the history of biology generally), what can be said of their profiles, particularly in their usage profiles, that can be gleaned for the levels concept? For one thing, doctrinal ideas typically exhibit familiarity within lay and scientific communities, as well as serve in a wide justificatory manner in scientific literature to support general theses about the nature of the scientific phenomena they countenance. Thus, their appearance is often a mainstay in introductory or outreach texts, a feature I return to below. Moreover, doctrine or doctrinal captures a status assigned to an idea by the scientific community that evokes codified principles or observations that in turn comprise core beliefs, inferences, and assumptions that belong within the conceptual foundation of a field of inquiry. In the following box, I collect a number of cursory (and surely incomplete) candidate postulates expressed by 'levels of organization,' seen as a doctrine (compare with Wimsatt, this volume). These items are not self-evident (although often treated as such) but rather reflect the work of earlier workers in biology that sought to articulate the idea of levels and apply it to the realm of biology (see especially Brooks, *unpublished data a*). These, then, were later taken up and incorporated by the scientific community as a kind of programmatic worldview that could do work for scientific purposes (Brooks, *unpublished data* b). Nonetheless, the veracity of levels as an overarching scientific doctrine in turn rests on certain key terms appearing in the doctrine's postulates, which appear bolded.

A Cursory Levels Doctrine

Principal Postulates

1. The natural world (or some part thereof) is separated into layers of **composition**.

2. These layers, or 'levels,' comprise classes of natural units grouped together based on some **criterion of similarity**.

3. Levels result from **generative processes** governing the formation of matter into new organizational forms.

4. Each level comprises **qualitatively different** and **projectible** generalizations.

Corollaries and Auxiliary Postulates

1. Investigation of biological phenomena comprises a **division of labor** that tracks one or more organizational levels.

2. Levels of organization preserve the **material continuity** of different forms of matter in the natural world.

This should already elicit some recognizability to the reader regarding the usage dynamics of 'levels.' Like other doctrines in biology, the idea of levels of organization is cited in many instances as inherently true or sufficiently well understood so as to not require further explication but is in turn capable of doing work by its mere mention. This, on the one hand, comports well with the initial intuitive appeal that facilitates the widespread usage of the levels concept. On the other hand, this demands further characterization as to what work, if any, the levels concept *in this doctrinal form* is doing for the scientific enterprise at large to earn such status, among other ideas.

Concerning the work performed by a doctrine, consider the germ theory of disease, particularly its historical expression by the Henle–Koch postulates (based on Evans, 1976, p. 177):

1. The agent occurs in every case of the disease in question and under circumstances that can account for the pathological changes and clinical course of the disease.

2. It occurs in no other disease as a fortuitous and nonpathogenic agent.

3. After being fully isolated from the body and repeatedly grown in pure culture, it can induce the disease anew.

These postulates express concrete, empirically based means not only in the service of positing a type of entity that becomes a basic phenomenon (i.e., a "germ," or pathogen), but it also provides guidance to experimental work toward inferring that a given agent is causally linked to clinical disease. Thus, although Robert Koch's earlier empirical work was linked to identifying agents causing specific diseases (e.g., anthrax and tuberculosis), the ability of his (later) formalized postulates to extrapolate widely to other pathogenic illnesses was parcel to the establishment and acceptance of germ theory as a widely successful scientific doctrine. That is, the Henle–Koch postulates were not only compellingly supported by existing empirical work but also formed the basis for further investigative study into pathology and epidemiology (Foster, 1970, chap. 1; King, 1983).

From this short discussion, three initial criteria for indicating an idea as a scientific doctrine can be gleaned.[3] First, the idea in question should be *basic*. That is, the idea should represent or appeal to a collection of well-established observations, facts, or inferences that are now treated as obvious or *basically true* as a point of departure for an area of inquiry. Thus, it is now common knowledge that cells form the basic unit of living systems (due to cell theory) or that pathogens (rather than humors or miasmatic airs) cause disease. It should be noted, however, that ideas that become scientific doctrines are anything but obvious and familiar in their early formulations. Rather, ideas gain doctrinal status after much deliberation, debate, improvements in techniques and instruments, revision, and further explication by other scientists, often requiring several generations of painstaking research (see especially Bechtel, 1984, for a discussion of the formation of cell theory).

Second, a scientific doctrine should also *guide scientific efforts*. That is, the idea in question should in some way provide direction or impetus to scientific activities (research oriented or otherwise). This may proceed, for example, by creating further avenues of empirical investigation, aiding to explain the phenomena, formulating standards of evidence, or characterizing what stands in need of explanation or investigation. Germ theory's postulates above serve as an example of this criterion.

Finally, scientific doctrines are *fallible* and should *openly admit limitations*, even major ones. This is a criterion of adequacy, as no examples of doctrinal knowledge are free of exceptions. Each of the examples I have mentioned here (cell theory, neuron doctrine, and germ theory of disease) admits or has admitted significant exceptions in their various iterations. In the case of germ theory, exceptions to the Henle–Koch postulates soon became clear, and irreconcilable, with germ theory's then current expression; asymptomatic carriers of disease (which violate the second postulate above) and viral and prion diseases (which violate the third postulate) posed scenarios that exposed the limitations to earlier iterations of germ theory (these are discussed in detail, and with a proposed revision to germ theory's postulates, in Evans, 1976). Other doctrinal notions have also shown limitations or failures (see, e.g., Bullock et al., 2005; Luyet, 1940). Thus, various cases within a doctrine's domain of application will predictably deviate, even extensively, from articulations of doctrine. However, whether the deviation constitutes an *exception* to the doctrine is an important issue to pose for the idea in question; too many exceptions, or too critical ones, can justify discarding a doctrine as false, useless, or antiquated.

2.3.1 Lines of Evidence for 'Levels' as a Doctrine

I turn now to providing evidence that 'levels of organization' in fact exhibits a doctrinal character. The strategy I will pursue will be to identify major patterns of usage involving 'levels' to substantiate its doctrinal profile. These patterns are indicative of the three identifying criteria I mentioned above.

Levels of organization in textbook literature

Perhaps most relatably, and directly implied by the notion of a doctrine, is that the idea in question figures into pedagogical efforts within the fields that feature the idea. This is probably the most visible contribution delivered by the levels concept and is supported by wide community consensus among textbook authors and their reviewers (see also Schneeweiß & Gropengießer, 2019). This makes sense, as established ideas that garner wide attention in textbooks become mainstays in scientific training oftentimes *due to* their doctrinal character. This exploits the fact that, in other words, *doctrinal knowledge frequently becomes textbook knowledge*. In table 2.2, I summarize the presence of 'levels' as a major organizing principle in five textbooks currently used in North American introductory college biology courses (four of these are aimed at majors, one at nonmajors).[4] Here the prominence of the concept is manifest in that it is not just point references but rather whole sections and even whole chapters that are dedicated to presenting the levels concept as warranting attention from the student of biology. Moreover, these dedicated spaces are almost always accompanied by large (i.e., full-page or centerfold) depictions of the concept. This further suggests the doctrinal status of the concept, as allotting limited space to an idea is indicative of its standing in the field.

Textbook uses of 'levels' exhibit several usage patterns salient to our purposes here. First, 'levels' is used to introduce the basic nature of biological phenomena as objects of study. In *Campbell Biology* (eleventh edition), "levels" is integral to "the study of life on Earth" and that "*as biologists*, we can" apply the levels concept to divide nature into major objects of study. Likewise, Mader and Windelspecht's *Biology* (twelfth edition) enumerates that any given level-bound object "builds upon the previous level and *is more complex*" than the things composing it. Solomon et al. (2019) and Simon (2015) also tellingly emphasize *organization* as a primary feature of living matter where "structure and function are precisely coordinated." Levels are also depicted and described as *stratified* into hierarchies, sometimes earning the synonymous moniker of the "hierarchy of life."

Another set of usage patterns includes the concept's use in initiating contact with basic themes and issues that found the study of biology. Particularly, several of the textbooks in table 2.2 apply 'levels' to introduce the themes of reductionism and emergence. Interestingly, these ideas are not applied as opponents but as independent and complementary ideas. The compatibility of the two is purchased by emphasizing the methodological importance of reductionism, while emphasizing the ontological importance of emergence.[5] For instance, Solomon et al.'s *Biology* and *Campbell Biology* each refers to reductionism as a means of "learning about a structure by studying its parts" (Solomon et al., 2019, p. 6) and a "powerful strategy" (Urry et al., 2016, p. 5) that is frequently applied in biological investigation. Emergence, on the other hand, is invoked as an idea whereby new features are exhibited by matter at different levels (Urry et al., 2016).[6]

These usage patterns exhibit directly the three criteria above for identifying a doctrinal character. For one thing, the presentations of 'levels' in these textbooks are unmistakably *basic*: the spaces dedicated to the levels concept are found at the beginning of the textbook as a means of introducing the study of biology as a whole. Moreover, the concept is clearly used to *guide scientific efforts*, being accompanied by thematic tasks to transfer not only empirical knowledge but more abstract skills to the reader, particularly those involved in posing questions in biology, how these questions are investigated, and, ultimately, how they have been solved in the past (I will return to this point in section 2.4). Thus, in transferring knowledge to novices, textbooks not only introduce facts and methods to students and beginners but also explicitly seek to teach how to think like a scientist.

The guidance afforded to scientific efforts by the levels concept is not one-sided, providing only the students with direction in navigating the broad field of biology. The authors, too, are guided by the levels concept, having chosen the concept to do the work

Table 2.2
Textbook usage of levels of organization

	Campbell Biology; Urry, Cain, Wasserman, Mivorsky, and Reece	Biology; Mader and Windelspecht	Biology; Mason, Losos, and Singer	Biology; Solomon, Martin, Martin, and Berg	Biology: The Core; Eric J. Simon
Year of publication and edition	2017 (eleventh edition)	2016 (twelfth edition)	2017 (eleventh edition)	2019 (eleventh edition)	2015 (first edition)
Containing chapter	Chapter 1: Evolution, the Themes of Biology, and Scientific Inquiry	Chapter 1: A View of Life	Chapter 1: The Science of Biology	Chapter 1: A View of Life	Chapter 1: An Introduction to the Science of Life
Contextualizing unit	Concept 1.1: The study of life reveals unifying themes	1.1 The Characteristics of Life	1.1 The Science of Life	1.3 Levels of Biological Organization	1.2 Life Can Be Studied at Many Levels
Thematic title introducing levels	New Properties Emerge at Successive Levels of Biological Organization	Life is Organized	Living Systems Show Hierarchical Organization	Levels of Biological Organization	The Levels of Biological Organization
Textual introduction	"The study of life on Earth extends from the microscopic scale of the molecules and cells that make up organisms to the global scale of the entire living planet. As biologists, we can divide this enormous range into different levels of biological organization."	"You should recognize from Figure 1.2 that each level of biological organization builds upon the previous level and is more complex. Moving up the hierarchy, each level acquires new emergent properties, or new, unique characteristics, that are determined by the interactions between the individual parts."	"Life forms a hierarchy of organization from atoms to complex multicellular organisms. Atoms are joined together to form molecules, which are assembled into more complex structures such as organelles. These in turn form subsystems that provide different functions. Cells can be organized into tissues, then into organs and organ systems such as the goose's nervous system pictured. This organization then extends beyond individual organisms to populations, communities, ecosystems, and finally the biosphere."	"Whether we study a single organism or the world of life as a whole, we can identify a hierarchy of biological organization (FIG. 1–6). At every level, structure and function are precisely coordinated."	"The study of life encompasses a very broad range of scales, from the microscopic world of cells to the vast scope of Earth's ecosystems. This figure summarizes some of the levels at which biologists study life on Earth, starting at the upper end of the scale."
Page in text	p. 4	p. 2	p. 3	p. 6	p. 4
Accompanying depiction—prominence in text	Yes—centerfold	Yes—full page	Yes—centerfold	Yes—full page	Yes—centerfold
Thematic task or lesson	Key Concepts: 1.1 The study of life reveals unifying themes.	Learning Outcomes: 1. Distinguish among the levels of biological organization.	Learning Outcomes: 3. Characterize the hierarchical organization of living systems.	Learning Objective: Construct a hierarchy of biological organization, including levels characteristic of individual organisms and levels characteristic of ecological systems.	Core Question: Which level of life's organization is the smallest one that can be considered alive?

it exhibits in these sources. This holds not only for the thematic lessons and tasks that educators would have of their students but also for how they structure the textbook themselves. Witness one prominent textbook author explain her motivations in writing her textbook:

We are all dedicated to the desire that students develop a particular view of the world—*a biological view*. When I wrote the first edition of Biology, it seemed to me that a *thorough grounding in biological principles* would lead to an appreciation of the structure and function of individual organisms, how they evolved, and how they interact in the biosphere. *This caused me to use the levels of biological organization as my guide*—thus, this edition, like the previous editions, begins with chemistry and ends with the biosphere. (Mader, 2010, preface, emphasis added)

Finally, depictions of 'levels' in textbooks *openly admit limitations*. Particularly, the meaning of 'levels' in these contexts is not just ambiguous but oftentimes *vague*. In any one of the textual instances of 'levels' given in table 2.2, there are multiple possible content fragment combinations and weightings that can be reconstructed (see section 2.2), but little or no guidance is given as to how exactly the concept should be understood. This leaves it quite open concerning how the concept should precisely be conceptualized. Balancing this, however, is the pedagogical spirit in which the concept is conveyed (the usage of concepts in textbooks, including the levels concept, is, after all, oriented toward novices). Textbooks, in effect, invite the reader to try out the concept in their own reasoning, first by introducing topical themes and tasks (such as reductive methodologies), then by citing examples where the levels concept is salient to understanding a particular lesson. With this in mind, the limitation of the levels concept in textbook literature is compatible with a more lenient reading; that is, the concept is capable of expressing many things simultaneously (Brooks, 2019).

Scientific statements of 'levels' as doctrine

A second collection of usages involves the appearance of 'levels' in professional scientific literature. These patterns reveal that 'levels,' invoked as a doctrine, is an attractive device not only for students learning the ropes of biological research but also for practicing scientists, who invest in the term's usage beyond pedagogical contexts and into research-organizing agendas. To illustrate this, I gather below a gallery of statements from prominent sources that directly invoke the principled, *doctrinal* character of the levels concept. This is particularly noticeable in the concept's presentation, where the authors apply the levels concept with an imperative tone and frequently exhort their audiences to acknowledge, or begin work using, the levels concept. A second feature concerns the concrete context of each passage: each instance has been chosen to represent strategic junctures in the scientific literature (in contrast to research and review articles) that have to do with planning or executing research by mobilizing scientific opinion surrounding one or another agenda concerning the direction research in biology should proceed.

Consider first a passage from the National Science Foundation's (NSF's) website for its "Big Questions" initiative calling for large-scale grant applications, which states directly the basic importance of 'levels' for both studying and conceptualizing living phenomena:

• "*Life on our planet is arranged in levels of organization* ranging from the molecular scale through to the biosphere. There exists a *remarkable amount of complexity* in the interactions within and between these levels of organization and across scales of time and space" (National Science Foundation, (n.d.), "Big Questions—Understanding the Rules of Life" website, accessed January 3, 2019, emphasis added).

This passage hearkens back to the presentation of 'levels' in introductory textbooks in that the existence of levels in nature is posited as a primitive fact, only this time, the concept is not directed toward novices but rather researchers interested in NSF funding. This presentation is also common in prominent monographs authored by advanced researchers who engage in theorizing efforts in biology. Theodosius Dobzhansky and Ernst Mayr, for example, both frequently referred to the existence of levels in their theoretical works, two passages of which emphasize quite nicely the doctrinal character the concept had already acquired:

• *One of the outstanding characteristics of living matter is complexity of organisation.* There is a *hierarchy of complexity* that runs from atoms and molecules, through cells, tissues, individual organisms, populations, communities and ecosystems, *to the whole of life on Earth. Different biological disciplines concentrate on the study of one or several levels of hierarchical organisation.* (Ayala & Dobzhansky, 1974, p. vii, emphasis added)

• The formation of constitutive hierarchies is *one of the most basic properties of living organisms.* At each level there are different problems, different questions to be asked, and different theories to be formulated. Each of these levels has given rise to a separate branch of biology: molecules to molecular biology, cells to cytology, tissues to histology, and so forth, up to biogeography and the study of ecosystems. Traditionally, the recognition of these hierarchical levels has been one of the ways of subdividing biology into fields. *To which particular level an investigator will turn, depends on [their] interests.* (Mayr, 1982, p. 65, emphasis added)

Like the NSF's initiative, these passages emphasize the basic importance of levels but go further by emphasizing that biological inquiry *follows* organizational levels in nature. This highlights the work being done by the levels concept in that different putative levels will house different experimental or theoretical investigations to be conducted.[7]

Doctrinal invocations of the levels concept are not relegated to point sources that reflect on the basic nature of biological phenomena; they are also found in community-oriented literature concerning the state and future of biological research, seen as a whole. Two especially telling contexts here include "New Biology" and "Grand Challenges" literatures. New Biology items identify or advertise a general research agenda characterizing how biology should develop in order to meet future research and practical demands, often laying out guidelines for what kind of research should be funded or what conceptual issues are at stake by refining the way that biology as a whole is conducting its investigations. Grand Challenges literature, similarly, appeals to "multi-sector, multi-disciplinary research" in the life sciences, as well as political and funding apparatuses (Efstathiou 2016, 48), to galvanize research agendas on large-scale problems facing individual or constellations of scientific fields, society, or human well-being. Both kinds of literatures prominently apply the levels concept to convey their message to a broad audience. For example, in Carl Woese's (2004) aptly titled article "A New Biology for a New Century," Woese invokes the levels concept as a central, assumptive feature of the biological realm, one that is instrumental to reconceptualizing biological phenomena *in general*:

• Approaching evolution of the cell with a *clean slate* requires establishing a perspective, a framework, and ground rules—not simply for this one problem *but for biology in general*. Let us … try to imagine a biology released from the intellectual shackles of mechanism, reductionism, and determinism. …

What, if anything, do these examples of presumed evolutionary critical points have in common? How might they have come about? *All of them, of course, involve the emergence of higher levels of organization*, which bring with them qualitatively new properties, properties that are describable in reductionist terms but that are neither predictable nor fully explainable therein. (Woese, 2004, pp. 179–180)

Interesting here is that although Woese rhetorically speaks of a "clean slate" with "a [new] perspective, framework and ground rules," he nonetheless ("of course") refers to levels as if the idea were immune to the sweeping revisions he envisions for biology's conceptual inventory and empirical agenda. In a similar manner, a pamphlet released by the National Research Council in 2009, entitled *A New Biology for the 21st Century: Ensuring the United States Leads the Coming Biology Revolution*, centrally features the levels concept in the authors' rendition of how their vision for a New Biology should unfold.[8] With a nod to Woese's article, the committee announces that

• the time has come for *biology* to enter the nonlinear world. … The practical ability to achieve Woese's vision is now beginning to emerge. Biologists are increasingly able to integrate information across many organisms, *from multiple levels of organization* (such as cells, organisms, and populations) and about entire systems (such as all the genes in a genome or all the cells in a body) to gain a new integrated understanding that *incorporates more and more of the complexity that characterizes biological systems*. (National Research Council, 2009, p. 41, emphasis added)

This passage is accompanied by a prominent depiction of levels, one that recalls the depiction of levels in textbooks. Once again, this investment in space and effort in adding an image to the pamphlet shows nicely that the concept of levels is far more than merely a pedagogical tool for novices; it appears in sources textually and in images to sway audiences who are invested in the future of science: policy makers, citizens, other scientists, and other stakeholders are influenced by the idea of levels, which, seen as a doctrine, is capable of guiding scientific efforts at its mere mention.

2.4   Understanding the Contributions of 'Levels'

In the preceding sections, I compiled an epistemic characterization of the notion of levels of organization in biology, focusing on the dynamics it exhibits in the hands of scientific users. I turn now to discussing the upshots of this characterization and where this leaves us in making progress on understanding the concept. In the introduction, I observed that a significant tension surrounds the levels concept: this tension stems, on the one hand, from the enthusiastic and ubiquitous appearance of the concept and, on the other, from the obscurity concerning what general features of the concept mark genuine contributions to science. I then proposed relieving this tension by answering three general questions concerning the character of 'levels of organization.' These questions concerned how and where the concept is used and what roles in biological thought it exhibits. Having now explored different profiles of the levels concept in scientific usage, what can be said about the concept's character such that this tension is resolved, or at least reduced, and the concept's productive qualities are better revealed?

2.4.1   How 'Levels of Organization' Is Used in the Biological Sciences

The characterization I offered countenanced two usage profiles attached to the concept of levels. Each of these profiles reveals distinct ways that the concept figures into scientific efforts. One profile encompassed the concept's use as a tool, which aids scientists in navigating the layout of biological systems or focusing their descriptions of the same. Here the expressed content of 'levels' was revealed as highly sensitive to the circumstances of its usage, becoming operationalized in part by the techniques, interests, and empirical setting in which it is applied (this is developed in detail in Brooks, *2020*). This "fragmentary" character attached to the concept's use as a tool reveals 'levels' to be a useful descriptor of systems or phenomena of interest by accumulating increasingly determinate, albeit context-sensitive, content that produces robust and reliable scientific claims and generalizations. The second usage profile encompassed the usage of the levels concept as a scientific doctrine, evoking a number of theses or postulates that have garnered wide acceptance as ideas that capture scientific thinking about biological phenomena. Although rarely explicated beyond a collection of principled statements when invoked as a doctrine, the ideas constituent to levels of organization are still capable of doing scientific work, particularly in guiding efforts in educating and training lay and novice audiences in scientific literacy but also figuring into appeals for large-scale research challenges facing biology as a whole.

2.4.2   Where the Concept Is Applied

Although the levels concept exhibits a staggering ubiquity and heterogeneity across the scientific enterprise, there appears to be some order to the way its characteristic uses are applied in scientific literature. Specifically, the characterization of each usage profile drew upon on key *kinds of texts* to elaborate on the different aspects of the character of 'levels': usage of the levels concept as a tool tends to appear more frequently in texts that provide detailed descriptions of particular systems or structures, such as research and review literature, while doctrinal uses typically appear in texts oriented toward larger, more general audiences in preparation for broader messages, like textbooks and Grand Challenges literature.[9] Structural considerations of the usage context, such as the intended audience or user community, with their attendant readership and intellectual orientations, provide critical interpretative information concerning how we should understand 'levels.' This information is often specified by the type of scientific text in which the concept is applied.

Both of these aspects to the character of 'levels of organization' (how and where it is used) already contribute to relieving some of the tension surrounding the concept. For one thing, they provide crucial information regarding how to specify, and thereby effectively contextualize, what precisely is contributed by the concept to scientific efforts. For instance, extracting structural features from the textual source in which the concept appears, such as the intended audience, allows us to more fully reconstruct details of the concept's circumstances of usage. In particular, capturing or reconstructing the task(s) that scientific users of 'levels' are working toward is central to understanding what exactly, according to the user, *should* be delivered by using the levels concept (see also Brooks, 2019). This, in turn, enables us to better articulate, and then assess, criteria concerning the concept's usefulness (or lack thereof) for that task (see especially Brooks & Eronen, 2018, §4). In this way, an analysis of 'levels' is not terminated by concluding that what is expressed by the levels concept is *merely* context sensitive, *tout court*. Instead, what emerges is an active interpretative framework that attends to, and recovers, the use (and usefulness) of 'levels' for different breadths of scientific activity.

2.4.3   The Roles of Levels in Biological Thought

The final question I posed of the character of 'levels of organization' concerns its overarching role in biological thought. Here I must offer a preliminary answer, as a fuller, more systematic analysis is far too ambitious for this chapter. In short, I maintain that 'levels of organization' gains currency in biological thought at least by figuring, where it is used, deeply into the way that biologists conceptualize, articulate, and interact with the scientific problems they engage in the course of their investigative efforts. That is, the foregoing characterization reveals an *erotetic core* to the levels concept, meaning that the concept appears as a potent source for posing, engaging, and solving biological problems.

This idea has already been alluded to in the foregoing characterization. Recall, for instance, Ernst Mayr's statement concerning levels in conceptualizing biological phenomena. There he qualified the importance of the concept with respect to its erotetic capacities, saying, "*At each level there are different problems, different questions to be asked, and different theories to be formulated. ... To which particular level an investigator will turn, depends on [their] interests*" (Mayr, 1982, p. 65, emphasis added). Although somewhat simplified, this representation portrays levels as sources for distinct problems, engendered by the qualitatively different structures and phenomena housed at each level. Of course, grant-attracting scientific problems are rarely contained within only one particular level and can be analyzed from different perspectives housed at or stemming from multiple levels. However, the idea here is that focusing attention on different levels will frequently require adopting new investigative strategies or terminologies, and also novel sets of interests or research purposes oriented by different professional training. Each of these comprises different elements for articulating (distinct) questions, that is, problems, of biological phenomena.

Erotetic capacities of the levels concept are especially prominent in its historical underpinnings, figuring into both its injection into and consolidation within the biological lexicon.[10] Indeed, one of the primary motivations behind early references to the levels concept was that it aided in articulating what makes biological questions biological, rather than physical or chemical. This is particularly visible in the context of the level concept's genesis within the mechanist-vitalist dispute of the early twentieth century. Here the concept was applied to mediate between the austere, reductionistic claims of the mechanists that life was nothing over and above the physicochemical and the extravagant vitalist claims that life required postulating nonmaterial vital forces to account for its phenomena. In the hands of organicist biologists such as Joseph Woodger and Joseph Needham, early advocacy and development of 'levels' focused on its ability to *make autonomous* the types of phenomena faced by biologists in their investigative efforts, thus forming a basis for the autonomy of biology as a whole. Needham (1943), reflecting on this point, had this to say:

This deadlock [between mechanism and vitalism], which in various forms had run through the whole history of human thought, *was overcome when it was realised that every level of organisation has its own regularities and principles*, not reducible to those appropriate to lower levels of organisation, nor applicable to higher levels, but at the same time in no way inscrutable or immune from scientific analysis and comprehension. (p. 18, emphasis added)

Biological phenomena, the organicists further argued, are segregated from one another and ultimately made distinct from physical and chemical ones due to their *organizational* features (see Nicholson & Gawne, 2015, especially §4.2). These features require distinct methods, techniques, and, most important, *questions* to adequately attend to explaining and accounting for them. The apparently hierarchical trend running through different biological phenomena, in the form of part–whole relationships, inspired them to postulate systematic differences *and* connections between these phenomena (such as cells, tissues, and organisms), and thus 'levels of organization' entered the scientific lexicon (see especially Woodger, 1929, pp. 292–293, quoted in Eronen & Brooks, 2018).

These themes are visible in both usage profiles characterized above. For instance, in section 2.2, I mentioned that the approximative descriptions of biological phenomena afforded by the levels concept, seen as a tool, aid directly in different kinds of problem scenarios: ill-structured or ill-defined problem scenarios gain a foothold into further research by using 'levels' to depict or describe a novel system of interest, while mature problem scenarios are aided by 'levels' to robustly represent what is known about a given system.

Further erotetic motifs of the levels concept are visible in its doctrinal profile. One particularly clear use of 'levels' in this regard is in pedagogical contexts of usage, where the concept is applied with the goal in mind of acquainting students with how to "*apply scientific skills to solving problems*" (Urry et al., 2016, p. xvii, emphasis added). Likewise, Grand Challenges literature and New Biology narratives invoke the problem-orientedness of 'levels.' Grand Challenges, as the term implies, deal directly with larger-scale scientific problems, but what about exhortations involved in calls for a New Biology? In section 2.3, I provided instances where the levels concept has survived into the new millennium as a basic notion in biological science. However, here too 'levels' figures centrally into introducing or characterizing questions and challenges (or prospective solutions) that various authors see facing biological science in the twenty-first century. For instance, in their edited volume *A Biology for a New Century*, editors John Kress and Gary Barret are explicit in the role of levels in conceptualizing the work to be done in future biological research. First, they write,

In the new century we have a better understanding of biology at both the lower (i.e., cellular and molecular) and the higher (i.e., ecosystem, landscape, and global) levels of organization. We are poised to reassemble into a synthetic biological construct the elemental parts that have been carefully dissected over the last fifty years by the molecular and cell biologists at one end of the biological hierarchy as well as the ecosystem and landscape ecologists at the other end. *Our next step is to develop a transdisciplinary science that integrates concepts, theories, and approaches across all levels of organization.* (Kress & Barret, 2001, Preface, emphasis added)

The approach to solving problems of twenty-first-century biology, following Kress and Barret, will follow an *integrative* strategy. Although the precise nature of "integration" is left a bit hanging, the spirit here is an antireductionistic attitude toward conceptualizing solutions that emphasizes constructing links between research areas perched at various levels of organization. The doctrinal character of 'levels' here is undeniable and figures overtly into the editors' erotetic thinking, as they explain in their choices for selecting the contributing authors to their volume. Witness:

The authors were selected to maintain (a) a balance among the levels of biological organization; (b) an emphasis on integrative processes among levels of organization; and (c) an organization that highlighted transdisciplinary accomplishments in all fields of the life sciences rather than narrow intradisciplinary work. These chapters thus represent a breadth of insight and experience across a wide range of the biological sciences. (Kress & Barret, 2001, Preface)

Thus, a collection of erotetic themes appears to surround the levels concept in its key contexts of application. In addition to its contributions toward arguing for the autonomy of biology and its phenomena from the physical sciences (by, for example, substantiating the presence of different kinds of questions due to organizational features of biological phenomena), the concept is also germane to the problem-solving tasks around which scientific texts are constructed and contributes materially to scientific reasoning by shaping the problem space shared by a community of researchers.[11]

2.5   Conclusion

In this chapter, I provided a characterization of the levels concept and sketched one overarching role it exhibits in biological thought. This characterization focused on the concept's usage in the hands of scientific users and revealed two distinct but salient profiles that together contribute to a coherent depiction of the character of 'levels of organization' in biology. The overarching role I identified of the concept concerns its ability to interact with scientific problems and their solutions. Here I claimed that 'levels of organization' finds currency throughout different contexts of usage by figuring, where it is used, into how scientists conceptualize and articulate the questions that they engage in during the course of their investigations.

In looking forward, a number of general upshots for 'levels' come to mind. For one thing, I believe the characterization offered in this chapter offers useful points of departure for considering the *nature* of levels (see especially Wimsatt, 1994/2007a). For example, one unarticulated motivation behind my characterization concerns a dialectical element between the two different usage profiles. Recovering this would entail demonstrating that users of 'levels' show interest in, reliance on, or a commitment to the ontological promise exhibited by one or another postulate expressed by the levels doctrine. Similarly, tool-centered uses of the concept can be seen as revealing local details of the doctrinal predictions of the concept, tailored to that instance of usage.

Moreover, I believe there is much here to be marshaled for advertising the term's *operational potential* in scientific usage (Brooks, *2020*). Particularly, the 'fragmentary' approach laid out here and elsewhere for actively interpreting the expressed content of levels in different instances should allow for a more nuanced and sophisticated articulation of the *openness* of 'levels' as a primary explanans for the term's longevity in scientific parlance (cf. Potochnik, this volume). The level concept's adaptability to myriad empirical settings and investigative interests, coupled with its overarching promises as a scientific doctrine, surely reinforce the concept's fidelity to substantiating the scientific image of the world.

Acknowledgments

This chapter benefited from multiple exchanges with James DiFrisco, as well as helpful feedback from Markus I. Eronen and Lena Kästner. Special thanks are also reserved for all the participants of the workshop, who collectively created an outstanding atmosphere of discussion.

Notes

1. I use single quotes throughout this chapter for technical terms whose meaning and significance are in question.

2. The characterization of 'levels' I develop here is meant to directly complement, and hopefully in some way enhance, William C. Wimsatt's corpus of work on the nature of levels (1976, 1981/2007b, 1994/2007a, this volume) by "getting at" levels from the opposite end from which Wimsatt begins: specifically, whereas Wimsatt recovers significant epistemic consequences of the idea of levels from a broadly ontological characterization, I seek to begin with a broadly epistemic characterization in light of its behavior as a concept and then recover ontological consequences from this along the way. I refer the reader especially to Jim Griesemer's discussion of the epistemic mirror image of Wimsatt's ontology (Griesemer, this volume).

3. Two caveats should be clear as I proceed: (1) I certainly do not mean for these criteria to be exhaustive, but rather suggestive, perhaps representative, and in any case spur further thinking on what we take to be doctrinal contributions to science. (2) Furthermore, although other criteria come to mind for better articulating what I take to be doctrinal knowledge (such as *unifying diverse phenomena*), space considerations compel me to focus only on those I disclose here. I nonetheless take them to be sufficient for establishing my proposal.

4. These textbooks were selected from a list compiled by an online wiki (https://wiki.ezvid.com/best-biology -textbooks; accessed October 17, 2019), and as such I can make no authoritative claims regarding their representativeness. As it turns out, locating statistics concerning which textbooks, for example, are most widely used in university classes is elusive.

5. It is clear upon investigation that these textbooks' authors clearly mean "emergence" in the innocuous (non-spooky, "weak") variety. Rather than speaking of, for example, in-principle irreducibility, they merely refer to the appearance of new features at each new level, which are not exhibited by the level below it.

6. Interestingly, these same sources further characterize reductionism as inadequate in isolation for answering all biological questions and introduce systems biology as an alternative means of study in biology. Systems biology in these passages emphasizes *integrative* approaches toward understanding biological systems as a complementary method to reduction (e.g., Solomon et al., 2019, p. 21; Urry et al., 2016, p. 6).

7. Although a strict correspondence between scientific *disciplines* with nature is now a troubling attribution, there is certainly something to be said of how questions in biology influence or are influenced by the levels concept.

8. The authors seem to directly implicate "levels" as a "core organizational principle" of biology, saying in a section on "The Fundamental Unity of Biology," "The great potential of the life sciences to contribute simultaneously to so many areas of societal need rests on the fact that biology, like physics and chemistry, relies on a small number of *core organizational principles*" (National Research Council, 2009, p. 40, emphasis added). They continue, "The *reality* of these *core commonalities*, conserved throughout evolution—[e.g.]…*that cells can be organized into complex, multicellular organisms*…means that any knowledge gained about one genome, cell, organism, community, or ecosystem *is useful in understanding many others*" (National Research Council, 2009, p. 40, emphasis added).

9. Of course, this division is not airtight, as individual research or review articles may also appeal to the doctrinal character of levels, and general texts like textbooks may focus on more detailed, tool-like uses of the concept.

10. These are discussed in greater detail in my emerging historical analysis of the concept, especially Brooks (*unpublished data a* and *b*).

11. This last erotetic currency of 'levels' actually recovers a primary rationale for applying the levels concept in scientific contexts (Brooks, 2019). Specifically, the *epistemic goal* (*sensu* Brigandt, 2010) of *structuring scientific problems* guides and motivates the level concept's usage in scientific contexts by serving to connect concrete scientific tasks concerning problems to their execution by investigative activities.

References

Ayala, F. J., & Dobzhansky, T. (1974). *Studies in the philosophy of biology: Reduction and related problems.* London, UK: MacMillan.

Bechtel, W. (1984). The evolution of our understanding of the cell: A study in the dynamics of scientific progress. *Studies in History and Philosophy of Science, 15*(4), 309–356.

Brooks, D. S. (2017). In defense of levels: Layer cakes and guilt by association. *Biological Theory, 12*(3), 142–156.

Brooks, D. S. (2019). A new look at "levels of organization" in biology. *Erkenntnis.*

Brooks, D. S. (2020). *Levels of organization: Lessons from a fragmentary concept.* Unpublished manuscript.

Brooks, D. S. *unpublished data a*. The organicist roots of the levels concept, 1910–1937.

Brooks, D. S. *unpublished data b*. Consolidation despite fragmentation: Initial dispersal of the levels concept in biology in the mid-20th century, 1937–1961.

Brooks, D. S., & Eronen, M. I. (2018). The significance of "levels of organization" for scientific research: A heuristic approach. *Studies in History and Philosophy of Biological and Biomedical Sciences, 68*, 34–41.

Brigandt, I. (2010). The epistemic goal of a concept: Accounting for the rationality of semantic change and variation. *Synthese, 177*(1), 19–40.

Bullock, T. H., Bennett, M. V. L., Johnston, D., Josephson, R., Marder, E., & Fields, R. D. (2005). The neuron doctrine, redux. *Science, 310*(791), 791–793.

Douglass, J. K., & Strausfeld, N. (1996). Visual motion-detection circuits in flies: Parallel direction- and non-direction-sensitive pathways between the medulla and lobula plate. *Journal of Neuroscience, 16*(15), 4551–4562.

Efstathiou, S. (2016). Is it possible to give scientific solutions to Grand Challenges? On the idea of grand challenges for life science research. *Studies in History and Philosophy of Biological and Biomedical Sciences, 56*, 48–61.

Egelhaaf, M., & Borst, A. (1993). A look into the cockpit of the fly: Visual orientation, algorithms, and identified neurons. *Journal of Neuroscience, 13*(11), 4563–4574.

Eronen, M. I. (2015). Levels of organization: A deflationary account. *Biology and Philosophy, 30*(1), 39–58.

Eronen, M. I., & Brooks, D. S. (2018). Levels of organization in biology. In E. N. Zalta (Ed.), *The Stanford encyclopedia of philosophy*. Retrieved from https://plato.stanford.edu/archives/spr2018/entries/levels-org-biology/

Evans, A. S. (1976). Causation and disease: The Henle-Koch postulates revisited. *The Yale Journal of Biology and Medicine, 49*, 175–195.

Foster, W. D. (1970). *A history of medical bacteriology and immunology*. London, UK: William Heinemann Medical Books Ltd.

Guttman, B. S. (1976). Is "levels of organization" a useful concept? *BioScience, 26*(2): 112–113.

Hausen, K. (1984). The lobula-complex of the fly: Structure, function and significance in visual behaviour. In M. A. Ali (Ed.), *Photoreception and vision in invertebrates* (pp. 523–559). New York, NY: Plenum Press.

King, L. S. (1983). Germ theory and its influence. *Journal of the American Medical Association, 249*(6), 794–798.

Kress, W. J., & Barret, G. W. (2001). *A new century of biology*. Washington, DC: Smithsonian Institution Press.

Love, A. C. (2012). Hierarchy, causation and explanation: Ubiquity, locality and pluralism. *Interface Focus, 2*(1), 115–125.

Luyet, B. J. (1940). The case against the cell theory. *Science, 91*(2359), 252–255.

Mader, S. S. (2010). *Biology* (10th ed.). Boston, MA: McGraw Higher Education.

Mader, S. S., & Windelspecht, M. (2016). *Biology*. New York, NY: McGraw Hill.

Mason, K.A., Loso, J. B., & Singer, S. R. (2017). *Biology*. New York, NY: McGraw Hill.

Mayr, E. (1982). *The growth of biological thought: Diversity, evolution, and inheritance*. Cambridge, MA: Belknap Press.

National Research Council. (2009). *A new biology for the 21st century: Ensuring the United States leads the coming biology revolution*. Washington, DC: National Academies Press.

National Science Foundation. (n.d.). Big ideas: Understanding the rules of life. Retrieved October 17, 2019, from https://www.nsf.gov/news/special_reports/big_ideas/life.jsp

Needham, J. (1943). *Time the refreshing river (Essays and addresses, 1932–1942)*. London: G. Allen & Unwin Ltd.

Nicholson, D. J., & Gawne, R. (2015). Neither logical empiricism nor vitalism, but organicism: What the philosophy of biology was. *History and Philosophy of the Life Sciences, 37*(4), 345–381.

Pierantoni, R. (1976). A look into the cock-pit of the fly: The architecture of the lobular plate. *Cell and Tissue Research, 171*, 101–122.

Reznikov, N., Bilton, M., Lari, L., Stevens, M. M., & Kröger, R. (2018). Fractal-like hierarchical organization of bone begins at the nanoscale. *Science, 360*(6388), 1–10.

Schneeweiß, N., & Gropengießer, H. (2019). Organising levels of organisation for biology education: A systematic review of literature. *Education Sciences, 9*(207).

Simon, E. J. (2015). *Biology: The core*. New York, NY: Pearson

Solomon, E. P., Martin, C. E., Martin, D. W., Berg, L. R. (2019). *Biology*. Boston, MA: Cengage.

Urry, L.A., Cain, M.L., Wasserman, S.A. Minorsky, P.V., Reece, J.B. (2016). *Biology*. New York, NY: Pearson

Waters, C. K. (2008). Beyond theoretical reduction and layer-cake antireductionism: How DNA retooled genetics and transformed biological practice. In M. Ruse (Ed.), *The Oxford handbook of philosophy of biology* (pp. 238–261). Oxford: Oxford University Press.

Weiner, S., & Wagner, H. D. (1998). The material bone: Structure-mechanical function relations. *Annual Review of Materials Research, 28*, 271–298.

Wimsatt, W. C. (1976). Reductionism, levels of organization, and the mind-body problem. In G. Globus, G. Maxwell, and I. Savodnik (Eds.), *Consciousness and the brain* (pp. 205–267). New York, NY: Plenum.

Wimsatt, W. C. (2007a). The ontology of complex systems: Levels of organization, perspectives, and causal thickets. In W. C. Wimsatt, *Re-engineering philosophy for limited beings: Piecewise approximations to reality* (pp. 193–240). Cambridge, MA: Harvard University Press. (Original work published 1994)

Wimsatt, W. C. (2007b). Robustness, reliability, and overdetermination. In W. C. Wimsatt, *Re-engineering philosophy for limited beings: Piecewise approximations to reality* (pp. 43–74). Cambridge, MA: Harvard University Press. (Original work published 1981)

Woese, C. R. (2004). A new biology for a new century. *Microbiology and Molecular Biology Reviews, 68*(2), 173–186.

3 Our World Isn't Organized into Levels

Angela Potochnik

Overview

Levels of organization and their use in science have received increased philosophical attention of late, including challenges to the well-foundedness or widespread usefulness of levels concepts. One kind of response to these challenges has been to advocate a more precise and specific levels concept that is coherent and useful. Another kind of response has been to argue that the levels concept should be taken as a heuristic, to embrace its ambiguity and the possibility of exceptions as acceptable consequences of its usefulness. In this chapter, I suggest that each of these strategies faces its own attendant downsides and that pursuit of both strategies (by different thinkers) compounds the difficulties. That both kinds of approaches are advocated is, I think, illustrative of the problems plaguing the concept of levels of organization. I end by suggesting that the invocation of levels may mislead scientific and philosophical investigations more than it informs them, so our use of the levels concept should be updated accordingly.

Levels of organization have featured prominently in a number of philosophical debates. They have also been invoked in a variety of scientific contexts, especially in biology, where textbooks tend to be organized with levels of organization as the organizing principle. In recent years, increased philosophical attention has been devoted to concepts of levels of organization and how these have been deployed in science and philosophy. Some of this work has been critical. Rueger and McGivern (2010) challenge whether the traditional conception of levels is apt in physics and propose an alternative. Potochnik and McGill (2012) aim to identify and challenge broad invocations of levels of organization that occur in both philosophy and biology. In our view, such invocations tended to anticipate a number of broad implications from levels that by and large fail to materialize and to rest on intuitions that, although widely shared, are difficult to justify philosophically. Eronen (2013, 2015) points out limitations to mechanistic conceptions of levels of organization in particular, which many see as the most promising levels concept currently on offer.

Some of these same discussions, as well as other work, have used the ambiguity of and associated difficulties with the broad, intuitive levels concept as an opportunity to refine the concept and how it is employed. Craver (2007) advocates a mechanistic conception of levels and distinguishes this from the more traditional conception of levels of organization based on compositional relationships of parts and wholes. Rueger and McGivern (2010) propose scale as a replacement for levels in physics. Potochnik and McGill (2012) also suggest the use of scale to articulate "quasi-levels" according to the bounds within which some causes tend to dominate. DiFrisco (2017) argues that temporal scale alone holds promise for achieving the goals of the invocation of levels. These are all attempts to disambiguate the levels concept or motivate a suitable substitute, giving the concept a precise, specific interpretation based on the relation used to articulate levels—composition, mechanism, spatial scale, or temporal scale. Some also explicitly attend to the relation used to group entities or processes onto the same level, which—as Eronen (2013) and DiFrisco (2017) make explicit—is a distinct question. I'll refer to this strategy as making the levels concept precise.

A different strategy has been to embrace the broad, intuitive levels concept, along with its limitations and difficulties. Many of our scientific concepts are imprecise and ambiguous. This has been shown to be so for the concept of the gene, for example, which might refer (at least) to a stretch of DNA or to the genetic basis for some identified physical or behavioral trait. (For a recent discussion of the ambiguous and changing meanings of the gene concept, see Rheinberger & Müller-Wille, 2017.)

Moreover, many concepts are usefully applied in some scientific contexts but not others. It's been crucially important to identify the gene variants related to some genetic diseases, but for other diseases, we have not identified identifiable genetic causes, and there's no reason to expect we will. Finally, imprecise and ambiguous concepts, like that of the gene, can be fruitfully applied in different ways in different scientific contexts. Perhaps the levels concept is like the gene concept in these regards.

This seems to be the tack taken by Wimsatt's (1972, 2007) classic and deeply influential treatment of levels of organization. He emphasizes that levels aren't found across nature and that, although some phenomena display a remarkable convergence of different features of levels, there may still be exceptions. And then, Brooks (2017) encourages philosophers and scientists to attend to the ways in which the levels concept varies in different applications and to accept that variability rather than see it as a weakness of the concept. Brooks and Eronen (2018) advocate the heuristic use of the levels concept in structuring scientific problems. In his chapter in this volume, Brooks argues for a "doctrinal" use of levels as structuring education and inquiry in the field of biology. This view posits the broad significance of the levels concept at least in biology, while also tolerating variability in application and in precise meaning. These treatments of levels of organization refuse to make the levels concept precise, as they see its multiple relevances and breadth of applicability as an important strength of that concept. They instead accommodate criticisms of the levels concept by granting variability in its use and limitations to its usefulness. I'll refer to this strategy as endorsing the heuristic use of the levels concept.

Here I explore these two strategies to clarifying the levels concept. I suggest that while each of these strategies is initially plausible, each also encounters its own significant difficulties. A specific, precise levels concept turns out to be remarkably difficult to characterize. One or another precise concept of levels may be useful in certain constrained scientific contexts, but none should be expected to apply generally or to preclude the usefulness of other related or opposed concepts. And then, heuristic uses of the levels concept may not prove to be of much scientific or philosophical value, and they also risk being taken too literally and may occlude the recognition of other, more productive heuristics. Moreover, the pursuit in philosophical and scientific literatures of both strategies—making levels precise and treating levels as a heuristic—compounds the difficulties facing each. Indeed, that both kinds of approaches are advocated is, I think, illustrative of the problems plaguing the concept of levels of organization. In section 3.1, I evaluate the strategy of making the levels concept more precise; in section 3.2, I evaluate the strategy of endorsing the heuristic use of the levels concept. In section 3.3, I suggest that the invocation of levels can mislead scientific and philosophical investigations just as much as it informs them. The limitations and ambiguity of levels concepts should be recognized, organization into levels should not be presumed, and scientists and philosophers should actively try to cultivate alternative heuristics for the organization of our world.

3.1   In Pursuit of a More Precise Levels Concept

Philosophical and scientific discussions of levels of organization have variously considered a number of different relations as relevant to levels. Classically, in what has been called the "layer-cake" view of levels developed by Oppenheim and Putnam (1958), compositional relationships among parts and wholes are the basis for levels of organization in our world. In his seminal work on complexity, Simon (1962) emphasizes the significance of evolved levels of biological composition in particular. Wimsatt (1972, 2007) largely grounds his analysis on compositional levels but also focuses on the causal significance of levels. In ecology, O'Neill et al. (1986) highlight the importance of time scale in distinguishing levels, although they variously appeal to spatial scale and composition as well. Craver and Bechtel (2007) and Craver (2007) conceive of levels as based on mechanisms and their components. Some influential philosophical discussions of levels of explanation have instead ordered levels by realization relationships and, relatedly, abstractness of description (e.g., Fodor, 1974; Putnam, 1975; Sober, 1999). Reductionists and antireductionists about explanation alike have tended to presume that these relations track compositional relationships (see Potochnik, 2010).

In this section, my aim is to explore the feasibility of a more precise levels concept that is immune to criticisms of ambiguity and related difficulties of the broad, intuitive concept of levels of organization. We can use these various ideas about levels to articulate a general problem space, a range of possibilities for where we might locate a workable levels concept. Options include the differentiation of levels with regard to part-whole composition, spatial scale, temporal scale, mechanistic composition, realization, and/or abstraction. Another approach, illustrated by Simon (1962), would be to use specific scientific theory—for instance, regarding major evolutionary transitions to more complex forms of life—to inform an articulation of levels that then has broader relevance beyond the scientific theory on which it is based.

It seems that an advocate of an unambiguous and generally applicable levels concept will need to choose among these ways of differentiating levels. When we look closely, we can see that these modes of differentiating levels don't always or even often yield the same results. Rather, composition, spatial scale, temporal scale, mechanism, realization, and abstraction relations define different organizational schemes that only partly and haphazardly overlap with one another. I briefly support this claim here; for more detail, see Potochnik (2017, §6.2.2).

Notice first that spatial scale and compositional levels are independent. While wholes and their proper parts are larger and smaller, respectively, this scale relationship need not hold for the parts of different wholes at the same compositional level. Further, different parts of a given whole may be at the same compositional level but need not be similar in size. Different parts of a single whole can differ radically in size: large, lumbering organisms and individual waste particles are both important parts of the same ecosystem. And an entity can be a proper part of wholes of radically different sizes: a sodium ion may be an independent part of elephant or in a simple saline solution. (Thomas Reydon has pointed out in correspondence that spatial scale seems to relate more closely to compositional relationships among subatomic particles.) Spatial and temporal scales also vary independently of each other. Creep in an individual solder joint of an old lead water pipe occurs very slowly (small spatial scale, large time scale), but the catastrophic failure of an entire water distribution system resulting from cumulative solder creep is a sudden event (large spatial scale, small time scale). For similar reasons, temporal scale also doesn't relate neatly to compositional relationships.

Now consider where mechanistic organization fits in all of this. As Craver and Bechtel (2007) and others have made clear, mechanistic composition is distinct from part-whole composition. Material parthood doesn't guarantee mechanistic parthood, and the same component can participate in different mechanisms. What about the relationship between mechanistic organization and spatial and temporal scale? Mechanisms and their components do, it seems, obey a spatial ordering in much the same way material parts and wholes do. Further, the kinds of processes parts undergo that contribute to directly to action of a mechanism must occur at a faster time scale than the action of the whole mechanism, insofar as they constitute elements of the mechanistic activity. But these are only some of the processes or changes that mechanisms and their parts undergo. Other changes need not obey any such temporal ordering. For example, a gene complex involved in some trait expression will evolve very slowly, much more slowly than the time scale on which the trait is expressed in any given organism. Also similar to the relationship between part-whole composition and scale, there is no basis for spatial or temporal comparison among components or across mechanisms (Eronen, 2013). One component may be much larger or smaller or act much more quickly or slowly than another; distinct mechanisms may be of virtually any size and act at any of a variety of speeds.

The realization relationship, as philosophers have discussed it, does not follow part-whole composition or scale relationships. Because of multiple realization, some properties of interest may be realized by any number of different kinds of entities of different sizes: using Fodor's (1974) classic example, monetary exchanges may be realized by a stack of dollar bills or the movement of electrons. And, because of what I have called complex realization (Potochnik, 2010, 2017), there's no reason to think that the properties of parts alone realize some property of a whole. That a fish is camouflaged, for example, is jointly realized by a number of properties of the fish, its immediate environment, and its potential predators.

Finally, unlike all the other organizational schemes I've surveyed here, level of abstractness or specificity is a representational rather than metaphysical relation. As such, it's fully distinct from all of the above concepts. It is true that if you represent only big things, long time periods, and general functional relationships, you omit lots of details. Abstract representations can be of high levels in one or more of these other senses. But you also omit

many details if you represent only immediate interactions among microphysical entities—
including, among other things, how those relate to big things, long time periods, and general
functional relationships. Ultimately, you can have more or less abstract representations of
anything: objects of any size, changes at any time scale, and properties structurally or functionally specified.

Because none of these different relations sometimes used to define levels relate to each
other in consistent ways, any attempt to identify a precise formulation of the levels concept
will need to identify one relation among these on which to base levels. To make the levels
concept precise, then, we could consider choosing one of these organizational schemes as
the sole basis for articulating levels. I'll next explore the prospects of that strategy. But
first I want to propose a few ground rules.

Any concept we select should do at least some of the basic work we expect of levels of
organization. First, I take it that this minimally includes providing a basis for levels comparisons across entities. For this, we need to be able to show how some entity should be
articulated into levels, and we also need to be able to show across different entities (or properties, processes, etc.) how these levels boundaries align with one another. Without the latter,
levels comparisons across individual entities are impossible, and this has been fundamental
to the invocation of levels. Second, it also seems like we want our revised levels concept to
make judgments that at least loosely correspond to how levels of organization have custom-
arily been understood. If this is a revision of our intuitive levels concept, it should work in
at least some ways similar to that intuitive concept. Third and finally, when we have a handle
on the features of this revisionary concept of levels, we'll need to set aside any anticipated
or desired implications that we can no longer reasonably expect to obtain in light of our
conceptual refinement. If a precise levels concept can be worked out, it has no claim to the
various significances our prior, ambiguous idea of levels was taken to have.

So, let's survey our candidate organizational schemes with an eye to (1) the prospects
for articulating levels across different entities, (2) how well the scheme coheres with
traditional invocations of levels, and (3) what implications of levels we'll need to give up
because of this conceptual refinement.

Part-whole composition provides an obvious candidate for the articulation of levels: any
whole is at a different level from its proper parts. But, a difficulty immediately emerges.
Are the waste particle and the elephant that (separately) partly compose the same ecosystem at the same level as each other? (See Eronen's chapter in this volume for additional
discussion of this question.) How about the sodium ion and liver that (separately) partly
compose the same elephant? If the answer to these questions is yes, then different waste
particles (and different sodium ions, etc.) are on different levels from one another depending on what they compose, since these can be freely participating parts of an entity or of
a part of that same entity (or something else entirely). If we articulate levels following
simple composition relationships in this way, then we run amok of suggestion (1) I made
above, that kinds of entities (or properties, processes, etc.) should be distinguishable into
levels. Individual sodium ions, waste particles, and all the rest can occur at different levels
depending on what sort of entity they happen to compose. An individual sodium ion (etc.)
may even change level over time. This, I take it, is a counterintuitive outcome.

If instead we don't hold that all proper parts of a whole are on the same level, one level
lower from the whole, then why not? What's the basis for distinguishing which parts

qualify as one level down? Perhaps we expect wholes on the same level to be composed of the same *kinds* of parts, or parts with similar properties to one another. Again considering the elephant, a common suggestion for the next compositional level down is organs, which are in turn composed of tissues, tissues composed of cells, and so on. But if wholes on a given level should be composed of the same kinds of parts, then because lots of multicellular organisms don't possess organs, the next level down from multicellular organisms simply would be cells. Organismal organization turns out to be a lot flatter than expected. And this problem recurs; for instance, are eukaryotic and prokaryotic cells on the same level? If so, organelles are not a distinct level from eukaryotic cells. This also violates common expectations. If not, then types of cells are partitioned into incommensurate systems of levels. Things quickly get messy and violate my suggestion (2) above by deviating significantly from traditional invocations of levels.

No matter which choices are made on inclusivity or exclusivity of level membership, an organizational scheme based on composition looks less like well-behaved levels than we might hope. Either the same types of entities occupy different levels and potentially change level over time depending on what they compose, or we are left with rather unintuitive, partial ordering systems with incommensurate levels in different kinds of systems. Part-whole composition is very good at yielding a nesting relationship by dividing a type of entity into the types of parts that compose it. It is less good at showing how these nesting relationships relate to one another, and so less good at yielding a leveling relationship.

One way out of this morass is to turn to scientific theory to help us articulate levels. For example, major transitions in evolutionary history have been suggested as a basis for articulating compositional levels, perhaps most famously by Simon (1962). In this case, prokaryotes are on the same level as organelles, eukaryotic single-celled organisms on the same level as cells in multicellular organisms, and multicellular organisms in a class of their own. I grant that this way of conceptualizing the types of lifeforms may be enlightening for some specific inquiries. But the general significance of this ordering—even in biology— is extremely limited. For example, it is useful for developmental biologists to employ an alternative ordering based on development processes, according to which tissues and organs are each distinct levels intermediate to cells and multicellular organisms. The obvious distinction between evolutionary levels and developmental levels is regularly missed and the significance of evolved levels for other phenomena regularly overstated. This includes by Simon (1962) himself, who stresses evolved levels even as he counts tissues, organs, and organ systems as levels.

Thus, the approach of basing compositional levels on particular scientific theory does not yield a single conception of levels of organization so much as a recipe for the construction of highly specific conceptions to suit particular inquiries. Such a highly specific levels concept is unproblematic and perhaps may be useful in the specific domain of the theory on which it is based. But, this limited scope of applicability seems (to me) to violate our expectations for a levels concept rather significantly. And, this also leads to high risk of overextension, tempting scientists and philosophers to violate suggestion (3) I made above, of giving up unearned expectations for a refined levels concept. The domain of relevance must be policed to avoid overextension and conflation of different levels concepts that give rise to different orderings, as illustrated with evolutionary and developmental levels. As things stand, the evolutionary levels concept in particular has regularly been overextended, generating scientific and philosophical error in the process. This risk of overextension also limits the predictive and explanatory value of these posited theory-specific levels. More than anything, they simply summarize one aspect of our theory in some domain.

Part-whole composition didn't fare well as the basis for a more precise levels concept. Now on to mechanistic levels, taken by many to be the most promising approach to levels currently available. The difficulties here mirror the problems with compositional levels. Eronen (2013) shows compellingly that mechanistic composition provides a way to distinguish nested subcomponents, components, and mechanisms but lacks a way to create levels comparisons across different mechanisms or even different components of a given mechanism. Consequently, similar to the plight of the sodium ion above, the same entities that participate in different mechanisms will be classified as being at incomparable or potentially different levels as themselves. Further, DiFrisco (2017) suggests that mechanistic composition runs into trouble distinguishing levels above the organismal level, where (I take it) the very idea of participating in a mechanism applies in a less clear-cut way. Much like classical composition, mechanistic composition is good at yielding nested hierarchies but insufficient for generating anything like traditional levels of organization. It's also the case that mechanistic composition is relative to a particular scientific investigation, perhaps more than or at least more explicitly than material composition is, as the research focus helps guide which causal processes in any given entity are under consideration. This compounds the difficulty of not being able to generalize categorization into levels across mechanisms, as a levels articulation in one domain may have no significance for another.

Let's turn next to temporal scale. DiFrisco (2017), taking inspiration from Wimsatt, systems theory, O'Neill et al. (1986), and others, advocates time scale as an alternative basis for the articulation of organizational levels. He emphasizes that hierarchical segregation is a way of preventing dynamical interaction between subsystems of a system, which stabilizes the different levels of subsystems and sub-subsystems within a system. This is evocative of Simon's (1962) discussion of levels, where modular levels are taken to be required for the evolution of complex forms. DiFrisco's organizational scheme applies only to biological systems. It also applies only to highly limited kinds of phenomena experienced by those biological systems, namely, the physiological processes that maintain homeostasis. It is thus most naturally interpreted as resulting in a hierarchical ordering of processes of those types, not of the entities that participate in those processes. As DiFrisco appreciates, this conception of levels is a significant departure from traditional invocations of levels, and it is highly circumscribed in its applicability. Indeed, he points out that "a process rate has no significance except in relation to other process rates within the same local interactive context."

Notice two things about this limitation. First, this is yet again a highly specific levels concept with a very limited domain of applicability, so it is again at risk of overextension. Recall my suggestion (3) from above, the need to recognize the intended implications we'll have to give up if they don't follow from our precise refinement of the levels concept. Second, that process rate is only significant in relation to other related process rates indicates that this intended levels concept again gives rise to nests rather than levels. Much in the way of mechanistic "levels," there is no grounds for comparison across different processes, even processes the very same entities undergo, thus falling short of suggestion (1).

Here is a candidate for a levels concept arising from spatial scale. Inspired by O'Neill et al. (1986) and more recent investigations in macroecology, Potochnik and McGill (2012) suggest articulating "quasi-levels," acknowledged to be dependent on what phenomena are under investigation and what features are of interest to the investigation. On this approach, the articulation into distinct levels across the gradient of spatial scale is based on the empirical discovery of boundaries for the dominance of different causal processes. That is, if one causal influence predominates at a given scale and another at a higher or lower scale, these are at different levels. An example is the different influences on species dispersal. McGill (2010) suggests that this is controlled by random chance at the scale of a few meters, by species interactions at a scale of roughly a kilometer or so, by habitat preferences at tens to hundreds of kilometers, and by climate at a scale approaching continental size. A similar approach to levels also could be employed across time scale rather than spatial scale (giving rise to different levels schemes, ones that are still dependent on what phenomena are investigated and which features are of interest).

This conception of levels is significantly pared down. It is explicitly dependent on focal phenomena and investigative context. Further, this articulation of levels cannot guide causal attribution but rather follows from causal attribution. One might reasonably wonder how much is to be gained from this beyond the simple recognition that causal processes vary across scales. That said, it is worth noting that something like this conception might help motivate more specific levels concepts, like evolutionary levels, developmental levels, and perhaps DiFrisco's suggested hierarchical ordering for homeostasis-preserving processes. Note that this attempt at levels again leaves us without an ability to articulate levels across entities or processes—in this case, this is a direct consequence of the ordering relating to the phenomena investigated and features of interest.

To complete our survey of potential levels concepts based on each of the different relations sometimes thought to be bound up with levels, let's briefly consider the prospects of realization and abstractness used to define levels of organization. Neither is, I think, a promising candidate. Because neither of these orderings correspond in any systematic way with compositional levels, as discussed earlier in this section, realization relationships or relative abstractness used alone yields nothing remotely approximating the levels concept as it has typically been invoked. As discussed above, realization relationships do not track compositional or scale relationships. At best, this scheme would also result in nesting relationships akin to (but not aligned with) mechanistic organization. But even that might be out of reach, for multiple realization and complex realization significantly complicate the relationship between realizers and what is realized. Quite many properties will jointly and variously contribute to realizing any given property, and once realization is dissociated from material composition, there is no reason to expect size ordering or containment of realizers and what they realize.

As for abstractness, I suggested above that levels of abstraction are representational rather than metaphysical. Floridi (2008) calls these "epistemological levels" and explicitly distinguishes them from levels of organization, arguing that although the latter are untenable, the former are useful. He considers Marr's three levels of analysis to be a prime example of levels of abstraction. Marr (1982) distinguishes among the computational, algorithmic, and implementational levels of description for cognitive processes. Notice that even this very different approach to levels has some of the same limitations as the others we surveyed above. Marr's levels scheme in particular applies only to a very specific kind of phenomena: cognitive or information-processing systems. And as Floridi details, they have regularly been conflated with other, very different levels concepts. Like evolutionary levels, then, this levels scheme is motivated by specific scientific theory, and its usefulness is correspondingly constrained by the applicability of the theory, as well as whether and in what form the theory is maintained. So, for example, it may turn out that computational models of cognition are of limited value, and this would undermine the significance of that representational level.

Most important for my purposes, a conception of levels in terms of abstraction bears little similarity to levels of organization as they have typically been invoked. We have already explored how abstractness varies wholly independently from composition and scale: anything large or small can be represented in more or less detail. Levels of abstraction may be a useful conceptual tool in some scientific pursuits, but that is of no help in making good on an ontological or metaphysical levels concept. And, as with Marr's levels of abstraction, it has been common to conflate representational levels of abstraction with compositional levels of organization. It's also worth pointing out that, like the other relationships I have surveyed, abstractness relationships tend to characterize nests rather than levels. This is because any given representation can be made more specific, less abstract, by incorporating any of a range of different details. It is thus quite easy to generate incommensurate levels of abstraction by adding or removing different ranges of details.

In this section, I first showed that all of the different relationships sometimes taken to be bound up with levels of organization systematically deviate from one another, thereby contributing to the ambiguity of the levels concept. I then explored the prospects for each of those relationships to be used alone as the basis for articulating levels of organization. I believe this survey has revealed no promising general conception of levels of organization. To be clear, it's not that there is a bit of messiness in levels divisions or an occasional failure but the wholesale inability to ground any systematic ordering scheme of levels. Clear themes have also emerged in the limitations of each of these candidates for the basis of a levels concept.

Recall my specification at the outset that we should attend to (1) the prospects for articulating levels across different entities, (2) how well the scheme coheres with traditional invocations of levels, and (3) what implications of levels we'll need to give up because of this conceptual refinement. It turns out that every relationship surveyed fails on (1), articulating levels across different entities, thus giving rise to an ordering scheme not of levels but with more resemblance to nests. This is a significant shortcoming, as it gets at the very heart of the levels concept. Moreover, if all attempts to render the levels concept precise suffer from this same difficulty, then any levels concept invoking several of these relationships will likely suffer from the same problem. As for (2), realization relationships and abstractness by themselves bear no similarity to traditional expectations for levels of organization. Other ordering schemes, such as levels in temporal scale, deviate less radically from traditional expectations but still deviate markedly. And, I suggested for part-whole composition that the only way of solving problems with (1) results in a very flat hierarchy of levels that departs significantly from traditional expectations.

And then there is the question, (3), of what implications of levels we'll need to give up due to a conceptual refinement in what levels of organization are. The most promising candidates we have seen for a levels concept are highly domain specific and thus interest relative. Such a highly local approach to levels is explicitly motivated by Love (2012), who focuses on the example of the four levels of protein structure. Similarly, one may emphasize evolved levels, for example, due to what we understand of the major evolutionary transitions. But, as I have suggested, domain-specific and interest-relative levels have little to no implications for other biological processes. Evolved levels should not be expected to have significance for present structural organization, development, nor even for what entities are now subject to evolutionary forces. Similarly, a conception of levels based on spatial scale will give rise to very different levels systems for different phenomena, and levels in the temporal scale of homeostatic processes have no implication for any other kinds of processes.

Furthermore, even for highly specific formulations of levels, articulating well-defined levels may still be a challenge. Protein folding, Love's (2012) example, may be an instance where there is clear uniformity in structure and domain of applicability such that levels can be distinguished across entities. But this is more challenging for other, similarly localized levels articulations. Structural organization in many other entities does not involve this manner of discrete stages. And evolved levels may be better understood as simply the historical introduction of novel forms of evolutionarily relevant integration rather than levels of structural organization at all. It may well be that for some circumscribed scientific pursuits, like protein folding, there is a workable conception of levels of organization. But, I suspect "levels" is not the most apt description of many instances of local organization. And, with any successful localized levels concepts, we must strenuously avoid conflation or usage slip among them or with a more general conception of levels, the promise of which we have not been able to make good on. This task is made more difficult by the fact that both of these forms of conflation and usage slip are historically and currently overwhelmingly common in appeals to levels.

3.2 Heuristic Use of the Levels Concept

We have explored the prospects for refining our concept of levels of organization into a more precise and defensible version and come up short. So, let's now consider the other kind of strategy to defending the usefulness of our levels concept. In the previous section, I distinguished among the various features levels have been taken to have, to show that they don't hang together and that none is by itself promising as a basis for well-founded levels of organization. But, you might think, this is common for our concepts, commonsense and scientific alike. There are often limitations in the domain of a concept's applicability, just as evolutionary and developmental levels are applicable to different biological phenomena. Concepts can also be usefully ambiguous, as the gene is defined in different, inequivalent ways in different scientific contexts that are nonetheless useful to lump together. Indeed, the philosopher Otto Neurath thought vague, ambiguous concepts—what he called *Ballungen*—were by necessity at the basis of scientific inquiry (Cartwright et al., 1996; Neurath, 1983; Potochnik & Yap, 2006). Perhaps, then, equivocation among distinct relationships is not a weakness of the levels concept but a strength. As many have emphasized, including Brooks in this volume, biology textbooks are organized around the concept of levels of organization. So, in line with this use, might levels of organization at least be a productive heuristic?

It seems Wimsatt has something like this in mind. He sees levels as related to most of the features I identified in the previous section: composition, evolved organization, spatial and temporal scale. But he also acknowledges deviation from the alignment of these features, and he anticipates phenomena, "causal thickets," that do not feature levels at all. Inspired in part by Wimsatt, some other recent levels proponents more explicitly advocate this view. Brooks (2017) demonstrates the variability of the levels concept and defends this as a strength of the concept. And, in another paper, Brooks and Eronen (2018) explicitly advocate the heuristic use of levels, a use that they think relies on the term's polysemy. They say, "We offer an approach that ... embraces the ambiguity of levels as vital to the flexibility the concept exhibits in expressing many distinct ideas." Brooks and Eronen hold that the levels concept is useful in structuring scientific problems by suggesting how a phenomenon can or should be studied.

This seems like a promising strategy to accommodate a scientific concept that eludes precise definition. It may be impossible to give a global definition of levels of organization, and the usefulness of the term may vary. But, there's no denying that many scientists and philosophers have invoked levels terminology in a variety of projects. And levels play a fundamental role in biology education. So accepting variability in the meaning of levels, along with some limitations and inconsistencies, may be the best way forward.

Unfortunately, though, I'm not convinced that advocating a heuristic levels concept has much better prospects than our search for a precise concept of levels of organization. I'll outline four concerns; the first regards the value of this heuristic. This concern follows from the dead end at which we ended up in the previous section's search for a precise levels concept. There I described how the different features the levels concept is expected to have in fact come apart from one another and how none of those features is particularly well suited as a basis for articulating general levels of organization. All of the relationships under consideration produced orderings more like nests than levels and were in other ways significantly out of step with what has traditionally been said about levels. Contrast this to gene concepts. These differ in their import and applicability across projects in biology, but gene concepts in one form or another have been central to a number of theoretical and empirical projects. There is some variability in what we call a gene, but genes are inarguably biologically central. I don't see evidence that such a thing can be said of any combination of the levels concepts we have surveyed. It's one thing for a heuristic to vary in its applicability and usefulness, but I am concerned that there is no scientific investigation in which this heuristic is more apt and useful than misleading.

The examples Brooks and Eronen (2018) provide of the usefulness of levels in structuring scientific problems are an opportunity for me to illustrate this point. One of their examples is the use of levels to separate a system into its component parts and then relate those back to the whole phenomenon to which they belong. They have in mind localization and decomposition of the kind Bechtel and Richardson (1993) emphasized. I agree that this can be useful for some scientific projects. But, I wonder, is any levels concept important for such a project? To the contrary, this simply seems to be a matter of exploiting part-whole or mechanistic composition when such composition is important to the phenomenon under investigation. Material or mechanical composition is certainly important to this kind of a project, but I fail to see how the articulation of levels is doing any work beyond this. One can, of course, investigate parts and wholes without delineating *levels* of parts and wholes.

Another of Brooks and Eronen's (2018) examples is the identification of levels to support moving scientific inquiry to another level in order to gain insight into a problem or to generate hypotheses for solving that problem. They suggest this can involve "looking to new things (i.e. natural objects) or looking to new epistemic resources (descriptions, methods, models)." Brooks and Eronen appeal to systems biology's consideration of dynamics in a whole system to illustrate this idea. I also grant the usefulness of this strategy: considering new influences or methods can be scientifically fruitful. But, here too, I don't see what work is done by levels per se. These authors emphasize the historical importance of appeals to levels in systems biology's resistance of reductionism. I agree that some of the appeal of a levels concept is this historically important role. But one can just as well resist reductionism—and be creative about where in nature we look and what methods we apply—without positing that nature is organized in levels. So, granting Brooks and Eronen (2018) the potential value of both of these strategies, that value does not provide evidence of the usefulness of the levels concept per se.

The first concern I identified with a heuristic levels concept is that there's not clear evidence of its value. My second concern is that I fear this heuristic invites overextension and overly literal interpretation. The recent history of scientific and philosophical work is littered with instances of scientists and philosophers conflating different levels schemes and presuming the strict reality of levels. Here are just a few examples. Simon (1962) emphasizes evolved levels even as he discusses development. Craver and Bechtel (2007) claim that mechanistic levels are a species of compositional, or part-whole, levels (although later works by both authors explicitly distinguish these levels schemes). And, evocative of earlier reductionist views of levels in philosophy, ecologist Lidicker (2008) describes the universe as "composed of hierarchically arranged systems" and biological systems as comprising the "pyramid of life." Examples also abound of scientists and philosophers overselling the significance of levels, anticipating metaphysical, epistemic, explanatory, and causal significance that fails to obtain (see Potochnik & McGill, 2012). In the face of the difficulties with making good on the levels concept identified earlier in this chapter, simply endorsing the heuristic use of the levels concept invites such overextension and overly literal interpretation.

Here's a third worry. Beyond limited value and risk of overextension and overly literal interpretation, I also think the levels heuristic may *systematically* mislead. Despite all the strong claims to the contrary, and all the smart people who part ways with me on this, I wonder whether our world might just not be aptly described as ordered into levels. I am supposed to be impressed by pockets of well-ordered behavior seemingly insulated from external fluctuations, by components of systems capable of largely independent action, and by the extent of similarities among complex forms and the implications of those similarities. But instead, I am shocked by how often external influences shape local behavior (and in ways that tend to escape our notice), how variable the action of components of a system often turn out to be and how subject to top-down control, and by the extent of difference among complex forms and the infinite variability produced by evolution and development. I realize that I am an outlier in my ideas about levels of organization (but see Thomasson, 2014). Perhaps it's best for me to leave this point with an invitation for responses to the following question: why *should* I believe that the levels heuristic is apt—that our world, or significant swaths of our world, is hierarchically structured?

This brings me to the last concern I want to raise with the heuristic use of the concept of levels of organization. I wonder whether there aren't other, better heuristics that looking through the lens of a levels heuristic has occluded. I suspect there is a broad human tendency, perhaps in part due to the nature of our sensory systems, to focus on the significance of factors that operate on our spatial and temporal scales and to look to the parts of a system to explain its behavior. If this is not a human tendency in general, it has at least been a tendency in much of the history of science to date. Particularly in light of the foundational role that the concept of levels of organization has played in biology education and elsewhere, the levels heuristic may contribute to reinforcing these expectations of our world. Yet these expectations are, I believe, regularly violated. If this is right, then maintaining levels in a heuristic use enables us to postpone the recognition of other heuristics, heuristics that are potentially more apt and more productive either across the board or in particular scientific domains.

Here are a few candidates to illustrate the idea. Perhaps the concept of networks would be a promising alternative heuristic to explore. This relates to familiar technological innovations that might be used to disrupt our expectation that local influence is the only form of important influence and to help us anticipate large-scale influences and top-down coordination. A related notion is that of causal complexity (see, among others, Potochnik, 2017). This is less familiar and less evocative, but it does anticipate open-ended lists of causal relationships traversing spatial and temporal scales, causal interaction, and nonlinear behavior. All of these are, I suspect, more common than causal relationships well ordered by levels of organization. Here's an idea for a heuristic organizational scheme—or rather two schemes—to replace the levels heuristic, inspired by Eronen (2015). The levels heuristic anticipates that scale and composition go together, but as we have seen, they do not. But these two customary ingredients of the levels heuristic might be used independently of each other and of levels to categorize the range of phenomena in our complex and variable world. Scientists describe relationships that are scale invariant and those that are scale dependent. Similarly, composition is sometimes of crucial relevance and other times nearly irrelevant. Exploring the extent to which phenomena are scale bound or not and compositionally dictated or not would be a radically different, open-ended approach to categorizing our world.

None of these alternative heuristics requires setting aside the levels heuristic. But it seems likely that viewing the world through the lens of levels—organizing textbooks and philosophical and scientific inquiry according to the expectation of levels—has led us to recognize and emphasize some features of the world to the exclusion of others. Scientific discovery and philosophical breakthroughs have, I think, increasingly made clear that deep features of our world deviate from the expectation of well-ordered levels. These features are rendered less visible and taken less seriously when they are merely viewed as exceptions or limitations to an overall architecture of levels.

My concerns with preserving the concept of levels of organization as a heuristic are thus the limited value of this heuristic, the likelihood that the heuristic is overextended and interpreted too literally, the possibility that the heuristic of levels systematically misleads us about the nature of our world, and filling conceptual space that, given all of these limitations, might be better filled by a competing heuristic. Science is filled with simplifications, polysemic concepts, and heuristics. The question I am urging is whether levels of organization is one that earns its keep. Perhaps the basis for our levels concept turns out to be simply our impulse to categorize the variety we encounter in our world and to identify similarities among the various resulting categorizations. If this is right, calling the levels concept a heuristic is not simply an acknowledgment of the necessary variability and ambiguity in scientific terminology but a fallback position adopted when one is faced with the failure to make good on the idea that our world is organized into levels.

3.3 What's to Lose with Levels?

In this chapter, I have surveyed two approaches to a concept of levels of organization: rendering it precise and advocating its heuristic use. I have worked through problems that I believe face each approach. It's worth pointing out that, because of the nature of these problems, the pursuit of both strategies simultaneously in philosophical and scientific literatures compounds the difficulties facing each. Attempts to identify a precise refinement of the levels concept tempts us to think there is a way to make good on the heuristic and thus to take the heuristic too literally. And then, the heuristic use of the levels concept tempts us to think that we don't need to fully work through a precise refinement of the levels concept, obscuring the conceptual difficulties it faces. That both kinds of approaches are advocated is, I think, evidence of the problems plaguing the concept of levels of organization.

Now is the point at which to ask, "So what?" What would I wish to be done? First and most basically, I urge wider recognition of the variety, dissociation, and limitations of levels concepts. If "levels" is polysemic, we should at the very least track the polysemy to avoid scientific error. This is made especially difficult given that some philosophers are advocating conceptual refinement while others are advocating broad heuristic use. Part of this project should be to carefully distinguish metaphorical from literal applications of the concept and to be aware of the domain within which any given use of the concept applies. For example, as I said above, that more complex forms have evolved at the major transitions of evolutionary history is not reason to expect those ancestral forms to have special significance for biological organization today, such as development. And yet, although it's now well appreciated that ontogeny does not recapitulate phylogeny, evolutionary levels are still invoked in developmental contexts. Close on the heels of this point is a second idea: scientists and philosophers should investigate the significance of levels in a given domain rather than draw conclusions about that domain from the presumed significance of levels. Some bits of our world may turn out to be organized into levels, at least in certain respects. But given the complex variability of our world, that form of organization cannot be presumed. Third and finally, scientists and philosophers should actively try to cultivate alternative heuristics—networks, causal complexity, scale invariance, and others. Try for a moment to imagine what organizational schemes are available for biology textbooks other than compositional "levels" and how these different schemes might differently educate our biologists. What about, for example, a textbook ordered by the spatial scale and temporal scale over which different influences tend to dominate? Or a textbook that begins with a figure depicting the interdependence of phenomena and is then ordered by the specific research questions that have motivated a focus on different features of these complex phenomena?

In my view, our adherence to the levels concept in the face of the systematic problems plaguing it amounts to a failure to recognize structure we're imposing on the world, to instead mistake this as structure we are reading off the world. Attachment to the concept of levels of organization has, I think, contributed to underestimation of the complexity and variability of our world, including the significance of causal interaction across scales. This has also inhibited our ability to see limitations to our heuristic and to imagine other contrasting heuristics, heuristics that may bear more in common with what our world turns out to actually be like. Let's at least entertain the possibility that the invocation of levels can mislead scientific and philosophical investigations more than it informs them. I suggest that the onus is on advocates of levels of organization to demonstrate the well-foundedness and usefulness of this concept.

Acknowledgments

Thanks to Daniel Brooks, James DiFrisco, and Bill Wimsatt for coordinating a productive and fun workshop and to the Konrad Lorenz Institute for hosting. I learned so much from the other participants, and I had a great time in the process. Thanks also to Thomas Reydon and Markus Eronen for helpful comments on an earlier draft of this chapter. The chapter also benefited from audience feedback at talks in the Department of History and Philosophy of Science at the University of Pittsburgh and the Department of Philosophy at the University of Colorado, Boulder.

References

Bechtel, W., & Richardson, R. C. (1993). *Discovering complexity: Decomposition and localization as strategies in scientific research*. Princeton, NJ: Princeton University Press.

Brooks, D. S. (2017). In defense of levels: Layer cakes and guilt by association. *Biological Theory, 12*, 142–156.

Brooks, D. S., & Eronen, M. I. (2018). The significance of levels of organization for scientific research: A heuristic approach. *Studies in History and Philosophy of Biological and Biomedical Sciences*. Advance online publication.

Cartwright, N., Cat, J., Fleck, L., & Uebel, T. E. (1996). *Otto Neurath: Philosophy between science and politics*. Cambridge, UK: Cambridge University Press.

Craver, C. F. (2007). *Explaining the brain: Mechanisms and the mosaic unity of neuroscience*. Oxford, UK: Oxford University Press.

Craver, C. F., & Bechtel, W. (2007). Top-down causation without top-down causes. *Biology and Philosophy, 22*, 547–563.

DiFrisco, J. (2017). Time scales and levels of organization. *Erkenntnis, 82*, 795–818.

Eronen, M. I. (2013). No levels, no problems: Downward causation in neuroscience. *Philosophy of Science, 80*, 1042–1052.

Eronen, M. I. (2015). Levels of organization: A deflationary account. *Biology and Philosophy, 30*, 39–58.

Floridi, L. (2008). The method of levels of abstraction. *Minds & Machines, 18*, 303–329.

Fodor, J. (1974). Special sciences: The disunity of science as a working hypothesis. *Synthese, 28*, 97–115.

Lidicker, W. Z., Jr. (2008). Levels of organization in biology: On the nature and nomenclature of ecology's fourth level. *Biological Reviews, 83*, 71–78.

Love, A. C. (2012). Hierarchy, causation and explanation: Ubiquity, locality and pluralism. *Interface Focus, 2*, 115–125.

Marr, D. (1982). *Vision: A computational investigation into the human representation and processing of visual information*. Cambridge, MA.: MIT Press.

McGill, B. J. (2010). Ecology: Matters of scale. *Science, 328*, 575.

Neurath, O. (1983). *Philosophical papers 1913–1946*. Dordrecht, Netherlands: Reidel.

O'Neill, R. V., DeAngelis, D. L., Waide, J. B., & Allen, T. F. H. (1986). *A hierarchical concept of ecosystems*. Princeton, NJ: Princeton University Press.

Oppenheim, P., & Putnam, H. (1958). Unity of science as a working hypothesis. In H. Feigl, M. Scriven, & G. Maxwell (Eds.), *Minnesota studies in the philosophy of science* (pp. 3–36). Minneapolis: University of Minnesota Press.

Potochnik, A. (2010). Levels of explanation reconceived. *Philosophy of Science, 77*, 59–72.

Potochnik, A. (2017). *Idealization and the aims of science*. Chicago, IL: University of Chicago Press.

Potochnik, A., & McGill, B. (2012). The limitations of hierarchical organization. *Philosophy of Science, 79*, 120–140.

Potochnik, A., & Yap, A. (2006). Revisiting Galison's 'Aufbau/Bauhaus' in light of Neurath's philosophical projects. *Studies in History and Philosophy of Science, 37*, 469–488.

Putnam, H. (1975). *Philosophy and our mental life*. Cambridge, UK: Cambridge University Press.

Rheinberger, H.-J., & Müller-Wille, S. (2017). *The gene*. Chicago, IL: University of Chicago Press.

Rueger, A., & McGivern, P. (2010). Hierarchies and levels of reality. *Synthese, 176*, 379–397.

Simon, H. A. (1962). The architecture of complexity. *Proceedings of the American Philosophical Society, 106*, 467–482.

Sober, E. (1999). The multiple realizability argument against reduction. *Philosophy of Science, 66*, 542–564.

Thomasson, A. (2014). It's a jumble out there. *American Philosophical Quarterly, 51*, 285–296.

Wimsatt, W. C. (1972). Complexity and organization. In K. F. Schaffner & R. S. Cohen (Eds.), *PSA: Proceedings of the biennial meeting of the Philosophy of Science Association* (pp. 67–86). Dordrecht, Netherlands: Reidel.

Wimsatt, W. C. (2007). *Re-engineering philosophy for limited beings*. Cambridge, MA: Harvard University Press.

4 Levels, Nests, and Branches: Compositional Organization and Downward Causation in Biology

Markus I. Eronen

Overview

The idea of compositional levels of organization is deeply rooted into contemporary biology and its philosophy. I argue that when we take a closer look at the structure of biological systems, it turns out that hierarchical organization in nature is more complex and messy than most accounts of levels assume. More specifically, biological components tend to be very heterogeneous, which results in branching and tangled hierarchies instead of neat levels. Consequently, the traditional idea of levels should be treated as a heuristic abstraction that is only useful in some specific biological contexts. I demonstrate this approach by applying it to the debate on downward causation, where assumptions about levels play an important role and should be made explicit.

Scientists and others tend to be quite fond of neat, clear-cut patterns. Nature is not. Nature is inherently messy.
—Denis Noble (2006, p. 52)

4.1 Introduction: Levels and Compositional Hierarchies

The idea that nature is structured into nested compositional levels is ubiquitous in biology and its philosophy. As early as 1929, Joseph Woodger wrote, "The organism is analysable into organ-systems, organs, tissues, cells and cell-parts. There is a hierarchy of composing parts or relata in a hierarchy of organizing relations. These relations and relata can only be studied at their own levels … and not simply in terms of the lower levels" (Woodger, 1929, p. 293). In a classic article, Lewontin (1970) argued that the principle of natural selection can apply "at all levels of biological organization" and gave as examples of such levels molecules, cells, gametes, individuals, and populations. In his seminal work on levels of organization, Wimsatt (1976, 1994/2007) argued that compositional levels are a deep and fundamental feature of the ontological architecture of the world. And the list goes on (see, e.g., Bunge, 1977; Eldredge, 1996; Novikoff, 1945; Simon, 1962/1996; Tëmkin & Eldredge, 2015; Umerez, 2016; Zylstra, 1992).

The core idea of nested compositional hierarchies is clear: most things studied in biology consist of parts or components, and those parts themselves also consist of parts. Thus, there are wholes (e.g., cells) that consists of parts (e.g., cell parts), and these wholes and their parts are "nested" within bigger wholes (e.g., tissues). This kind of organization is particularly salient in multicellular organisms. An organism such as a human is composed of tissues, tissues are composed of cells, cells are composed of cell parts, cell parts are composed of molecules, and so on.

The approaches to these compositional levels fall into two broad categories: global and local (see, e.g., Craver, 2007, 2015; Love, 2012). In global approaches to levels (e.g., Oppenheim & Putnam, 1958), the idea is that levels span the whole of nature, or at least a substantial part of it (i.e., not just one organism or mechanism). Thus, it is thought that there is a global level of molecules, level of cells, individuals, populations, and so on, and that there is some clear and consistent sense in which, for example, cells in fruit flies and cells in humans are "at the same level." Depictions of such global levels are very common in biology, for example, in textbooks (Eronen & Brooks, 2018).

In local approaches, no general framework of this kind is assumed. Instead, the focus is simply on compositional hierarchies as they appear in a specific (type of) system or mechanism (Bechtel, 2008; Craver, 2007, 2015; Love, 2012; Winther, 2006). For example, there is a certain compositional structure, with certain hierarchical levels, in the spatial memory mechanism of mice and a different hierarchy with different levels in the human respiratory system (Craver, 2007). With such a local approach, there is no need to assume that levels span the whole of nature or even a whole organism: they can be determined on a case-by-case basis and can be extremely local and context specific. The most famous and influential local approach is "levels of mechanism," proposed by Bill Bechtel and Carl Craver, where the idea is that a mechanism as a whole forms the highest level, its components form a lower level, the subcomponents of those components form the next lower level, and so on (Bechtel, 2008; Craver, 2007, 2015; Craver & Bechtel, 2007).

In this chapter, I will point out certain features of biological compositional hierarchies that make them messier and more tangled than is usually assumed by philosophers and scientists. Although I focus on local levels, my arguments apply *a fortiori* to the more general global levels. However, they are not aimed at levels that are defined by other criteria than composition (e.g., scale-based levels or levels of processing). After critically discussing compositional levels, I provide a more positive viewpoint by outlining various ways in which the notion of levels can still be useful for biological research. Finally, I apply these insights to clarify the debate on downward causation.

4.2   Branching Hierarchies

Biological compositional hierarchies are typically downward-branching. To see what I mean by this, it is useful to contrast biological hierarchies with physical ones. In many physical systems, there is a neat hierarchical organization (see, e.g., Simon, 1962/1996): for example, a volume of gas consists of gas molecules, and when we decompose those gas molecules, we find similar things, even if the gas molecules are different (e.g., nitrogen and oxygen molecules): the gas molecules are composed of atoms, the atoms are composed of electrons and protons, and so on. Thus, even after several rounds of decomposition, we find similar things in the different branches.

In systems like this, we can delineate very neat compositional levels, where each level is occupied by similar kinds of things, and these similar kinds of things compose higher-level things, which are also similar to one another. This idea of levels is very intuitive and widespread and is particularly evident in "layer-cake" conceptions of levels (stemming from Oppenheim & Putnam, 1958; see Brooks, 2017, for more).

In biology, there are also some cases that come close to this picture, most prominently the levels of (multicellular) organisms and cells. Multicellular organisms are arguably similar in important ways: for example, they are capable of reproduction and have a similar basic metabolism. Furthermore, they are composed of cells, which are also similar to each other in important ways (e.g., all eukaryotic cells have a membrane). Thus, we seem to have a level of similar things (cells) that make up higher-level things (organisms) that are also similar to one another, leading to two neat levels of organization.

However, when we take a closer look at this picture, cracks begin to appear (see also Potochnik, this volume). First of all, cells make up organisms only in a very indirect and abstract sense. Rather, it would be more accurate to say that cells make up tissues, which make up organs, which make up organ systems, which (in combination with many other things) make up multicellular organisms (see also the quote from Woodger, 1929, at the beginning of this article). If we adopt any notion of composition that does justice to the complex and nested nature of biological organization, it is a vast oversimplification to say that organisms are composed of cells (see also Brooks, this volume).

To make this more concrete, I will assume here (and in the rest of this chapter) that when we analyze the compositional structure of a biological system, we should look for components that are biologically relevant "working parts": that is, parts that play a non-redundant role in the functioning of the higher-level whole (Bechtel, 2008; Craver, 2007). In addition, what are relevant for analyzing levels are the *direct* components or working parts—for example, the cell organs are direct components of cells, but the components of cell organs are components of the cell only in an indirect way, as there is another compositional "step" in between.[1] Let us then look at the components of, say, a mouse. This will include things like the reproductive system, the cardiovascular system, and the nervous system. Although these are all organ systems in some broad sense, they are very different kinds of systems. And things get even messier when we look at the next lower level, formed by the components of these systems. One component of the mouse nervous system is the brain. One component of the mouse reproductive systems is an ovary. Ovaries and brains are very different kinds of things regarding their scale, function, and organization. Does it make sense to claim that they are at the same level, just like atoms in the gas example, because they are both two rounds of decomposition away from the mouse as a whole? Probably not.

What this example illustrates is that when we take a biological organism or a system and look at what it is composed of, we are likely to find very heterogeneous components, very much in contrast to the gas example. This becomes even more evident when we consider the stereotypical organizational unit in biology: the cell. In some sense, it is true that the components of cells are molecules, just like it is true in some sense that the components of organisms are cells, but again, this is an unhelpful oversimplification. If we look at the biologically relevant components or "working parts" of a (eukaryotic) cell, we find things such as the cell membrane, the nucleus, mitochondria, liposomes, ribosomes, and so on.[2] These are very different kinds of things, with different properties, and on different scales. For example, a mitochondrion is an organelle that has its own DNA and is a relatively large structure, whereas liposomes are much smaller and have a simple container-like structure. When we then look at the subcomponents of these two components, this heterogeneity is amplified. The components of the mitochondrion include things like the inner membrane, the outer membrane, and the matrix that contains the mitochondrial DNA. The components of liposomes do not include any of these things but rather things such as lipid molecules and water molecules. It is evident that water molecules and the mitochondrial matrix are very different kinds of things and at very different scales. Thus, in contrast to the gas example, after just a few rounds of decomposition, we find wildly heterogeneous things in the different "branches" in the compositional structure of the cell.

This kind of downward-branching is a very common feature in nature. Although I have focused on downward-branching, analogous observations can also be made regarding upward-branching: for example, the same kind of protein can be a component in many different kinds of wholes (see also Wimsatt, 1994/2007, especially figure 10.2). Branching is also not an exclusive feature of the biological world but can in principle also occur in nonliving systems, such as complex artifacts. All it requires is sufficiently heterogeneous components, as then this heterogeneity will be amplified when it comes to subcomponents. Branching also occurs in the standard example of mechanistic levels, namely, a mouse navigating a water maze (Craver, 2007). One of the levels in this mechanism is the "cellular-electrophysiological level," where, according to Craver (2007), we find most importantly the synaptic long-term potentiation (LTP) mechanism. The components of this mechanism include things such as NMDA receptors, synaptic vesicles, cell membranes, magnesium and sodium ions, and so on. Just as in the case of the cell, it is clear that if we decompose these heterogeneous components, we do not find similar things that would form a coherent level but rather many different branches with wildly different components.

Although branching is often briefly mentioned in the literature on levels (e.g., Bechtel, 2008; Craver, 2007; Kim, 2002; Wimsatt, 1976, 1994/2007), its consequences have not been sufficiently acknowledged. Supporters of the mechanistic approach to levels have correctly emphasized the local and case-specific nature of levels, but as the examples above show, even in local and specific contexts, compositional organization can get extremely messy. Downward-branching implies that biological compositional hierarchies rarely form neat levels, *even when we focus on a single mechanism or phenomenon.* Instead, they form branching structures, where usually after just two rounds of decomposition, we find very heterogeneous things in the different branches.

This does not necessarily mean the end for the notion of compositional levels or levels of mechanisms. One can also bite the bullet, accept downward-branching, and stipulate that each set of (direct) components that we find is worthy of the title "level."[3] For example, the (direct) components of a nerve cell (i.e., cell organs, etc.) could be said to form a "level" in this sense. With this approach, compositional levels are in essence reduced or deflated to sets of (direct) components (see also Eronen, 2013, 2015; Krickel, 2018). Thus, it is debatable whether these sets of direct components should be called "levels" or whether the term just invites misleading intuitions here (Eronen, 2013, 2015; Potochnik, 2017 and this volume).

4.3   Scale-Based Levels and Heuristics

One reaction to the problems in defining levels has been to abandon the whole idea of levels in biology as too confused and misleading (e.g., Guttman, 1976; Thalos, 2013; see also Potochnik, 2017 and this volume). However, it is important to keep in mind that conceptions of levels serve various purposes and scientific goals (Brooks & Eronen, 2018; Eronen, 2015), and for different goals we may need different notions of levels. If one notion of levels, or a set of assumptions underlying the use of levels, turns out to be deeply problematic, this does not imply that other notions or uses of levels are problematic as well. In particular, downward-branching and other considerations against levels discussed above only apply to *compositional* levels and not to other kinds of levels. Most important, they have no implications for scales or scale-based levels, which are also central in biology.

The role of scales in understanding biological organization has been extensively discussed by other authors (e.g., DiFrisco, 2017; Noble, 2012; Potochnik & McGill, 2012; see also Simon, 1962/1996). The scale that is probably the most intuitive one is the size scale, which is based on how big things are. The size scale is a key element in Wimsatt's (1994/2007) approach to levels, and different ways of measuring size (e.g., length, mass, or volume) and the associated scales play an important role in fields such as ecology or allometry. Another biologically crucial scale is the time scale, which is based on the rate at which interactions or processes take place. For example, interactions between neurotransmitters and receptors are much faster than interactions between neurons, which again are faster than interactions between organisms (e.g., reproduction). Scales have two key advantages over levels of organization (Eronen, 2015; Potochnik & McGill, 2012). First, scales are continuous, not discrete. Sometimes scales can separate so that interactions at a higher scale become to a large degree independent from interactions at lower scales (e.g., Simon, 1962/1996; Wimsatt, 1994/2007), but in general, scales allow for continuity and do not require cutting organisms or systems into discrete layers (see also Green & Batterman, 2017). Second, scales are relatively well defined: when the relevant quantitative property of an entity or process is measured (e.g., mass, volume, or rate of interactions), it is straightforward to determine where that entity or process falls on the scale.

It is also be possible to formulate more substantive (noncompositional) accounts of levels based on scales. For example, Wimsatt (1994/2007) suggests that levels can be seen as local maxima (i.e., peaks) of regularity and predictability when plotted against a (size) scale.[4] Potochnik and McGill (2012) argue that scale-based quasi-levels can be defined based on causal relationships. DiFrisco (2017) puts forward a dynamical approach, where levels are conceptualized in terms of time scales. These kinds of scale-based levels and their role in research and theory is undoubtedly an important field for future work on levels, but it is important to emphasize that they are different from composition-based levels and should not be conflated with them.

Another, complementary approach to levels is to take a *heuristic* perspective to them (Brooks & Eronen, 2018). Instead of seeing levels as ontological categories of levels of nature, they can be understood heuristically, in much the same way as Bechtel and Richardson (1993/2010) treat decomposition and localization as heuristics that work in some contexts (and for some purposes and goals) but not in others. As an example, consider the intuitive idea of levels as "similar things made up of similar things." As we have seen above, there are very few if any cases in biology that *exactly* fit this idea. However, there are cases and specific contexts where it can be a good approximation. For instance, consider the protein network underlying the circadian clock mechanisms in mammals, as summarized by Bechtel (2017). The most important components in the cellular circadian clock mechanism are different kinds of proteins (e.g., PER, CRY) and the genes that transcribe these proteins. Both proteins and genes are composed of monomers (amino acids and nucleotides, respectively). In this context, it is perhaps useful (for some purposes) to conceptualize a higher level that is formed by similar things (genes and proteins) and a lower level that is formed by their components (monomers), which are also similar to one another. This two-level framework is obviously an abstraction, as the actual mechanisms are vastly more complex and include many more components (see, e.g., Skillings 2015). Nevertheless, this two-level picture can, for example, provide structure to the scientific debate or help in transporting the basic ideas of the organization of the mechanism to a context where the details matter less. A key question in this case as well as in others is whether the benefits of the levels-abstraction outweigh the drawbacks of distorting and losing details. The answer to this question will depend on the specifics of the case and the purpose or goal of the use of the notion of levels (see also Brooks & Eronen, 2018; Potochnik, this volume).

4.4   Levels and Downward Causation

Above I have argued that the traditional idea of compositional levels of organization is problematic but that there are nevertheless biologically relevant ways of making sense of levels. What is important to keep in mind is that these different notions of levels (e.g., minimalistic compositional levels, scale-based levels, heuristic idealizations) come with different assumptions and implications. In most scientific or philosophical contexts, these differences cannot be overlooked but are crucially important. One example of this is the debate on downward or top-down causation.

The concept of downward or top-down causation is intertwined with the concept of levels: in downward causation, higher-*level* states or processes cause lower-*level* states or processes. Many scientists and philosophers consider downward causation to be an important feature of nature (e.g., Campbell, 1974; Ellis, 2008; Noble, 2012), whereas others approach it with skepticism, for reasons having to do with causal exclusion and the nature of causation (e.g., Craver & Bechtel, 2007; Kim 1999, 2005). In this section, I will argue that there are many distinct forms of downward causation, corresponding to different notions of levels, and that arguments in favor or against one type of downward causation do not automatically apply to others.

First, the term "downward causation" often refers to causation where a change in a higher-level whole results in a change in the components of that whole. This kind of downward causation is the target of the now-classic paper by Craver and Bechtel (2007), who discuss downward causation in the framework of levels of mechanisms. Craver and Bechtel argue that what appears to be downward causation can be explained away as regular same-level causation that then has "mechanistically mediated effects" downward to the components in the mechanism. One of the examples provided by the authors is Hal's tennis match. Hal is playing tennis, and this results in changes in the glucose metabolism of cells in Hal's body. This looks like downward causation: the activity of the mechanism as a whole (i.e., Hal playing tennis) causes changes in the activity of the components of the mechanism (the cell metabolism). However, according to Craver and Bechtel, seeing this as downward causation would violate many of the core ideas that we associate with causation, in particular, that causes must precede their effects and that causes must be distinct from their effects. As an alternative, Craver and Bechtel propose to analyze these cases in terms of the constitutive (noncausal) relationship between the mechanism and its components: when the activity of the mechanism as a whole changes, the activities of some of its components change as well, but simply because these activities of components make up the activity of the mechanism as a whole (see also Romero, 2015). There is no causation going from the higher to the lower level; rather, the higher level is constituted by the lower-level entities and activities.[5]

Craver and Bechtel (2007) are thus targeting *compositional* downward causation, where a system (or mechanism) as a whole exerts causal influence on its *own* parts. It is important to understand that their reasoning against downward causation does *not* carry over to contexts where downward causation is understood in a *noncompositional* way. One prominent example of such a context is the debate on mental causation. In this context, "downward causation" refers to cases where a mental state is a putative cause of physical behavior (Baumgartner 2010; Kim, 1999, 2005; Raatikainen, 2010). For example, John's desire to drink beer is a putative cause for John walking toward the fridge (and the question then is whether or not this mental cause is excluded by a neural cause). Importantly, the physical behavior (e.g., walking toward the fridge) is *not* a component of the mental state (e.g., desire to drink beer) but is at a lower level in some more general sense.[6] In Kim's (1999) terminology, this kind of downward causation is *nonreflexive*, as the higher-level state or process is not influencing its own constituents. The sort of downward causation that Craver and Bechtel discuss, that is, where a whole influences its own parts, is *reflexive.* In nonreflexive downward causation, causes clearly precede their effects and are distinct from them, so the arguments of Craver and Bechtel do not apply.

As a biological example of (putative) nonreflexive downward causation, consider the following case discussed by Love (2012). When explaining protein folding, scientists appeal to chaperone molecules in the cellular environment. Chaperones interact with the primary structure of the folding protein (the amino acids in a sequence), and this causal impact is due to the three-dimensional overall structure of the chaperones. This seems to be a case of downward causation in the sense that the higher-level properties of chaperones affect the lower level of primary amino acid structure. However, as the amino acids in the folding proteins are *not* components of the chaperones, the (putative) downward causation here is not reflexive: the causal relationship is not between a whole (mechanism) and its own parts. Consequently, this kind of downward causation does not face the metaphysical problems raised by Craver and Bechtel.

In the context of this book chapter, a key question is how to understand the "levels" in this kind of downward causation. In my view, the heuristic approach characterized at the end of the previous section is perfectly suited for this purpose. As chaperones are proteins themselves and thus also made up of amino acids, they are similar to the folding proteins in important ways: both are proteins and composed of amino acids. Thus, in this limited context, the entities of interest seem to fit the idea of "similar things made up of similar things" sufficiently well to make the levels notion useful. For this reason, it is justifiable to label this case as downward causation, as long as it is appreciated that this kind of downward causation is very different from the reflexive downward causation discussed above. Moreover, as the levels here are not conceptualized ontologically but understood heuristically, the sense in which causation goes "downward" in these cases is also heuristic and not ontological. This suggests that in contexts where levels are understood heuristically, also the question of whether there is downward causation or not becomes heuristic in nature.

There are also further types of downward causation where a *scale-based* notion of levels seems to be more appropriate than a compositional or heuristic one.[7] Consider the following examples from the downward causation literature. Bishop (2008) argues that downward causation occurs when convection cells in a fluid constrain the behavior of the individual molecules in the fluid. According to Green (2018), properties of tissue structure (e.g., resistance) set boundary conditions for the propagation of action potentials and this amounts to downward causation. Based on his seminal work on cardiac rhythm, Noble (2006, 2012) points out that the behavior of ion channels in a cell depends on the cell potential, which is a system-level property of the cell. In all of these cases, the relevant levels do not seem to be compositional but scale-based levels, as is also explicitly stated by Green (2018) and Noble (2012). For example, there is no sense in which an ion channel is a component of the cell potential. Thus, the worries raised by Craver and Bechtel (e.g., that causes need to be distinct from their effects) do not apply, at least not in the same form as with reflexive downward causation (see also Green, this volume; Woodward, this volume, for more on downward causation). On the other hand, nonreflexive forms of downward causation (both scale based and heuristic) potentially face the causal exclusion problem, which has been intensely debated in recent years (see, e.g., Baumgartner, 2010, for a clear statement of the problem and Eronen & Brooks, 2014, and Woodward, 2015, for proposed solutions).

A more detailed analysis of the plausibility of downward causation in its different forms is beyond the scope of this chapter. I hope to have shown that it is crucially important to pay attention to the notion of levels underlying claims of downward causation, because different notions of levels correspond to different forms of downward causation. As arguments in favor or against one form do not automatically transfer to another form, it is important to clearly distinguish between them and to give each form a separate treatment.

4.5   Conclusions

The idea of compositional levels of organization figures prominently in contemporary biology and its philosophy. In this chapter, I have analyzed the complex relationship between compositional organization and levels of organization. The upshot is that compositional organization may result in neat layer-cake style levels when components are sufficiently homogeneous, but the more heterogeneous the components are, the more branching and less layered the organizational structure will be. In biology, components tend to be extremely heterogeneous, which leads to branching structures and highly local compositional levels that do not even span across one mechanism or system. The idea of levels as similar things that are composed of things that are similar to one another can at best be seen as a heuristic abstraction that may be useful in specific biological contexts. Finally, in the debate on downward causation, it is particularly important to make assumptions about levels explicit, as different notions of levels correspond to different types of downward causation.

Acknowledgments

I am very grateful to Dan Brooks, James DiFrisco, and Bill Wimsatt for organizing the Konrad Lorenz Insitute workshop "Hierarchy and Levels of Organization in the Biological Sciences" that this book is based on, and for their work in editing this chapter. I would also like to thank Laura Bringmann, Angela Potochnik, and James Woodward for their very helpful comments on an earlier draft of this manuscript. I also presented the main ideas of this chapter at the aforementioned workshop and would like to thank the audience for their valuable comments.

Notes

1. More technically, direct components can be defined as the components of a system that are not components of any other component of the system (Eronen, 2015).

2. This is just a simplified "textbook"-style picture of the components of the cells—the actual compositional organization is far more complex and heterogeneous (see, e.g., Skillings, 2015).

3. In some cases, it may also be possible to delineate compositional levels that are more general. The cases that I have in mind are those where *all* (or most) compositional branches in a system eventually involve certain types of components. As an example, let us assume that in all compositional branches in the spatial memory mechanism in a mouse brain, we will at some point end up with synapses as components. In this case, it can be useful and informative (for various purposes) to think of these synapses as being "at the same level." This would then amount to a synaptic level that is not just specific to a certain branch of the mechanism but cuts across the whole mechanism.

4. Wimsatt (1994/2007), however, argues that these levels are also compositional.

5. See, however, Bechtel (2017), who now argues that downward causation occurs when a mechanism as a whole constrains or controls the behavior of its parts. A similar argument has been made by Kistler (2009). Another way of saving compositional downward causation is to conceptualize mechanisms as temporally extended entities, in which case higher-level causes *can* be distinct from their lower-level effects; see Krickel (2017) for more.

6. It is far from clear in what sense behavior is at a lower level than mental states. In a closer analysis, it may well turn out that levels talk is not very suitable or helpful here.

7. See also Ellis (2008), Emmeche et al. (2000), Moreno and Umerez (2000), and especially Hulswit (2005) for overviews of yet further types of downward causation.

References

Baumgartner, M. (2010). Interventionism and epiphenomenalism. *Canadian Journal of Philosophy, 40*, 359–384.

Bechtel, W. (2008). *Mental mechanisms: Philosophical perspectives on cognitive neuroscience.* London, UK: Routledge.

Bechtel, W. (2017). Explicating top-down causation using networks and dynamics. *Philosophy of Science, 84*(2), 253–274.

Bechtel, W., & Richardson, R. C. (2010). *Discovering complexity: Decomposition and localization as strategies in scientific research.* Cambridge, MA: MIT Press. (Original work published 1993)

Bishop, R. C. (2008). Downward causation in fluid convection. *Synthese, 160*(2), 229–248.

Brooks, D. S. (2017). In defense of levels: Layer cakes and guilt by association. *Biological Theory, 12*(3), 142–156.

Brooks, D. S., & Eronen, M. I. (2018). The significance of levels of organization for scientific research: A heuristic approach. *Studies in the History and Philosophy of Biological and Biomedical Sciences, 68–69*, 34–41.

Bunge, M. (1977). Levels and reduction. *American Journal of Physiology: Regulatory, Integrative and Comparative Physiology, 233*(3), R75–R82.

Campbell, D. T. (1974). Downward causation in hierarchically organised biological systems. In F. J. Ayala & T. Dobzhansky (Eds.), *Studies in the philosophy of biology: Reduction and related problems* (pp. 179–186). London, UK: MacMillan.

Craver, C. F. (2007). *Explaining the brain: Mechanisms and the mosaic unity of neuroscience.* Oxford, UK: Oxford University Press.

Craver, C. F. (2015). Levels. Retrieved from www.open-mind.net

Craver, C. F., & Bechtel, W. (2007). Top-down causation without top-down causes. *Biology and Philosophy, 22*(4), 547–563.

DiFrisco, J. (2017). Time scales and levels of organization. *Erkenntnis, 82*(4), 795–818.

Eldredge, N. (1996). Hierarchies in macroevolution. In D. Jablonski, D. H. Erwin, & J. H. Lipps (Eds.), *Evolutionary palaeobiology* (pp. 42–61). Chicago, IL: University of Chicago Press.

Ellis, G. F. (2008). On the nature of causation in complex systems. *Transactions of the Royal Society of South Africa, 63*(1), 69–84.

Emmeche, C., Køppe, S., & Stjernfelt, F. (2000). Levels, emergence, and three versions of downward causation. In P. B. Andersen, C. Emmeche, N. O. Finnemann, & P. V. Christiansen (Eds.), *Downward causation: Minds, bodies and matter* (pp. 13–34). Aarhus, Denmark: Aarhus University Press.

Eronen, M. I. (2013). No levels, no problems: Downward causation in neuroscience. *Philosophy of Science, 80*(5), 1042–1052.

Eronen, M. I. (2015). Levels of organization: A deflationary account. *Biology and Philosophy, 30*(1), 39–58.

Eronen, M. I., & Brooks, D. S. (2014). Interventionism and supervenience: A new problem and provisional solution. *International Studies in the Philosophy of Science, 28*(2), 185–202.

Eronen, M. I., & Brooks, D. S. (2018). Levels of organization in biology. In E. N. Zalta (Ed.), *The Stanford encyclopedia of philosophy (Spring 2018 edition).* Stanford, CA: Metaphysics Research Lab. Retrieved from https://plato.stanford.edu/archives/spr2018/entries/levels-org-biology/

Green, S. (2018). Scale-dependency and downward causation in biology. *Philosophy of Science, 85*(5), 998–1011.

Green, S., & Batterman, R. (2017). Biology meets physics: Reductionism and multi-scale modeling of morphogenesis. *Studies in History and Philosophy of Science Part C: Studies in History and Philosophy of Biological and Biomedical Sciences, 61*, 20–34.

Guttman, B. S. (1976). Is "levels of organization" a useful concept? *BioScience 26*(2): 112–113.

Hulswit, M. (2005). How causal is downward causation? *Journal for General Philosophy of Science, 36*(2), 261–287.

Kim, J. (1999). Making sense of emergence. *Philosophical Studies, 95*(1–2), 3–36.

Kim, J. (2002). The layered model: Metaphysical considerations. *Philosophical Explorations, 5*, 2–20.

Kim, J. (2005). *Physicalism, or something near enough.* Princeton, NJ: Princeton University Press.

Kistler, M. (2009). Mechanisms and downward causation. *Philosophical Psychology, 22*(5), 595–609.

Krickel, B. (2017). Making sense of interlevel causation in mechanisms from a metaphysical perspective. *Journal for General Philosophy of Science, 48*(3), 453–468.

Krickel, B. (2018). *The mechanical world: The metaphysical commitments of the new mechanistic approach.* Cham, Switzerland: Springer.

Lewontin, R. C. (1970). The units of selection. *Annual Review of Ecology and Systematics, 1*(1), 1–18.

Love, A. C. (2012). Hierarchy, causation and explanation: Ubiquity, locality and pluralism. *Interface Focus, 2*(1), 115–125.

Moreno, A., & Umerez, J. (2000). Downward causation at the Core of Living Organization. In P. B. Andersen, C. Emmeche, N. O. Finnemann, & P. V. Christiansen (Eds.), *Downward causation: Minds, bodies, matter* (pp. 99–117). Aarhus, Denmark: Aarhus University Press.

Noble, D. (2006). *The music of life.* Oxford, UK: Oxford University Press.

Noble, D. (2012). A theory of biological relativity: No privileged level of causation. *Interface Focus, 2*(1), 55–64.

Novikoff, A. B. (1945). The concept of integrative levels and biology. *Science, 101*(2618), 209–215.

Oppenheim, P., & Putnam, H. (1958). The unity of science as a working hypothesis. In H. Feigl, M. Scriven, & G. Maxwell (Eds.), *Concepts, theories, and the mind-body problem* (pp. 3–36). Minneapolis: University of Minnesota Press.

Potochnik, A. (2017). *Idealization and the aims of science.* Chicago, IL: University of Chicago Press.

Potochnik, A., & McGill, B. (2012). The limitations of hierarchical organization. *Philosophy of Science, 79*(1), 120–140.

Raatikainen, P. (2010). Causation, exclusion, and the special sciences. *Erkenntnis, 73*, 349–363.

Romero, F. (2015). Why there isn't inter-level causation in mechanisms. *Synthese, 192*(11), 3731–3755.

Simon, H. A. (1996). The architecture of complexity: Hierarchic systems. In H. A. Simon (Ed.), *The sciences of the artificial* (3rd ed., pp. 183–216). Cambridge, MA: MIT Press. (Original work published 1962)

Skillings, D. J. (2015). Mechanistic explanation of biological processes. *Philosophy of Science, 82*(5), 1139–1151.

Tëmkin, I., & Eldredge, N. (2015). Networks and hierarchies: Approaching complexity in evolutionary theory. In E. Serrelli & N. Gontier (Eds.), *Macroevolution: Explanation, interpretation and evidence* (pp. 183–226). Dordrecht, Netherlands: Springer.

Thalos, M. (2013). *Without hierarchy: The scale freedom of the universe.* Oxford, UK: Oxford University Press.

Umerez, J. (2016). Biological organization from a hierarchical perspective: Articulation of concepts and inter-level relation. In N. Eldredge, T. Pievani, E. Serrelli, & I. Tëmkin (Eds.), *Evolutionary theory: A hierarchical perspective* (pp. 63–85). Chicago, IL: University of Chicago Press.

Wimsatt, W. C. (1976). Reductionism, levels of organization, and the mind-body problem. In G. G. Globus, G. Maxwell, & I. Savodnik (Eds.), *Consciousness and the brain* (pp. 205–267). New York, NY: Plenum Press.

Wimsatt, W. C. (2007). The ontology of complex systems: Levels of organization, perspectives, and causal thickets. In W. C. Wimsatt, *Re-engineering philosophy for limited beings: Piecewise approximations to reality* (pp. 193–240). Cambridge, MA: Harvard University Press. (Original work published 1994)

Winther, R. G. (2006). Parts and theories in compositional biology. *Biology and Philosophy, 21*(4), 471–499.

Woodger, J. H. (1929). *Biological Principles.* New York, NY: Harcourt, Brace.

Woodward, J. (2015). Interventionism and causal exclusion. *Philosophy and Phenomenological Research, 91*(2), 303–347.

Zylstra, U. (1992). Living things as hierarchically organized structures. *Synthese, 91*(1/2), 111–133.

5 Levels, Perspectives, and Thickets: Toward an Ontology of Complex Scaffolded Living Systems

James Griesemer

Overview

Complex systems appear to be hierarchically organized, and the idea of compositional levels of organization is one classic strategy to capture this type of complexity. However, compositional levels are inadequate for capturing the complexity of complexity. Wimsatt's ontology for complex systems—comprised of compositional levels, theoretical perspectives, and causal thickets—is mobilized in an argument that living developmental systems are typically scaffolded in ways that not only make them "interactionally" as well as descriptively complex, but with the effect that their system/environment boundaries change dynamically through the developmental process.

5.1 Introduction

Herbert Simon (1962, 1966/1981, 2002; cf. Callebaut 2007) offered a model and a parable to explain the intuition that evolved natural and designed artificial systems should be hierarchically structured in nearly decomposable, compositional levels of organization. I argue that Simon's model does not adequately represent the phenomena of his parable and that his parable does not appropriately illustrate his model. There is a missing ingredient: *scaffolding* explains how hierarchical order appears to arise out of complex nonhierarchical order, whether built up from fully decomposable aggregations or instead originating in interactionally complex, nondecomposable structures—"causal thickets"—that are "pruned down" (Wimsatt, 1974, 1997, 2007). In other words, there is a dynamical aspect of the parable that Simon did not consider that explains plausibly how the parable works better than what Simon says about it. My explanation for the appearance of hierarchy is epistemic: we tend to ignore developmental scaffolds in the environment when we describe the structure of scaffolded systems and thus see them (from some perspective or other) as neatly hierarchical, when in fact they are far more complex.

At the same time, the role I propose for scaffolding in Simon's parable *undermines* the applicability of Simon's model to his parable and therefore its applicability to his competitive evolutionary explanation favoring hierarchy, so we are back to scratch in explaining the appearance of hierarchical order in nature. In complicating Simon's picture in the parable, then, we are also seeking a clue as to how to revise or substitute for the model he applied to evolution and thus a clue as to how to formulate an extended evolutionary theory (Laland et al., 2014, 2015; Müller, 2017; Pigliucci & Müller, 2010; Wray et al., 2014). Before we can get there, however, we will need to understand more about what Simon says and, importantly, what Wimsatt says in response to Simon about the complexity of complexity. I suggest the idea of scaffolding lends credence to Wimsatt's intuition that complex systems are often such tangles, such concrescences (Winslow, 2017) of causal processes that we tend to recognize hierarchical structure in those cases in which human interpreters have "pruned" the multiple perspectives in a causal thicket to reveal hierarchy as *simplifications* of actual causal order.

5.2   Simon's Argument for Hierarchical Levels of Organization

Simon says that real systems are not truly decomposable into strict compositional hierarchies. Instead, they are *nearly* decomposable. In so-called hierarchical systems, interactions among components across subsystems are weak relative to interactions of components within subsystems. If the former were zero, the system would be strictly decomposable and the behavior of the whole a mere aggregate of the behaviors of its components (see Wimsatt, 2007 pp. 274–312, on aggregativity).

In most physical systems, however, there is at least *some* interaction across subsystems. Covalent bonds *within* organic polymers are much higher energy than hydrogen bonds linking atoms *between* polymers, yet many hydrogen bonds can hold polymers together. Temperatures in corners of closed rooms of a building equilibrate faster than temperatures for each floor of many rooms or the whole building of multiple floors because gas molecules within rooms interact much more frequently and with higher energy than do those among rooms (via heat transfer through dividing walls or under closed doors). Cells of a multicellular organism or bacterial mat carry on their metabolic processes somewhat in isolation from their neighbors due to their bounding membranes, but there is exchange of material across membranes as well, regulated by various mechanisms.

We can think of near-decomposability as a criterion for *relational* dynamical specification of boundaries of objects or processes or as a criterion for biological units, "modules," or even "individuality." The ratio of among-to-within (sub)system interaction is a measure of degree of modularity and dynamical independence and thus a measure of the extent to which real systems are ordered hierarchically. Simon's insights about hierarchy and near-decomposability have gained renewed currency in discussions of modularity in evolutionary developmental biology (Callebaut & Rasskin-Gutman, 2005; Schlosser, 2004), where there is great need to conceptualize units of developmental construction as well as units of adaptive evolution.

At the heart of Simon's essay is a parable of two watchmakers, Hora and Tempus. They each make watches of 1,000 parts, but Hora's are put together in ten-part sub-subassemblies, which in turn are assembled into ten 100-part subassemblies, which in turn are assembled into a watch of 111 subassemblies comprising $10 \times 10 \times 10 = 1,000$ parts. Each component subassembly, like the whole watch, is stable (on the time scale of the assembly process). Tempus, on the other hand, puts 1,000 parts together and only the whole is stable. When the phone rings from a customer to order a watch, the watchmakers have to put down their partially built watches to answer the phone, and the work under construction falls to pieces. Hora thrives and Tempus goes out of business because Hora only loses a subassembly-in-the-making while Tempus loses a whole watch in-the-making each time the phone rings.

While Simon's parable is clever, it does not support the mathematical model he offers to interpret it, nor is it an apt model of developmental construction. Most commentary focuses on what Hora's procedure adds compared to Tempus's. I ask a developmental question to challenge Simon's assumptions: how can either Hora or Tempus make watches *at all* given just the conditions Simon describes?

The argument is roughly that the homogeneous Poisson model Simon uses to describe mathematically how phone calls disrupt the work carries with it the assumption that the step-by-step processes of Hora and Tempus are the same. Yet they cannot be the same due to the very dynamical argument Simon makes because Hora's subassemblies of up to ten parts are internally stable while Tempus's are not. This means that Tempus must be doing something different from Hora when he puts parts together, so the formal model *of the individual steps* for Hora and Tempus should differ as well as the overall structure of the resulting watches. But if that's right, Simon can't apply the same Poisson model to both watchmakers step by step. I think that Tempus must be applying more sophisticated and extensive external *scaffolding* than Hora, either in how they hold parts with their hands or how they use external scaffolding jigs and clamps (which is what real watchmakers do). Without fancy scaffolding, it is hard to imagine Tempus could build a watch at all, phone call or no phone call.[1]

Thus, while Simon's formal model (cf. Simon & Ando, 1961) may be an appropriate heuristic null model for some purposes, developmental construction is not one of them. What Simon should have said about his parable is that "scaffolding" complicates, but may answer, the developmental question about construction dynamics for complex systems. Unfortunately, what Simon should have said about scaffolding (if he had recognized the concept)— that it leads to "interactional complexity" (i.e., causal interaction across different perspectives on the descriptive decomposition of a system into parts at different levels)—violates the assumptions of near-decomposability. Scaffolding undermines Simon's argument for nearly decomposable hierarchy as the core notion individuating entities of development (and evolution). This is a result due to Wimsatt (1974, 2007, chap. 9).

I suggest that typical living systems, the kinds Wimsatt (1974) identifies as "interactionally complex," are constructed or develop by means of exogenous scaffolding or by means of self-scaffolding events in the developmental process. A different approach to complexity and organization than one based on what Simon says, taking scaffolding interactions in the developmental and evolutionary origins of modules into account, is needed to articulate concepts of complexity and modularity suited to evolutionary developmental biology and extended evolutionary synthesis.

It is important to keep in mind that Simon framed his argument as a *contrast* between competing aggregative (Tempus) and hierarchical (Hora) systems of construction. Wimsatt's (1974, 2007, chap. 9) critique looks at how hierarchy emerges from failures of aggregativity in the complex interplay between epistemic descriptions of systems as comprising parts (Wimsatt's "K-decompositions") and ontic descriptions of systems of causally interacting parts (Wimsatt's "S-decompositions"). Interactional complexity involves causal interactions that "cross-cut" system descriptions in ways not recognized by Simon.

5.3 Wimsatt's Three-Tiered Ontology of Complex Systems

I propose a shift of perspective: that hierarchical organization be viewed as an *achievement* from the (ontological) top down, not something built from elemental building blocks from the bottom up. In my view, what is given, as William James said, is how nature appears to a baby lacking experience: "one great blooming, buzzing confusion" (James 1918, p. 488). Even human design doesn't proceed from the bottom up. To make a watch, car, or computer, you design from the top down in that designing and building must conjure a context and not just building blocks or at least a vision of system function that can be realized in a configuration of material parts.

Even without design, evolution works with *preexisting* organisms and environments, which form the "given" context in which evolution operates. Scientific investigations of nature must begin with humans entering into relationships with nature, which always already creates "causal thickets" in Wimsatt's sense, even if we have a theory by which to work some descriptive simplification on the phenomena. A theory may guide an observer, collector, or experimenter to move quickly from thickety experience to a more orderly, interpretable interaction by resolving, pruning, or reducing. However, the increased speed and confidence in how to approach a new problem or phenomenon when armed with a theory does not make "nature" any less of a thicket than it appears at the start of a new experience.

Where Simon might have started with his parable of two watchmakers would have been to set the scene of his investigation, just as he started his mathematical modeling by announcing presuppositions. To tell the parable, Simon might have imagined walking into the two watchmakers' shops to see how they make watches. There would be a mass of clutter and confusion (to those uninitiated in the habits of watchmakers) of jigs and clamps, tools and workbenches, the accoutrements of watchmaking. These would be noticed before the tiny watch parts. Look at a building or bridge under construction and the first thing you notice is the scaffolding, not the obscured building-in-formation. Watchmaking is as much about the scaffolding as it is about watch parts. A story or model that does not tell us *how* to put two watch parts together, only *that* they are, cannot tell us about the (dynamical) organization of watchmaking, only about the structure of (finished) watches. But the whole point of Simon's argument was supposed to be to open up the black box of watch *making* in order to consider why the finished products tend to have the structures they appear to. Simon just didn't open the box wide enough.

I think Simon's parable would be better articulated if it started from consideration of the workshop, not just of watch parts. In the same spirit, I draw from Wimsatt's work an articulated framework of concepts about complex systems to interpret dynamic adjustments of system/environment boundaries in developmental scaffolding processes. I think that most or all developmental systems, just as most or all systems of manufacture of artifacts, are in the class of systems Wimsatt calls interactionally complex.

In his 1974 critique, Wimsatt pointed out that there are two kinds of complexity relevant to Simon's argument (as Simon himself noticed in his discussion of descriptive vs. system complexity in the original paper). Wimsatt gives an elegant and penetrating account of a distinction between descriptive and interactional complexity (Wimsatt, 1974, 2007, chap. 9). Systems are descriptively complex when they admit of different descriptive "K-decompositions" (see Kauffman, 1971) of a system into parts from different theoretical perspectives. "Decompositions of the system into parts whose boundaries are not spatially coincident are more descriptively complex than systems whose decompositions under a set of perspectives are spatially coincident" (Wimsatt, 1974, p. 176, cf. 2007, p. 181). In nested hierarchical organizations, the boundaries of parts are contained in or coincident with those of more inclusive parts (at higher levels of composition). If different descriptive decompositions of a system give noncoincident spatial boundaries, the system is complex in a way that Simon's story of temperature equilibration of rooms in a building or parable of watches as systems of solid state parts is not. Wimsatt compares a piece of descriptively simple granite with a descriptively complex fruit fly (Wimsatt, 1974, see his figure 1; 2007, figure 9.1, p. 183).

A system can also be decomposed into subsystems of *causal* interactions ("S-decompositions"). Wimsatt describes interactional complexity in terms of the ratio of the strength of causal interactions among elements *between* interacting subsystems (as bounded by different parts decompositions) relative to the strength of causal interaction among elements *within* interaction subsystems. A system is interactionally simple if the interactions *within* subsystems tend not to cross the boundaries of K-decompositions and interactionally complex to the extent that they do (Wimsatt, 1974, figure 2; 2007, figure 9.2, p. 185).

Wimsatt revisited his distinction among three ontological notions or tiers in an ontological hierarchy: compositional levels, theoretical perspectives, and causal thickets, which he argues are required for "a full accounting" of the phenomena investigated by most sciences (Wimsatt, 1994, 2007, chap. 10, p. 194). He understands the relations between these categories of causation in terms of robustness of the entities that occupy them. One way to interpret Wimsatt's concept of object robustness across perspectives is by analogy with his views on perceptual robustness (Wimsatt, 1981). Objects detectable by means of different sensory modalities are more robust than objects detectable by fewer or a single modality. A mirage may be detected by sight, but perhaps not by touch, whereas a real pool of water can be detected by both. Wimsatt's perspectives are analogous to sensory modalities—both are "ontic" and both also deeply anchor representational successes and cognitive experiences relevant to our knowledge of causal systems. Perspectives and thickets reflect the messiness of ontology when robustness of the "objects, properties, events, capacities and propensities" (Wimsatt, 1981) at levels of organization weakens or fails. He characterizes compositional levels this way: "Levels of organization can be thought of as local maxima of regularity and predictability in the phase space of alternative modes of organization of matter" (Wimsatt, 2007, p. 209). Differently put, levels are maxima of detectability across sensory or perspectival modalities.

For Wimsatt, what it means for objects in this phase space to lack robustness is for the levels themselves to break down, to cease to be a mode of organization. Because robustness is both ontic and epistemic—a property of things in nature *and* a feature of our description and modeling of nature—Wimsatt understands levels, perspectives, and thickets as both ontic and epistemic. The two are hand in glove, but it makes little sense to think of one as the hand and the other as the glove. "Things are robust if they are accessible (detectable, measurable, derivable, definable, producible, or the like) in a variety of independent ways" (Wimsatt, 2007, p. 196), where independence is understood as independent probabilities of failure of the different means of access (Wimsatt, 2007, p. 197). To be an object is to be knowable robustly (Wimsatt, 2007).

Wimsatt characterizes his ontological categories in terms of these ideas about complex organization. By levels of organization, Wimsatt means compositional part–whole relations in which wholes at one level function as parts at all higher levels (Wimsatt, 2007, p. 201). The "all" in this characterization expresses the sense in which decomposability or near-decomposability captures the relevant causal interaction aspect of the levels concept. His view of levels is something Simon could agree to. "I urge a view that Simon would share: that levels of organization are a deep, non-arbitrary, and extremely important feature of the ontological architecture of our natural world" (Wimsatt, 2007, p. 203).

But we can *describe* entities as residing at multiple levels of organization, so if the entities we can most easily identify are the most robust ones, "relations must hold between descriptions of the same object at different levels" (Wimsatt, 2007, p. 205). Because entities at higher levels of organization will have more parts with more degrees of freedom and emergent properties, the former will have "increased richness of ways ... of interacting with one another" (Wimsatt, 2007, p. 205). By the same token, regularities at higher levels of organization will have more exceptions due to this increased number of ways of interacting. The "robustness" of objects begins to break down at higher levels as the exceptions overtake the rules. Because levels pick out maxima of regularity and predictability, their robustness "tends to make them stable reference points that are relatively invariant across different perspectives," and hence explanation tends to be "level-centered" (Wimsatt, 2007, p. 214).

When phenomena seem to occur "between levels," for example, Brownian motion (where entities of substantially different sizes interact), we tend not to recognize the clusters of objects (e.g., a dust mote plus the disturbing particles that cause it to display Brownian motion) "as entities" at all (Wimsatt, 2007, p. 215). We may give functional descriptions of such things, but "they" seem not to have "intrinsic properties," "since only levels have the intensity of different kinds of interactions among entities to fix unique sets of intrinsic properties as being causally relevant" (Wimsatt, 2007, p. 216). These points bear on the status of scaffolds: when a scaffold and what it scaffolds interact (combine), even when they are roughly of the same size, their functional roles are so different that we tend not to recognize the pairings as entities in their own right. We also tend to underestimate the strengths of interaction between them.

Wimsatt continues with the story of the breakdown of regularity: "As the richness of causal connections within and between levels increases, levels of organization shade successively into two other qualitatively different kinds of ontological structures that I have called, respectively, 'perspectives' (Wimsatt, 1974) and 'causal thickets' (Wimsatt 1976a)" (Wimsatt, 2007, p. 205). In particular, when interactions become complex in Wimsatt's sense, the "neat compositional relations break down," so "levels become less useful as ways of characterizing the organization of systems" (Wimsatt, 2007, pp. 221–222) and "other ontological structures enter, either as additional tools or as replacements," including perspectives. Indeed, "perspectives cannot be ordered compositionally relative to one another" as levels could (Wimsatt, 2007, p. 231).

Wimsatt's notion of perspectives recognizes that the state variables describing causal interactions of entities, organized as parts of a whole, may be salient to a class of observers because of the ways those observers (and, I would add, scaffolds), in turn, causally interact with complex systems (Wimsatt, 2007, p. 227). Moreover, the variables are not taken to be complete, so other perspectives are possible. However, for the problems and interests of a class of observers (or scaffolds), a perspective may describe a complex system well enough for their particular purposes, while another perspective may describe the same system differently but well enough for other purposes, so more than one perspective might be "applied" to the behavior of a system at a time (Wimsatt, 2007, pp. 227–228). Thus, scientific practices for studying complex systems tend toward a form of pluralism, rather than forms of (apparent) monism with a reductionistic strategy applicable to the more robust entities occupying (typically lower) levels of a hierarchical organization. Similarly, a theory of the engagements of scaffolds in facilitating the development (or maintenance) of a complex system must tend toward a form of theoretical pluralism, because the dynamics of different sorts of scaffolding interactions will call for different "calculi" of combinations and permutations afforded by the many ways scaffolds might interact with their "targets."

With still further interactional complexity, even perspectives begin to fail: empirical problems become "cross-perspectival" (Wimsatt, 2007, p. 234). And when the boundaries between perspectives begin to break down, Wimsatt describes the situation as a "causal thicket." For example, what counts as a genetic perspective and what counts as a physiological perspective in phenomena of gene expression become entangled with questions of the nature of hereditary transmission (e.g., in research into transgenerational epigenetic inheritance). Is the methylation reaction leading to so-called epigenetic marking of DNA a chemical reaction, a physiological interaction, or a hereditary transmission? Or is it a causal interaction crossing all three of these perspectives on atoms, molecules, cells, and organisms? Another example is when biological entities participate in "culture": it becomes challenging to distinguish organisms as cultural or as biological phenomena and thus subject to sociocultural versus biological theories and perspectives other than by relying on reductionistic theories promising to map all phenomena to a lowest level of organization (e.g., genes), despite the many causal and compositional relations inherent in such "thickety" problems. Wimsatt motivates the notion of a causal thicket by appeal to those biological systems that have emergent psychologies, ecologies, socioecologies, and sociocultures (see figure 10.2 and its legend in Wimsatt, 2007, pp. 233–234).

5.4   Pruning Wimsatt's Causal Thickets: A Scaffolding Perspective

Most approaches to thinking about complexity are bottom-up, from some base notion of simple parts or simple systems. But if Wimsatt is right, evolved living systems are highly complex, entangled in (and as) causal thickets. Looked at from the top down, from top-tier causal thickets, we might consider the work done to understand these systems as a form of "pruning," of idealizing or abstracting away from some ontological perspectives in order to reveal the hierarchical structure "within" the system rather than only recognizing complexity in failures of aggregativity or robustness of the organized wholes on some account of hierarchical organization of the lowest tier of the ontology.

Wimsatt does *not* think that there is a natural (e.g., evolutionary) trajectory from levels to perspectives to thickets. He takes the ontological ordering of complexity and the breakdowns of regularity and robustness that form perspectives and thickets to be taxonomic rather than temporal (Wimsatt, 2007, p. 222). I am not so sure. Breakdowns and buildups of regularities can have a temporal dynamic in real systems. In my view, developmental scaffolding presents such a type of case. Scaffolding presents situations in which the orderliness of compositional levels is disrupted dynamically by the association of scaffold and scaffoldee. The disruption is critical to taking the scaffoldee through developmentally unstable states to new ones affording further developmental possibilities not previously available (Bickhard 1992).

Viewed temporally, Wimsatt's ontology can be read as supporting a dynamical account of "developmental transition," following the pattern of accounts of evolutionary transition (Maynard Smith & Szathmáry, 1995, 1999; Szathmáry, 2015). A developmental transition is a contingently irreversible change in the way development can proceed. Some developmental transitions may mark what classical embryology calls "stages," although not in as theoretically principled a way as I have in mind for scaffolded developmental transitions (see Griesemer, 2016, 2018).

No matter how one characterizes complexity, our tendency to characterize origins in terms of simple, minimally complex systems entails that complexity can only increase over time. Moreover, if evolutionary transition is a process that creates new levels of organization with new units of selection, by composing biological entities at one level as parts of emergent wholes at a new level (Maynard Smith & Szathmáry, 1995; Szathmáry, 2015), then it would seem that Wimsatt's characterization of the increased degrees of freedom and emergent properties of wholes at the new level should afford more ways of interacting compared to the components at lower/earlier levels. I think it likely, although I cannot argue it here, that evolutionary transitions and what I am calling developmental transitions increase the degree of intersystem, cross-perspective interaction.

It would seem to follow, as new levels emerge in transition processes, that scaffold–scaffoldee pairs are just the sorts of entities that mark a breakdown of levels of organization into organisms and groups. Scaffolding bonds may be physical or social, but they are more intimate and serve developmental functions more fundamental than ordinary interactions among members of a group. Groups formed from scaffolding processes should look more like "individuals" than mere aggregates but less like individuals than organisms.

Differently put, the problem of temporal order in the ontological structure of nature looks different in light of evolutionary transition theory than from the perspective of a multilevel selection theory fixed by a static compositional hierarchy such as Richard Lewontin's (1970) interpretation of Darwin's principles and the units of selection. Note, however, that Hull's account of units of selection (Hull, 1980, 1988; see also Griesemer, 2005) recognized that levels of selection are dynamic and "wander" among the compositional levels as entities *at* levels dynamically come to satisfy or cease to satisfy Darwinian criteria. I think Hull's observation points to the dynamism of levels of evolution inherent in Wimsatt's story of "breakdown," not merely of descriptions, but of the hierarchical order of nature itself. What I add here is the thought that this same dynamism applies to the scaffolded development of organisms as well.

Because Wimsatt's notion of interactional complexity involves both epistemic (K-decompositions) and ontic (S-decompositions) aspects, one can see how his ontological ordering in terms of complexity might be read as taxonomic rather than temporal. Interactional complexity requires there to be multiple K-decompositions, and scientists may only produce those for systems already organized hierarchically. I do not, however, think it follows that there cannot be local causal progressions or trajectories between these ontological "states" of organization.

Wimsatt's view that the ordering is taxonomic only may be an artifact of his account having been formulated in terms of departures from Simon's account of hierarchy and near-decomposability, which Simon built on a physical theory of the organization of inactive matter. Simon only argues that hierarchically organized developmental systems (Hora-like modular constructions) should outcompete nonhierarchical developmental systems (Tempus-like aggregative constructions). Neither Simon nor Wimsatt offers an origin story of the two forms of variant organization. An origin story would include imagining the context in which construction can occur, such that the two kinds of systems might compete.

Simon, for his part, never seemed to inquire into the causes of the fixity of the starting conditions for his parable of the watchmakers or thus for the fixity of parameters of the model purported to describe it. I don't think Simon would have thought to ask how Tempus's workbench might be organized to make watches at all under the severe requirements of Simon's aggregative picture. Anyone who troubles to go into the "back room" of a workshop where such work gets done or into the yard of Wimsatt's "backyard mechanic" will find a causal thicket of parts, tools, scaffolds, and systems in various states of assembly, disrepair, or decay.

It may be that when scientists approach new problems or phenomena, they have a highly constrained, theory-guided perspective on how to make sense of what they experience, but that doesn't mean their *engagement* isn't messy and interactionally complex, at least at first, or that "the system" isn't so messy that it must be given an idealized description by the scientist as containing just the building blocks of interest while backgrounding everything else as context or environment or mere conditions.

Philosophers idealize this engagement when they neglect to consider the mess of multiple perspectives of science in practice: getting the grant, negotiating research permissions, organizing equipment and personnel, arranging to get into the field, setting up the lab in the first place, and whatever other "external," "social," "bureaucratic," or "merely pragmatic" conditions enable scientists to *justifiably* engage with "the phenomenon." That socioecological or sociocultural thicket is part of the phenomenon (no matter whether the discipline in question is anthropology, sociology, psychology, biology, chemistry, or physics). Until the mess is tamed and the thickety tangle of causes is pruned in an attempt to simplify the system and its description into a working empirical phenomenon from the perspective of the work to be done, the theory to be used as a guide will be no match for the phenomenon. Nor can a guiding theory succeed in targeting a focal level in "the" compositional hierarchy of elementary particles, atoms, molecules, and so on, until the thicket is pruned into a manageable shape (Wimsatt, 2007, figures 10.1, p. 224 and 10.2, p. 233). For a physics lab, the vibrations shaking the equipment and the trucks in the street outside the lab that are doing the shaking are all parts of the system until physicists intervene to impose physical barriers to vibration that "isolate" the phenomenon they are studying from its "surroundings."

That is merely an epistemic view of Wimsatt's ontology from the top down: while nature might be "organized" in levels, perspectives, and thickets, our human investigations of nature tend to progress from thickets to perspectives to levels: that is what it means to pursue a reductionistic research strategy (see also Bechtel & Richardson, 1993).

I aim to make a Wimsattian ontic claim as well: living things in general encounter "nature" as a process of entering causal thickets whose complexity they resolve, manage, and indeed impose causal order upon by their actions and interventions. The picture here is that to enter a causal thicket without taming, mastering or coping with it is simply to be swept up by it and to survive by mere luck or perish in bewilderment, like William James's proverbial baby.

Perhaps, abstractly speaking, one might view interventionist accounts of causation, such as Woodward's (2003), as imposing conceptual order on causal networks first faced as causal thickets. The interventionist view can be used not only to interpret the causal character of a network of interactions but also to simplify it. An intervention "sets" the value of a variable in order to assess what relations change or not as a result of the intervention and can be used to infer causation in terms of that changed network structure. By the same token, intervention *simplifies* the network. To intervene, to "do" (Pearl, 2000; Pearl & Mackenzie, 2018), as opposed to infer or calculate, is to cut off part of a thickety network, insinuating an investigator-controlled cause as a substitute for a "natural" or "observed" input. If that natural input is at the head of a complex subnetwork, then the intervention has the effect of simplifying the network under investigation by pruning away the subnetwork upstream. To understand causes in this interventionist way is to simplify a causal network's thickety structure to the point where one can take a perspective or find a level at which to interact with "it."

The epistemic picture of Wimsatt's ontology, viewed from the top down, can be deployed as part of a heuristic strategy to look for how "ontic" developmental transitions can happen in the opposite direction to Wimsatt's description of breakdown from the bottom up. Breakdown occurs epistemically from levels to perspectives and perspectives to thickets, as scientists realize that their descriptions fail. But ontically, or we might say, "in practice," as scientists engage phenomena and intervene in order to understand and interpret, they prune causal thickets so that they can gain perspectives (plural) on phenomena and perhaps trace causes further to discover the sort of regularity Wimsatt thinks is characteristic of levels. At each of these transitions, we can look at them the other way and consider how interactional complexity might be exploited to impose functional order to achieve a form of dynamical autonomy, "as if" operating in more simplified causal circumstances—how, in other words, *additional* complexity in the form of a (developmental) scaffold might *simplify* causal circumstances of the developing entity, so that the developing system may develop effectively.

A parent might help a child learn to ride a bicycle by *adding* training wheels (or running alongside while loosely holding the back of the seat). In this way, the parent scaffolds the child so that the uncontrolled minor fluctuations moving the rider out of balance do not lead to a catastrophic fall as the child shifts her weight and attention from street to handlebars, to legs, to parent, and back again in a trial-and-error attempt to maintain balance while pedaling forward. The addition of the parent to the "system" of child and bicycle creates a whole thicket of interactions that cross-cut perspectives on the mechanical, interpersonal, infrastructural, and organismic aspects of a child-operating-a-machine-in-the-street. At the same time, some of those interactions add constraints in the bicycle-child-parent system that screen off or prune some causes, freeing the learner from some thickety features that may not even be apparent to an observer (e.g., the choice of which street the parent guides the child down so as to avoid that dog on a leash in the next block that is likely to bark and scare the child into a dangerous wobble).

What I am gesturing at here is the notion that reading Wimsatt's ontology from the top down also suggests an *epistemic* perspective on the conduct of inquiry focused on "preparing" phenomena by pruning nature to fit perspectives. This reading suggests looking at causal thickets and tracing certain specific interactions pointing toward those causes or constraints that have "scaffolding effects" (i.e., consequences for delimiting or changing the boundaries between what is system and what is environment). Developer and scaffold join together to achieve a developmental effect for the developer, but in the process, their interaction causes a developmental transition to a new, "hybrid" compositional level *of the duo* that persists at least temporarily while the developmental process operates.

My example was a case of learning. Learning can be interpreted as a kind of developmental process. I view development quite broadly as any process in which new capacities are acquired, refined, or actively maintained. In other work, I characterize *biological* development as a process in which a capacity to *reproduce* is acquired (Griesemer, 2000a, 2014a, 2014b, 2016). The theoretical perspective I seek considers the role of scaffolding in biological development. In a sense, I offer a refined idealization of what Waddington called the "*epigenetic* environment"—but rather than picking out *all* environmental factors operating in development, I pick out just those that play a scaffolding role.

5.5   Dynamic System/Environment Boundaries

The problem with treating system/environment boundaries as fixed that is raised by the phenomenon of scaffolding has to do with ways in which interactions between a developing system and its scaffolds can themselves develop and evolve. In my view, developmental scaffolds, to do their work of facilitating development, *change* the system/environment boundary *while they are in interaction*, which means we cannot use static representations of causal relations and fixed structure of complex systems to capture the dynamism of system/environment relations. Scaffolding systems are cases, like Brownian motion, of phenomena that seem to occur *between* levels (Wimsatt, 2007, p. 214). We can interpret these changing system/environment boundaries as dynamical transitions to a new, higher degree of complexity and level of developmental organization. The combinations of scaffolds and scaffoldees are at least temporarily "at" a new level of developmental organization. These may only be transient, and the release of a developing system from a scaffold may represent the return of the system to a "lower" compositional level marked by the regularity we associate with system "autonomy" (see Moreno & Mossio, 2015).[2] Because understanding transient levels of organization is itself a conceptual boundary problem in which the orderliness of interlevel relations used to characterize things and their environments breaks down into conceptual thickets, we must view the conceptual landscape from multiple perspectives that track system/environment divides differently.

Students of phenotypic plasticity often endorse a "norm of reaction" view of phenotypes to express the idea that phenotype results from interactions with environment rather than by gene determination, so variation among environments as well as variation among genotypes (and phenotypes) must be tracked to express phenotypic outcomes. On this view, the range of possible environments an organism might experience in its development must be considered. But in development, organisms experience *sequences* of "developmental environments," and "genomes" (in a properly expanded sense) experience *sequences* of scaffolding interactions (or their containing systems do). What conventional norm of reaction diagrams plot as "locations" on x- and y-axes are, in my view, *vectors* of changing quantities or qualities in development (Griesemer, 2016, 2018; compare figure 5.2A and 5.2B in this chapter). We need something akin to Waddington's full epigenetic landscape picture, or more, to represent it (figure 5.1A and 5.1B).

Nevertheless, norm of reaction diagrams are still genotype-centered descriptions, tracing genotype effects across alternative environments and into alternative phenotypic outcomes.[3] A more dynamic representation of developmental reaction norms is needed to represent these sequences of environments, phenotypes, and genomes as vectors in open, contingent, dynamic landscapes. Even Waddington's epigenetic landscape took genes to be pegs, working via gene expression and interaction in ways that render the *landscape* dynamic but the foundation as static: as pegs fixed in some static, abstract foundation so the landscape could change against this fixed scaffold as an organism developed. So long as the scaffolding is abstracted away, the calculus of population genetics holds the promise of a complete, if highly idealized, dynamical theory, in the sense that all "nongenetic" causes impinging on development are encapsulated (and marginalized) in the "error" or "environment" term of its equations.

One thing Waddington got right is that the temporality of the epigenetic environment in interaction with genes over the course of development is crucial, although there are (at least) two different relevant senses of environment: epigenetic and selective environments. He speculated that these concepts could be brought together by someone with greater mathematical ability than he had (Waddington, 1974, 1975, chap. 26). What I reject, though, is the notion that there *is* any fixed foundation to anchor genes in their interaction with environments: *all* of the elements of a developmental system can change, at different rates relative to one another, and what counts as the boundary of the system may also change.

5.6   A Developmental Scaffolding Reaction Norms Perspective

I propose a shift of perspectives on norms of reaction to bring the concept of scaffolding into the picture of interactionally complex systems. The perspective seeks to integrate traditional norms of reaction representations of genotype × environment interactions without the gene-centrism. I start from a Waddington-like epigenetic landscape perspective on development but without the abstraction of separate developmental and selective environments and without the abstracted fixed foundation from which genes may act.

I view biological development as the acquisition of a capacity to reproduce (Griesemer, 2000a, 2000b). Typically, this involves multiple stages of acquiring developmental capacities in a sequence until a capacity to reproduce is acquired. To become reproductive, for a metazoan, for example, may entail gastrulation, which in turn entails blastulation, which in turn entails cleavage from a fertilized egg. Each of these "stages" of development involved acquiring new capacities that resulted in part from changes in gene expression but also cell movements, tissue differentiation, and much more that affords the capacity to get to the next stage in the sequence.

Reproduction, on my view, is progeneration (propagule generation) of material entities with parts carrying a capacity to develop (Griesemer, 2000a). This functional, capacity-centered account is intended to be sufficiently abstract as not to be constrained to any particular mode of development evolved in extant life or constrained even to cellular life. Some molecular systems must have been developmental systems if cellular life was to have *evolved* (through a Darwinian process entailing entities that can reproduce) from "proto-biological" molecular systems (Griesemer, 2014b; Griesemer & Szathmáry, 2009).

As I have argued previously,

Most developmental processes are facilitated rather than autonomous in the sense that the aid of an "external" process interacting with the developing system makes easier, or more likely, or with lower fitness cost, the acquisition, refinement or exercise of a new developmental capacity (Caporael et al. 2014). I say "external" because on traditional accounts of compositional levels of organization, the scaffold is one thing, the thing scaffolded is another thing, and the pair in interaction may or may not be treated as a thing at a higher level. (Griesemer, 2018, p. 155; see also Caporael et al., 2014a; Griesemer, 2014a, 2014b; Wimsatt & Griesemer 2007)

Scaffolds are typically treated as parts of the environment that facilitate development. They add constraints, guidance, coordination, timing, and (in some cases) safety or "fitness modulation" to a developmental process, sending it down a different and more (or less) felicitous developmental pathway than it otherwise would have gone without the scaffold. Think of scaffolds as specific triggers, switches, models, templates, or constraints that operate at branch-points of developmental trajectories in Waddington's epigenetic landscapes of developmental possibilities. Temperature triggers that drive development toward male versus female adult forms in turtle species with temperature-dependent sex determination are scaffolds set up by mother turtles, since the depth at which eggs are laid in beach sand cavities are determined by parental digging activities together with ambient environmental temperatures (Valenzuela & Lance, 2004).

Waddington imagined whole epigenetic environmental contexts of developing systems represented by balls rolling down landscapes, shaped by gene expression ("chemical tendencies which the genes produce") (figure 5.1A). The three dimensions of such landscapes are X: phenotype, Y: inverse of developmental probability given gene expression, and Z (into the plane of the figure): time. In some images (figure 5.1B), the landscape was portrayed from "underneath" as a system of gene pegs pulling gene expression guy-wires shaping the landscape into a (relatively stable) surface (Waddington, 1957, figures 4 and 5, pp. 29 and 36).[4] To depict developmental scaffolding, I combine Waddington's two representations (figure 5.1C) and add the "pull" of developmental scaffolds as abstractly anchored environmental factors that interact with genes to shape the epigenetic landscape. In the example above of temperature-dependent sex determination, mother turtle and ambient temperature are both pegs pulling from the sides of figure 5.1C, just as the baby turtle's genes are pegs pulling from below to create a window of environmentally regulated developmental opportunity for steroid hormonal control of gene expression, in this case.

Now we can idealize Waddington's landscape to consider not the whole landscape and trajectory but just those points in the landscape at which scaffolds mark and guide developmental trajectories (figure 5.1D). In this image, we can represent Waddington's notion of a selective environment as similarly temporally extended and dynamic, operating over the whole epigenetic landscape (rather than only on the end-state adult phenotypes) as a sequence of fitness functions representing a changing fitness surface overlying the epigenetic landscape. This representation idealizes the selective environment by depicting it as operating only through the fitness-modulating effects of scaffolding interactions. That is,

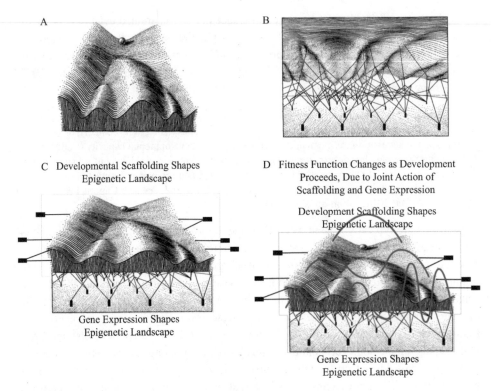

A

B

C Developmental Scaffolding Shapes
Epigenetic Landscape

D Fitness Function Changes as Development
Proceeds, Due to Joint Action of
Scaffolding and Gene Expression

Development Scaffolding Shapes
Epigenetic Landscape

Gene Expression Shapes
Epigenetic Landscape

Gene Expression Shapes
Epigenetic Landscape

Figure 5.1
Waddington's epigenetic landscape. (A) Waddington's original depiction of the epigenetic landscape. The ball represents an organism at the beginning of its ontogeny. As it develops, it rolls down the landscape following a trajectory of favored paths (troughs). Alternative paths are possible, although more or less probable. (B) The underside of Waddington's landscape is underpinned by genes (pegs) in an undefined abstract foundation. The genes pull the epigenetic landscape into a shape due to gene expression (guy-wires), which manifests "chemical tendencies which the genes produce." (C) Waddington's figures (1A, 1B) fused together, with the addition of developmental scaffolds (pegs to the sides), which also pull the landscape into shape by facilitating development through interaction with the developing system. (D) Fitness functions superimposed on Waddington's landscape, representing time slices of a fitness surface extending over the whole epigenetic landscape. Figures 5.1A and 5.1B (and elements of 1C and 1D) are copyright 1957 from *The Strategy of the Genes* by Conrad Hal Waddington. Reproduced by permission of Taylor and Francis, a division of Informa plc.

it does not represent the full complexity of the fitness surface but rather locates *developmental transitions* in fitness at developmental scaffolding "events."

From this perspective, developmental processes can be narrated as sequences of scaffolding interactions through which systems acquire new developmental capacities. The exercise of these new capacities advances developing organisms through epigenetic landscapes in their trajectories from birth to death. Along the way, scaffolds may modulate the fitness functions that describe natural selection operating among variant developmental systems in selective environments (figure 5.1D). These might be modeled as actual selective environments operating among variants in a population or as potential selective environments defined over the developmental possibilities of a given developing system, where the near-future developmental outcomes (of the rolling ball at $t+1$, given that it was at t) may afford different selective values to the developmental states that would be reached with or without a given scaffolding interaction.

Waddington's epigenetic landscapes represent the space of developmental possibilities for each state at each time but only incompletely insofar as they do not explicitly represent scaffolds that also have "guy-wires" pulling the landscape into a shape or constraint system, as Tavory et al. (2014) point out in their characterization of "social-developmental" landscapes. In figures 5.1C and 5.1D, I have added these scaffolding "guy-wires" as additional causal interactions crossing from the theoretical perspective of ecology to that of development.

Importantly, in considering the place of scaffolds in the epigenetic, social-developmental, or any other causal thicket-like landscape of causal factors that cross perspectival boundaries between ecology and development (e.g., if organisms' active behavioral choices co-construct their environments as their environments co-construct them), we can consider two ways of understanding and interpreting the implications. One mode relies on the kind of abstraction Waddington used to represent genes as conceptually *outside* the developing system, pulling on the landscape to shape developmental trajectories. In that mode, scaffolds are just an additional, interacting source of developmental constraint coming from the environment rather than from internal constraints. Waddington's abstraction reflected his desire to integrate development into the then-current neo-Darwinian gene-centric picture rather than to present an alternative theory incompatible with neo-Darwinism. Another interpretation recognizes the interaction of a scaffold with a developing system as generating a new developmental system—one with the scaffold as a *part*. Two balls (organisms) rolling down the (possibly transformed) landscape may temporarily join to form a hybrid developing duo and then perhaps separate.

In many cases of developmental scaffolding, the short temporal duration of the interaction relative to the duration of embryogenesis or the whole life trajectory of a developing organism may lead us to treat the scaffold as "environmental," especially if the resulting acquisition of a new developmental capacity of the system leaves it morphologically largely unchanged after the scaffolding interaction has ceased. People rarely judge gradual or slight phenotypic or parts change to constitute numerically distinct individuals. But for theoretical purposes, this might be a mistake if the change in developmental *capacity* marks a necessary condition for further development, or a seemingly slight deflection in capacity has the potential to produce a substantially different developmental outcome, as Waddington's landscapes were meant to illustrate.[5]

Examples of gradual change, in which the scaffold is likely to be treated as environmental, rather than as part of the developing system, might include the use of a fixed substrate as anchor for a stage in a developmental process. A rocky surface can scaffold pupation in insects. Inanimate triggers or conditions may scaffold a developmental response at an appropriate time in development (e.g., when a temperature profile in the sand triggers temperature-dependent sex determination of developing turtle offspring). These triggers or environmental conditions function as scaffolds when they are set up as triggers through the effects of scaffolding agencies that select among environmental variations.

The picture I have in mind is one in which organisms develop in *sequences* of developmental environments (the time steps in Waddington's diagram) rather than in *alternative* environments as described contrastively in norms of reaction diagrams in terms of, typically, a single parameter of measurable environments, for example, temperature or altitude (compare figures 5.2A and 5.2B). The point is that environments are processes, just as are developing systems and as is natural selection. All are distributed over the whole life

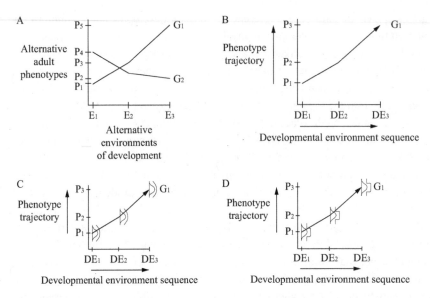

Figure 5.2
Conventional and developmental norms of reaction diagrams. (A) Conventional norm of reaction diagram showing the adult phenotypes (vertical axis) of organisms of genotypes G1, G2 as they would develop in alternative environments, E1, E2, and E3 (horizontal axis). (B) Developmental norm of reaction showing the developmental trajectory of an organism with genotype G1 through a sequence of developmental environments marked by developmental scaffolding events DEi = {DE1, DE2, and DE3} (horizontal axis). The phenotype also displays a trajectory through states Pj = {P1, P2, and P3} (vertical axis). (C) Same as B with phenotype distributions around Pj and (Gaussian) fitness function superimposed. (D) Same as B, but with a fitness function modulated (changed in shape to be more permissive of phenotypes farther from Gaussian optimum) due to developmental scaffolding.

trajectories of living organisms. They can, for very specific purposes, be idealized as occurring at points in life trajectories, but these are idealizations that may yield a misleading picture of system/environment boundaries as static. By not taking for granted what counts as system and what counts as environment when we first encounter a new phenomenon, we might look upon some aspects as scaffolding that changes the boundary *between* what might otherwise be identified as system and what as environment.

In this vein, recognition that a scaffold is interacting with a developing system amounts to identifying a mosaic or hybrid system. This "new" system would have to be recognized as creating a new system/environment relation as well, since a piece of the environment has become a part of the system. Conventional norm of reaction diagrams cannot (fully) capture what happens in the genotype × environment interaction of this developmental hybrid system. What counts as "environment," including both developmental and selective environments, changes at different points in scaffolded development. What is system at one point in development becomes a part later when a scaffold "joins" it. How selection operates may also change as scaffolds shift developmental trajectories from what they otherwise would have been (Griesemer, 2016, 2018). As different scaffolding interactions take place over the course of development, what count as system and environment depend on whether one takes the perspective of the scaffold or that of the developer or, rather, of the duo. On this reading, conventional norm of reaction diagrams are "adultocentric," in that they display adult phenotypic outcomes of development in possible alternative environments (Griesemer, 2014b, 2019; Minelli, 2011). An alternative "developmental reaction norm diagram" would take the environments of a norm of reaction to be the sequences of environments experienced and created by scaffolding interactions: a developmental sequence or vector in an epigenetic landscape. In figure 5.2C, time slices through Waddington's landscape are represented to show developmental trajectories of phenotypes subjected to local fitness surfaces (developmental transition probabilities). In figure 5.2D, the fitness surfaces of figure 5.2C are *modulated* due to the action of scaffolds at those points in development.

Genes are developmental constraints that interact in complex ways with scaffolds to change what counts as the developing system. The upshot is that the *units* of development may change with each successive scaffolding interaction. Scaffolds work by forming mosaic or hybrid developmental systems on some time and size scale defined by the interactors who form them. The salient processes during a scaffolding interaction cannot be reduced to effects of genes, even if the results of interaction can be calculated mathematically as "marginal" effects of genes in one (or more) contributors to the process. Marginal effect values cannot predict future developmental results because scaffolds, even if they are organisms, do not necessarily follow Mendelian or Darwinian dynamics in their functional-causal roles as scaffolds. We know from cultural evolutionary models that modes of cultural transmission and selection (e.g., social learning) need not follow standard gene dynamics, and scaffolding could be viewed as one generalization of "culture" in the context of development extending to many forms of developmental interaction inside organisms as well as to interorganism "ecological" interactions.

If the principles governing the dynamics of these mosaic/hybrid developmental systems are neither strictly Mendelian nor Darwinian, it is an open question whether there is a coherent theory to be had, assuming theory requires a "calculus" that is mathematically simple to operate, like population genetics. Moreover, it is not obvious that extended inheritance theories on my view of reproduction and development will look anything like an "expansion" by merely adding terms to a standard population genetics equation. This mirrors my understanding of the state of play in the debate over extended evolutionary synthesis: proponents charge that the modern, neo-Darwinian theory is inadequate, while skeptics wonder what form a proper theory *could* possibly take if the aim is to model a causal thicket where causal variables cross genetic, developmental, and ecological perspectives (Laland et al., 2014; Wray et al., 2014).

Perhaps recognizing the character of the conceptual challenge can provide insight into shaping a solution or at least guide us to instructive phenomena from which to prune the eco-devo-geno-evo thicket[6] into some semblance of explanatory, multiperspectival order. Or, alternatively, perhaps we will be forced to recognize that thickets by their nature, resist the regimentation of algebraic equation theories-as-usual, and conclude that thicket scientists must seek some other form of theoretical order.

The challenge can be sharpened considerably by diving deeper into the technical apparatus of Wimsatt's ontology of complex systems. He has offered hints, such as suggesting that we study the phase space represented by his epsilon, measuring the degree of interactional complexity, but against what other parameter? What would a Wimsattian phase plot for scaffolded development look like? Once we have the dimensions of a suitable phase space for the temporal dynamics of developmental transition, I think we will be on the path to an expanded and hopefully more explanatory configuration space for living systems that might answer Simon's question: why do the systems we see in nature tend to be (or look like they are) hierarchically organized?

5.7 Conclusion: Toward a Theory of Developmental Transitions

I sketched a developmental scaffolding perspective on phenomena that seem to display a breakdown of perspectival order, forming causal thickets in Wimsatt's sense. I have been pursuing an epistemic project, deploying the notion of causal thicket to reflect on the problem of identifying units or components of perspectival order from the top down rather than from the bottom up as failures of hierarchical order. The project is to find and understand patterns among scaffolding interactions, classify them into kinds, and explore whether new forms of theory can describe, predict, and explain these patterns. The units of development become units of interaction; a "process ontology" for them would take interactions to be primary. The material realizations of these developmental interactions are what I have been calling mosaics or hybrids.

New theoretical possibilities open up by embracing the scaffolding perspective resulting from the top-down epistemic strategy. Maynard Smith and Szathmáry characterize evolutionary transition in terms of changes in the way genetic information is transmitted before and after a transition, noting that the relevant sorts of changes involve "contingent irreversibility," that is, a low likelihood of return to the earlier mode of transmission (Maynard Smith & Szathmáry, 1995, 1999; updated in Szathmáry, 2015). I characterized their account in terms of transitions of reproducers rather than replicators (Griesemer, 2000b). Here, I suggest that the ways in which scaffolds contribute to mosaic/hybrid developmental systems look like "nongenetic" "parents" of those systems in a sense analogous to the very different *developmental* contributions of egg and sperm in sexual reproduction. An extended genetics on my view would need to count not just other organisms as developmental "parents" but also the diverse scaffolds that play developmental roles. This may stretch the notion of inheritance beyond credulity, but inroads can be made by a turn to process (e.g., Jaeger et al., 2012). In some cases, scaffolds cause "developmental transitions," marking contingently irreversible changes in the way developmental capacities can be acquired. Developmental transitions display analogous contingent irreversibility.

The value of describing transition as a matter of contingent irreversibility is to underscore that not just any sort of scaffolding event plays the relevant kind of role in a new or extended Darwinian theory: developmental transitions may be relatively infrequent in development, just as evolutionary transitions are presumably infrequent in evolution, or they may be so common as to occur multiple times in each life trajectory. Not every adaptive change or major evolutionary innovation such as the evolution of flight, or terrestriality, marks an evolutionary transition, nor does every developmental scaffold have the dramatic effects of, say, gastrulation.

What I seek is an account of developmental scaffolding that can classify, describe, and maybe explain eco-devo-evo patterns of developmental transitions across a wide range of developmental systems. Some of those systems may be ones we presently consider to operate in causal thickets that resist evolutionary explanations, like many aspects of socioculture, sociodevelopment, and sociopsychology. It may be that they will always resist because they genuinely are not evolutionary phenomena. But it is well recognized to be hard to answer that question using tools developed by a theoretical biology designed to handle hierarchies of biological organization grounded in a gene's eye perspective. That perspective begins to lose its grip in the face of Wimsatt's ontology and the new perspectives proposed by extended evolutionary synthesizers and has little to say in the face of causal thickets, other than to offer an in-principle reductionism that gets us nowhere.

Acknowledgments

This essay is dedicated to the memory of Werner Callebaut: philosopher, manager, editor, troublemaker, friend. I thank James DiFrisco and Dan Brooks for inviting me to their excellent conference at KLI. I thank James, Jan Baedke, and Alan Love for helpful criticism. I thank Sebastian Schreiber and Jun Otsuka for discussions of Simon's model. I thank Bill Wimsatt and Linnda Caporael for extensive discussion and critique of many of the ideas presented here. I thank Gillian Dickens for copyediting. I thank the people of California for financial support.

Notes

1. It is an interesting question whether Simon's analysis applies to "soft" materials such as living cells or only to solid-state machines and artifacts. As well, it is important to consider whether Simon's analysis applies equally to development and evolution. I do not have space to address these questions here. My argument here is that Simon's model and parable do not sit well together, so neither can be a good guide to thinking about the complexity of complex living systems even if there are disanalogies between solid-state and soft materials, assembly and biological development, or development and evolution. Thanks to James DiFrisco for pressing these points.

2. Jan Baedke wonders whether this account, reversing Wimsatt's "tiers" epistemically as pruning thickets rather than emergence from failing levels, applies to situations other than compositional levels. I think it may, for example, apply to scaling phenomena but would require a very different analysis beyond the scope of this essay.

3. Sonia Sultan (2015, 2017) has recently called for a reassessment of the genotype concept in light of her views on phenotypic plasticity.

4. As Jan Baedke points out, Waddington's pegs could be generalized from genes to developmental factors, but I will not characterize them as such here in order to distinguish exogenous environmental scaffolds from genes as kinds of developmental factors and to take special note of the shift of perspective on the relation between genetic factors and environments that I propose on norms of reaction.

5. It should be emphasized that merely necessary conditions for the possibility of development, like environmental oxygen for aerobic metabolisms, are not developmental scaffolds because their presence/absence does not per se mark transitions in developmental capacities (although in evolutionary history, they may once have done).

6. The problem is worse than this since behavior is yet another relevant perspective, so the thicket is really etho-eco-devo-geno-evo.

References

Bechtel, W., & Richardson, R. C. (1993). *Discovering complexity: Decomposition and localization as strategies in scientific research*. Princeton, NJ: Princeton University Press.

Bickhard, M. H. (1992). Scaffolding and self scaffolding: Central aspects of development. In L. T. Winegar and J. Valsiner (Eds.), *Children's development within social contexts: Volume 2: Research and methodology* (pp. 33–52). Hillsdale, NJ: Erlbaum.

Callebaut, W. (2007). Herbert Simon's silent revolution. *Biological Theory, 2*(1), 76–86.

Callebaut, W., & Rasskin-Gutman, D. (Eds.). (2005). *Modularity: Understanding the development and evolution of natural complex systems*. Cambridge, MA: MIT Press.

Caporael, L., Griesemer, J., & Wimsatt, W. C. (2014a). Developing scaffolds: An introduction. In L. Caporael, J. Griesemer, & W. Wimsatt (Eds.), *Developing scaffolds in evolution, culture, and cognition* (pp. 1–20). Cambridge, MA: MIT Press.

Caporael, L., Griesemer, J., & Wimsatt, W. (Eds.). (2014b). *Developing scaffolds in evolution, culture, and cognition.* Cambridge, MA: MIT Press.

Griesemer, J. (2000a). Development, culture and the units of inheritance. *Philosophy of Science, 67,* S348–S368.

Griesemer, J. (2000b). The units of evolutionary transition. *Selection, 1,* 67–80.

Griesemer, J. (2005). The informational gene and the substantial body: On the generalization of evolutionary theory by abstraction. In M. R. Jones & N. Cartwright (Eds.), *Idealization XII: Correcting the model, idealization and abstraction in the sciences* (pp. 59–115). Amsterdam, Netherlands: Rodopi.

Griesemer, J. (2014a). Reproduction and the scaffolded development of hybrids. In L. Caporael, J. Griesemer, & W. Wimsatt (Eds.), *Developing scaffolds in evolution, culture, and cognition* (pp. 23–55). Cambridge, MA: MIT Press.

Griesemer, J. (2014b). Reproduction and scaffolded developmental processes: An integrated evolutionary perspective. In A. Minelli & T. Pradeu (Eds.), *Towards a theory of development* (pp. 183–202). Oxford, UK: Oxford University Press.

Griesemer, J. (2016). Reproduction in complex life cycles: A developmental reaction norms perspective. *Philosophy of Science, 83*(5), 803–815.

Griesemer, J. (2018). Individuation of developmental systems: A reproducer perspective. In O. Bueno, R. Chen, & M. B. Fagan (Eds.), *Individuation, process, and scientific practices* (pp. 137–164). New York, NY: Oxford University Press.

Griesemer, J. (2019). Towards a theory of extended development. In G. Fusco (Ed.), *Perspectives on evolutionary and developmental biology: Essays for Alessandro Minelli* (pp. 319–334). Padova, Italy: Padova University Press.

Griesemer, J., & Szathmáry, E. (2009). Gánti's Chemoton model and life criteria. In S. Rasmussen, L. Chen, N. Packard, M. Bedau, L. Chen, D. Deamer, D. Krakauer, N. Packard, & P. Stadler (Eds.), *Protocells: Bridging nonliving and living matter* (pp. 481–512). Cambridge, MA: MIT Press.

Hull, D. L. (1980). Individuality and selection. *Annual Reviews of Ecology and Systematics, 11,* 311–332.

Hull, D. L. (1988). *Science as a process.* Chicago, IL: University of Chicago Press.

Jaeger, J., Irons, D., & Monk, N. (2012). The inheritance of process: A dynamical systems approach. *Journal of Experimental Zoology B (Molecular Development and Evolution), 318B,* 591–612.

James, W. 1918. *The principles of psychology, Volume 1.* New York, NY: Henry Holt.

Kauffman, S. (1971). Articulation of parts explanation in biology and the rational search for them. In R. Buck & R. Cohen (Eds.), *PSA 1970* (pp. 257–272). Dordrecht, Netherlands: Reidel.

Laland, K., Uller, T., Feldman, M., Sterelny, K., Müller, G. B., Moczek, A., Jablonka, E., & Odling-Smee, J. (2014). Does evolutionary theory need a rethink? Yes, urgently. *Nature, 514,* 162–164.

Laland, K., Uller, T., Feldman, M., Sterelny, K., Müller, G. B., Moczek, A., Jablonka, E., & Odling-Smee, J. (2015). The extended evolutionary synthesis: its structure, assumptions and predictions. *Proceedings of the Royal Society of London B, 282,* 20151019.

Lewontin, R. C. (1970). The units of selection. *Annual Review of Ecology and Systematics, 1,* 1–17.

Maynard Smith, J., & Szathmáry, E. (1995). *The major transitions in evolution.* Oxford, UK: W. H. Freeman Spektrum.

Maynard Smith, J., & Szathmáry, E. (1999). *The origins of life: From the birth of life to the origin of language.* Oxford, UK: Oxford University Press.

Minelli, A. (2011). Animal development, an open-ended segment of life. *Biological Theory, 6,* 4–15.

Moreno, A., & Mossio, M. (2015). *Biological autonomy: A philosophical and theoretical enquiry.* New York, NY: Springer.

Müller, G. B. (2017). Why an extended evolutionary synthesis is necessary. *Interface Focus, 7,* 20170015.

Pearl, J. (2000). *Causality.* New York, NY: Cambridge University Press.

Pearl, J., & Mackenzie, D. (2018). *The book of why: The new science of cause and effect.* New York, NY: Basic Books.

Pigliucci, M., & Müller, G. B. (Eds.). (2010). *Evolution: The extended synthesis.* Cambridge, MA: MIT Press.

Schlosser, G. (2004). The role of modules in development and evolution. In G. Schlosser and G. Wagner (Eds.), *Modularity in development and evolution.* Chicago, IL: University of Chicago Press.

Simon, H. A. (1962). The architecture of complexity. *Proceedings of the American Philosophical Society, 106*(6), 467–482.

Simon, H. A. (1981). *The sciences of the artificial* (2nd ed.). Cambridge, MA: MIT Press. (Original work published 1966)

Simon, H. A. (2002). Near decomposability and the speed of evolution. *Industrial and Corporate Change, 11*(3), 587–599.

Simon, H. A., & Ando, A. (1961). Aggregation of variables in dynamic systems. *Econometrica, 29*(2), 111–138.

Sultan, S. E. (2015). *Organism and environment: Ecological development, niche construction and adaptation.* Oxford, UK: Oxford University Press.

Sultan, S. E. (2017). Developmental plasticity: Re-conceiving the genotype. *Interface Focus, 7,* 20170009.

Szathmáry, E. (2015). Toward major evolutionary transitions theory 2.0. *PNAS, 112*(33), 10104–10111.

Tavory, I., Ginsborg, S., & Jablonka, E. (2014). The reproduction of the social: A developmental system approach. In L. Caporael, J. Griesemer, & W. Wimsatt (Eds.), *Developing scaffolds in evolution, culture, and cognition* (pp. 307–325). Cambridge, MA: MIT Press.

Valenzuela, N., & Lance, V. A. (Eds.). (2004). *Temperature dependent sex determination in vertebrates.* Washington, DC: Smithsonian Books.

Waddington, C. H. (1957). *The strategy of the genes: A discussion of some aspects of theoretical biology.* New York, NY: Macmillan.

Waddington, C. H. (1974). A catastrophe theory of evolution. *Annals of the New York Academy of Sciences, 231*(1), 32–41.

Waddington, C. H. (1975). *The evolution of an evolutionist.* Ithaca, NY: Cornell University Press.

Wimsatt, W. C. (1974). Complexity and organization. In K. F. Schaffner & R. S. Cohen (Eds.), *PSA 1972* (pp. 67–86). Dordrecht, Netherlands: Reidel.

Wimsatt, W. C. (1981). Robustness, reliability and overdetermination. In M. Brewer & B. Collins (Eds.), *Scientific inquiry and the social sciences* (pp. 124–163). San Francisco, CA: Jossey-Bass.

Wimsatt, W. C. (1997). Aggregativity: Reductive heuristics for finding emergence. *Philosophy of Science, 64* (Proceedings), S372–S 384.

Wimsatt, W. C. (2007). *Re-engineering philosophy for limited beings: Piecewise approximations to reality.* Cambridge, MA: Harvard University Press.

Wimsatt, W. C., & Griesemer, J. R. (2007). Reproducing entrenchments to scaffold culture: The central role of development in cultural evolution. In R. Sansom & R. Brandon (Eds.), *Integrating evolution and development: From theory to practice* (pp. 227–323). Cambridge, MA: MIT Press.

Winslow, R. (2017). *Organism and environment: Inheritance and subjectivity in the life sciences.* Lanham, MD: Lexington Books.

Woodward, J. (2003). *Making things happen: A theory of causal explanation.* Oxford, UK: Oxford University Press.

Wray, G., Hoekstra, H. E., Futuyma, D. J., Lenski, R. E., Mackay, T. F. C., Schluter, D., & Strassmann, J. E. (2014). Does evolutionary theory need a rethink? No, not at all. *Nature, 514,* 162–164.

6 Integrating Composition and Process in Levels of Developmental Evolution

James DiFrisco

Overview

Hierarchical organization in biology is usually conceptualized in terms of compositional levels of structurally defined entities. However, composition in biological systems is inescapably a dynamic affair concerning what a component *does* and not only what it *is*. This chapter examines how the dissociability of structure and function in complex evolved systems leads to ambiguities in the placement of things at levels, which makes "levels" a less useful concept. To address this difficulty, I develop a framework in which general levels of organization are restricted to levels decomposed under a specific causal process. In this approach, levels include only components and processes contributing to a focal process. After defining interlevel and intralevel relations in the process-relative framework, I explore a proposed decomposition of the process of developmental evolution. The resulting levels of developmental evolution are contrasted with levels of selection and replicator–interactor hierarchies.

6.1 Introduction

This chapter develops an approach to compositional levels in which composition is viewed as being relative to causal processes. I motivate this approach by first considering the roles that levels of organization play in biological theorizing as well as the difficulties of satisfying these roles. A major source of difficulties is that the same kind of component can have structural features that place it at one level and functional features that place it at a different level. Structural, functional, and hybrid structural–functional perspectives on levels each have limited resources to deal with this sort of problem. As an alternative, I explore a new approach based on functions understood as activities or processes, which contrasts with the standard understanding of functions as effects or outcomes. In the proposed approach, systems are "decomposed under a process," so that a given hierarchy only includes the entities and activities that contribute to the same causal process. The central conceptual strategy for excluding causally irrelevant structure and function is to rely on a notion of constitution between processes to define placement into levels.

I then explore a proposed decomposition of the process of developmental evolution, the focus of evolutionary developmental biology. The levels of developmental evolution exhibit the special property of "variational independence," which grounds substantive insights about causation and generalization across levels, in line with the main roles for levels distinguished in the beginning. I close by showing how the processual framework improves on the effect-functional frameworks of levels of selection and replicator–interactor levels and sketch some directions for future work.

6.2   Theoretical Roles of Levels of Organization

The concept of levels of organization has many important roles in the life sciences. The investigation of how these roles should be fulfilled issues in corresponding tasks or problem areas for an account of levels of organization. Among these roles we can distinguish three that have broad significance:

(R1) *Composition*. To identify and describe general types of entities in terms of the types of parts they have as well as the types of wholes they compose.

(R2) *Causal segregation*. To demarcate types of entities that are such that (1) causal interaction occurs exclusively or primarily within types rather than between types, and/or (2) variables measured on those entities are causally independent of variables measured on other entities.

(R3) *Generalization*. To identify types of entities that cluster together in their scalar or qualitative properties such that these types support projection or inductive generalization.

In standard framings, levels of organization are populated by objects, individuals, or substances, but we can also ask about how other categories of items such as properties and processes are hierarchically ordered. (R1) can accordingly be expanded to be categorially inclusive: how should we characterize constitution or "making-up" relations between types of items across the wide variety of biological systems (see Gillett 2013a, this volume)? Some connections between categories will be explored in section 6.4.

(R2) refers to an idea that routinely arises in discussions of levels—that causal interaction occurs within a level or not between levels (Craver & Bechtel, 2007; Eldredge & Salthe, 1984; Wimsatt, 1976, 2007; see also Gillett, this volume). A similar idea is that levels track relations of conditional independence between variables, so that causal models of variables at one level of organization can safely ignore or background variables at other levels (see Woodward, this volume; Batterman, this volume). Different empirical and conceptual factors have been invoked to explain this "causal segregation" across levels of organization. One is that parts and wholes cannot stand in causal relations because they are not suitably distinct and causation is irreflexive (e.g., Craver & Bechtel 2007; Gillett, this volume; see Introduction, this volume). Another is that causal interactions are more frequent and stronger between items of similar space, time, and force scales (DiFrisco, 2017; Salthe, 1985; Wimsatt, 2007). Finally, evolution under selection may lead to modular body plans in which interaction across modules is limited (Simon, 1962). Nonetheless, the life sciences are full of examples of causal interaction crossing levels, and so causal segregation is more plausibly interpreted as a general pattern rather than a law (DiFrisco, 2017). This has the consequence that causal segregation, even if it is an important feature of levels, will not be useful as a criterion for assigning items to levels.

The combination of (R1) and (R2) puts multiple ontological categories at play. Even if levels of organization are primarily serially ordered collections of *objects*, these collections are supposed to ground certain kinds of interactions, causal processes, activities, or dynamics. In turn, regularities deriving from general typologies of composition (R1) and causation (R2) support inductive generalizations about levels (R3).

The most in-depth discussion of (R3) can be found in Wimsatt's work (1976, 2007), where levels are pictured as "meshes" of different sizes that converge on groups of entities that are clustered together in spatial scale and other properties. These levels can be regarded as loci in type space where densest similarities in structural and causal properties are to be found. Otherwise expressed, levels are "local maxima of regularity and predictability" in the space of different modes of organization of matter (Wimsatt, 1976, 237–238; 2007, 209; this volume).

In general, the more roles or purposes there are for the same concept, the less likely it is that one and the same version of that concept will be able to fill all of the roles. Prioritizing (R2) may promote criteria for distinguishing levels other than compositional criteria (R1), such as space, time, or force scale (DiFrisco, 2017; O'Neill et al., 1986; Rueger & McGivern, 2010; see Batterman, this volume). These "causal" levels are then unlikely to track composition relationships between types very closely. Similarly, pursuit of (R3) may pull in a different direction than (R2) and especially (R1). This would occur when the "maxima of regularity and predictability" at different scales pick out entities that do not stand in neat causal and compositional relationships with one another. Some examples of this divergence between (R3) and (R1) in the context of developmental evolution are examined in section 6.5.

Whenever a single concept is enlisted for many heterogeneous investigative roles, an attractive strategy is to give a pluralistic reinterpretation of the concept. Implementing this strategy in the present context might yield different concepts of levels of organization for each different role (R1) to (R3). While it helps to deflate the problem, the pluralistic strategy is objectionable for sacrificing the unity of the concept and with it the linkages between its diverse roles. In the case of levels of organization, these linkages—such as between compositional and causal levels—are some of the most interesting and rich aspects of the problem area. A further limitation of the pluralistic strategy, as stated, is that it assumes that the roles are all equally valid or important, rather than being subject to revision in light of the conceptual and empirical development of an account of levels.

Once we ascertain the reasons why (R1) to (R3) are dissociable, a different strategy presents itself. Broadly speaking, the main reason has to do with a common feature of evolved complex systems, which is the decoupling between structure and function, or between the intrinsic properties of a thing and the activities it participates in. If this is the case, there are better prospects for ordering things into levels if we refrain from considering all kinds of structures and activities at once. Instead, the strategy would be to partition a system into levels *relative to* a theoretical perspective or to one kind of causal process. This "decomposition under a process" includes only functions that align in contributing to the process and includes only the structural parts that perform those functions. I show how this might work in section 6.5 for the process of developmental evolution. In the perspectival or process-relative approach, a pluralistic splitting occurs between perspectives or kinds of processes rather than between the roles (R1) to (R3), which are maintained across processes or perspectives on levels.

While this strategy can reduce the dissociation between roles, it only goes so far. The further step is to take a constructive rather than descriptive approach to levels. Instead of treating each role that scientists and philosophers have assigned to the concept as a static target, these roles are allowed to change and to be only partially satisfied. The desirability of this stance derives from a sense that the concept of levels of organization has been overburdened with roles of great scientific and metaphysical significance, even though work on the topic arguably hasn't delivered the resources commensurate with this burden. In view of this situation, a constructive approach to levels pursues the tasks (R1) to (R3) as far as the landscape of the problem actually affords, in order to find the most "stable conformation" addressing all of (R1) to (R3).

6.3   Compositional Levels and Structure–Function Dissociation

One of the major obstacles to formulating a general account of levels that satisfies (R1) to (R3) is the following: for the entities that theorists would like to place into levels, their structure and their functioning often do not covary. The same structure can perform multiple functions, and the same function can be performed by multiple structures. Given the additional premise that structure and function are both important for representations of biological hierarchy, we have the situation that structural aspects will sometimes place an entity at one level while functional aspects place it a different level. This creates a problem, because if the same entity is at multiple levels, then the inferences we want to make from level placements will become indistinct and confused. In this section, I illustrate the above argument by briefly outlining some of the limitations of mainstream accounts of levels based on structure, function, and combinations of the two.

6.3.1   Structural Levels

Most existing thought on (R1) is guided by the idea that certain structural types of objects identified by different sciences stand in orderly stepwise composition relationships. A classic expression of this idea can be found in Oppenheim and Putnam's (1958) "Unity of Science as a Working Hypothesis" (see Brooks, 2017). Here the aim is to map a hierarchy of levels that enables stepwise "micro-reductions" of things at one level (l_i) to things at the next lower level (l_{i-1}). From the aim of securing stepwise micro-reductions, Oppenheim and Putnam extract a number of "conditions of adequacy" for a theory of levels. The most important of these in the present context is the following: "(4) Any thing of any level except the lowest must possess a decomposition into things belonging to the next lower level" (Oppenheim & Putnam, 1958, p. 9). I will refer to this thesis as "type composition," by analogy to "type identity":

> **Type Composition**: objects of type T_i at level l_i are composed exclusively of objects of type T_{i-1} at level l_{i-1}, unless l_i is the lowest level of objects.

Molecules at the molecular level are entirely composed of atoms at the atomic level. Atoms are entirely composed of elementary particles, which are perhaps not further decomposable. Along with this claim about types, Oppenheim and Putnam (1958) make the following claim about individuals: "Any whole which possesses a decomposition into parts all of which are on a given level, will be counted as also belonging to that level. Thus each level includes all higher levels" (p. 10). The idea that one and the same object can be identified at the level appropriate to its kind, as well as at each lower level, implies that the object is identical to the set of its components. This thesis is known as "token identity":

Token Identity: for every object x at level l_i, there is an object y at level l_{i-1} such that x is identical to y, unless l_i is the lowest level of objects.

Oppenheim and Putnam claim that these two conditions—type composition and token identity—are satisfied in the following empirical levels: elementary particles, atoms, molecules, cells, (multicellular) living things, and social groups.

Both type composition and token identity are problematic when viewed in light of straightforward assumptions about composition and identity in biological systems. The problem with type composition is that most objects above the level of molecules have components from multiple lower levels and not just from the immediately adjacent lower level. By analogy to multiple realizability, the systems studied by biology are characterized by "multiple type composition": their components can belong to different structural levels due to their belonging to different types. For example, assuming Oppenheim and Putnam's empirical levels, cells are composed of molecules but also of atoms and ions that are not components of molecules. Multicellular organisms are composed of ions, atoms, and molecules, and these need not be parts of cells (for example, in interstitial fluids and extracellular matrix). Social groups of animals can be composed of multicellular and single-celled organisms in symbiotic relationships. In short, type composition will fail whenever the same type of component can exist at many levels in a higher-level system. The most general reason why multiple type composition occurs is that, above the level of macromolecules, the entities of interest to science are typically individuated, at least in part, by their modes of functioning, rather than by their material composition. In order to be part of a functioning system, it is not necessary that a component belong to a particular structural level, such as the next lower level, but only that it function in a certain way within the system.

The problem with token identity is that biological systems and sets of their components tend to have different conditions of individuation and persistence. Sets are individuated extensionally, so that sets are the same when they have the same members. But cells, organisms, and social groups can gain and lose parts while still remaining the same. Further, it is often not possible to individuate a higher-level entity starting from lower levels of organization. When we examine scientific descriptions of "the same" object at different levels, we find that lower levels include more parts and more detail, some of which is irrelevant for the aspect of functioning that characterizes the higher-level object. In such cases, the higher-level object cannot be identified with the sets of parts we can pick out using the descriptive resources of lower levels (DiFrisco, 2018).

These problems feed into the same core difficulty for structural approaches to compositional levels. If we say that l_{i+1} entities are composed of l_i entities, we at once include too much and too little in l_{i+1}: too much, because some parts from l_i may not be functional parts of l_{i+1}; too little, because parts from l_{i-1} and l_{i-2} may be functional parts of l_{i+1}. Purely structural approaches to compositional levels yield causally irrelevant and arbitrary objects, such as organisms without their noncellularized parts or organisms whose parts include physical entities that are too minute to actually affect their functioning.

A hierarchy of structural types defined without reference to functioning inevitably exhibits a divergence between compositional levels (R1) and causal levels (R2). Will the situation be improved if we define compositional levels starting from function, process, and causal interaction?

6.3.2 Functional Levels

The notion of function as it appears in the structure–function dialectic of biology can be understood in multiple ways (Brigandt, 2017; Gillett, 2013b; Griesemer, 2006; Love, 2007). The function of an entity can refer to its effects, relations, roles, goals, or outcomes, but it can also refer to its functioning—to activities, processes, or behaviors. Effect functions have been a central analytical category in traditional evolutionary theorizing about levels, for example, in work on levels of selection (Lewontin, 1970; Okasha, 2006) and replicator–interactor levels (Hull, 1980; Salthe, 1985). Activity functions or processes have more recently begun to play a more central role in evolutionary theorizing from a developmental perspective (Amundson, 2005; Griesemer, 2000, 2006; Jaeger, 2019; Jaeger & Monk, 2014; Love, 2007), but without yet issuing in an account of *levels* of developmental evolution.

The main rationale for a shift from effect function to activity function is based on the idea that understanding evolution requires "opening the black box" of developmental processes and examining their causal structure. As Griesemer (2006, 347) points out, analyses based on effect functions "depend on tacit reference to a process in which a goal is served, but they do not offer an account of those processes." This is why effect-functional analyses of levels are typically uninformative until they are mapped onto levels of structure or process, which provide the structures and causes that *realize* the effects (Griesemer, 2005). To say that l_i is a level of selection if entities at l_i engage in fitness-affecting interactions (Okasha, 2006) does not tell us which entities are at that level, nor does it describe the processes and interactions that affect fitness at that level. It simply gives an extra condition that some entities—whatever they are—must satisfy in order for some occurrence—whatever it is—to result in selection.

For now, I will only consider activity-functional analyses as candidate accounts of levels. There are two broad kinds of activity-functional approaches to levels: one classifies objects and their functioning or activity into levels, and the other only classifies functionings or activities themselves into levels. Probably the most well developed of the first form is Craver's (2007, 2015) account of levels of mechanisms. This account is concerned with understanding levels in the context of mechanistic explanation. The basic idea is that a component and its activity are at a lower level than a mechanism and its activity if these entity–activity pairs stand in two relations: a part–whole relation and additionally a relation of "mutual manipulability." For the part and whole to be mutually manipulable means that in intervening on the component or activity, one can observe changes in the mechanism and its activity and vice versa. As for the same-level relation, components and their activities are at the same level if they are manipulable parts of the same mechanism and neither is part of the other (Craver, 2007, p. 192).

By substituting a causal relevance criterion for structural types, the mechanistic view allows classification into levels without appeal to type composition and provides a way of limiting the lower-level components to only the causally relevant ones.[1] However, the advantages of abandoning structural types for understanding interlevel relations come at the expense of a well-defined *intralevel* relation (for detailed argument, see DiFrisco, 2017; Eronen, 2013, 2015). This issue has been addressed by restricting the scope of levels to local token mechanisms and denying that the intralevel relation is transitive and thus that it has much significance (Craver, 2015). Without a stable intralevel relation and without types, however, there is little possibility of reaching generalizations about composition (R1), causation (C2), or other qualitative features (R3).[2]

A second functional approach is to shift from placing entities at levels and instead to focus on hierarchical relations between dynamical categories—processes, activities, behaviors, functionings, events, and so on (Allen & Starr, 1982; Baedke & McManus, 2018; DiFrisco, 2017; Goodwin, 1963; O'Neill et al., 1986; Seibt, 2014). In DiFrisco (2017), I argued that this is the most generally reliable way of satisfying (R2), but this required abandoning typological and compositional criteria for level placements in favor of a dynamical one—namely, time scale. Like the mechanistic account, this sort of approach escapes the problems with Oppenheim and Putnam's (1958) scheme, but without losing a well-defined intralevel relation, since sameness of level is simply construed as similarity of time scale within an interactive context. As a consequence of the latter, time scale hierarchies of processes need not be restricted to tokens and to highly local contexts and thus give a more general answer to (R2). However, this approach only advances toward (R2) by setting aside (R1) altogether and also to some extent (R3). This approach does not help us to identify and describe the variety of noncausal "making-up" relations between types—neither between types of objects, *nor* between types of processes. Accordingly, this second functionalist approach fits into the pluralist strategy where different accounts are developed specifically for different roles (R1) to (R3) of levels of organization. In keeping with the limitations of the pluralist strategy noted earlier, we gain precision but lose the important linkages between the different roles (i.e., between composition, causation, and generalization over types).

6.3.3 Hybrid Structural–Functional Levels and Cluster Kinds

A different line of approach to levels combines structure and function concepts. Wimsatt's (1976, 2007) writings on levels of organization target each of (R1) to (R3) (among other tasks) while avoiding the above problems with Oppenheim and Putnam's (1958) structural hierarchy as well as functionalist hierarchies. He does this by reinterpreting type composition (R1) and causal levels (R2) as providing generally reliable *characterizations*, as opposed to *definitions* in terms of necessary and sufficient conditions for being at a level (Wimsatt, 2007, p. 204). This redirects attention away from finding definitions that avoid counterexamples and toward characterizing the many empirical linkages between composition, scale, causation, and scientific explanation. Although it is not explicitly conceptualized as such, Wimsatt's levels can be viewed as "cluster kinds," where membership is determined not by essential properties but by having any of a number of properties that tend to cluster together. Both structural and functional aspects can be included in the cluster of characteristic features of a level. The flexibility of this conception of levels allows for wide-ranging and hybrid characterizations like the following:

One thing is at a higher level than something else if things of the first type *are composed of* things of the second type [(R1)], and at the same level with those things it interacts most strongly or frequently with or is capable of replacing in a variety of causal contexts [(R2)]. (Wimsatt, 1976, p. 215)

The theories at different levels might be thought of as sieves of different sizes, which sift out entities of the appropriate size and dynamical characteristics [(R2)].... If the entities at a given level are clustered relatively closely together (in terms of size, or some other generalized distance measure in a phase space of their properties) it seems plausible to characterize a level as a *local maximum of predictability and regularity* [(R3)]. (Wimsatt, 1976, pp. 237–238)

These claims can be illustrated with many rich examples from the sciences (Wimsatt, 1976, 2007). The major advantage of adopting a nonessentialist view of levels is that one can capture interesting linkages between different features of levels without worrying about exceptions or counterexamples. Salthe (1985) offers another framework on levels of this same kind.

Understanding levels in terms of property clusters can also provide a straightforward way of comparing different empirical levels as to their degree of uniformity. When the properties of the cluster closely covary, there are highly uniform levels. These would include the atomic, molecular, and perhaps the cellular—the most lawlike of biological levels. When the properties of the cluster only loosely covary, for reasons such as the decoupling of structure and function, then levels are less uniform and less "level-like." Wimsatt (2007, p. 221) describes a situation where, as we ascend to systems with greater descriptive and interactional complexity, levels "break down" and are replaced by "perspectives" and "causal thickets." Levels tend to break down above the cellular level, yielding to multiple perspectives on multicellular organisms and, higher still, the "bio-psychological thicket" of the human brain, social institutions, and culture (Wimsatt, 2007, 233).

One way of explicating the idea of a "breakdown" of levels is to say that, as structure and function get decoupled in certain complex systems, level placement becomes *nondiscrete*. Discreteness of level placement means that the same thing is not at different levels in the same hierarchy. Discreteness is commonly assumed to be a necessary feature of levels (Bechtel, 2012; Potochnik & McGill, 2012), because without it, placement at a level becomes noncontrastive and uninformative. The discreteness requirement will be harder to meet as properties from a cluster covary more loosely. This is because membership in cluster kinds or levels tends to be quite permissive: any property in the cluster is sufficient, and none is individually necessary, for membership in the kind. Accordingly, an entity can have a structural property that is sufficient to place it at one level, while also having a functional property that is sufficient to place it at a different level. For example, colonies of eusocial insects are groups of metazoans, but their behavior, metabolic scaling, and other features are quite similar to that of solitary organisms (Hou et al., 2010). Should they be counted as belonging to the organism level or the level of groups or populations of organisms?

One could decrease the occurrence of nondiscreteness by stipulating that membership in a level requires surpassing a certain *threshold* of properties from a cluster. But this move would undercut the primary motivation for shifting to cluster concepts from essential properties. In setting a threshold of clustered properties for membership in a level, the risk is that many individuals would fail to surpass the threshold at all. Rather than being placed at *multiple* levels, they would be placed at *no* level. These level-less individuals would be excluded from generalizations (R3) and causal inferences (R2) about levels. Another possibility is that entities failing to surpass the threshold would be counted not as individuals at the focal level (e.g., a multicellular organism) but as *groups* of individuals at the next lower level (e.g., a collection of cells).[3] This reconceptualization may be appropriate when the individual really is more like a group of lower-level entities—for example, slime molds during the foraging phase of the life cycle. But such borderline cases of multicellularity— slime molds and obligately eusocial insect colonies—are precisely cases that do not easily fit into discrete levels of "cells" or "organisms." In a sense, *any* classification into one or the other levels is misleading. Hence, it seems that access to the wide diversity of generalizations about levels is only granted at the cost of nondiscrete or misleading level placements. When level placements are nondiscrete, it is difficult to single out the appropriate units of generalization to determine which phenomena instantiate the predictions and regularities, and thus it is difficult to apply the levels scheme in specific cases. When level placements are misleading, the generalizations and inferences may do more harm than good (see Potochnik, this volume).

Wimsatt's framework may capture the most that a monistic approach can say about levels of organization viewed as robust points of convergence across all scientific domains at once. In order to gain in precision about placements at higher levels without losing what is interesting about levels, one has to move from a general discussion of levels to a specific scientific contextualization. I propose to do this in section 6.5 in examining levels of developmental evolution. This move involves the following reorientation: instead of looking for unique levels placements that hold across diverse scientific perspectives, we consider only one perspective tracking one kind of causal process (e.g., metabolism, development, evolution) and decompose systems into levels relative to that perspective or process. In this approach, it should be the case that levels break down at higher scales than before, and complex entities such as organisms or traits can be ordered into levels without the same ambiguities native to *general* levels of organization.

6.4   Decomposition under a Process, Constitution, and Levels

From the preceding discussion of compositional levels, we can see how the nature of scientific composition constrains how levels of organization should be understood. Functional and Wimsattian approaches to levels depart from structural approaches on the basis of a different picture of scientific composition. They recognize that the criteria of individuation used in different sciences and at different scales pick out objects that do not necessarily stand in neat compositional relationships to one another. I will refer to this phenomenon as *relativistic composition*. Relativistic composition undermines the structuralist theses of type composition and token identity found in Oppenheim and Putnam's (1958) framework. More generally, relativistic composition creates the following challenge for accounts of compositional levels: how can a hierarchy include the entities that are targeted in actual scientific explanations, while also putting those very entities together into some form of unifying compositional relationship? This section will explore how the notions of decomposition under a process and constitution relationships between processes can address this challenge.

To start, it will be helpful to consider why relativistic composition occurs in science. Two key papers by Kauffman (1971) and Wimsatt (1972) develop the idea that different decompositions of the same system result from our bringing different *perspectives* to scientific investigation. From here we get the rich notion of *descriptive complexity*. The descriptive complexity of a system is the degree to which different perspectival decompositions yield nonoverlapping sets of parts. In biology, perhaps the best examples of descriptively complex systems are organisms and phenotypic traits, and these are also the entities for which levels start to break down. In the case of organisms, recent work on biological individuality has generally aligned in showing that there are irreducibly multiple, partially overlapping boundaries we can draw around organisms and that the different individuals that have these different boundaries figure into different perspectives and styles of explanation. For example, ecological perspectives tend to be more inclusive than evolutionary ones: in a given community of symbionts, there might be one large cohesive ecological individual but many distinct evolutionary individuals (Clarke, 2016; see DiFrisco, 2019).

As the language of "perspectives" suggests, relativistic composition is partly a consequence of the plurality of aims, interests, and cognitive strategies that we bring to bear on the world. But relativistic composition is also a reflection of the way the natural world is. Ecological perspectives differ from evolutionary perspectives not only because they have different investigative aims but also because they track different kinds of natural processes and discover different regularities and patterns. Importantly, these different processes have different participants. Animals together with their symbiotic microbes might jointly participate in the same processes of growth, migration, predation, and adaptation but in different processes of reproduction and developmental differentiation. If we ask whether the "animal plus microbes" system is on the organism level without specifying a reference process, the answer is likely to be ambiguous between decompositions of the system under different processes.

The idea of a causal process as an individuator of objects has been proposed in biology before.[4] In examining the individuation of organisms, Laubichler and Wagner (2000) maintain that "functionally relevant biological objects can be defined only within the context of a specific biological process" (p. 293). von Dassow and Munro (1999), in discussing the notion of modularity in evo-devo, write,

The way we conceptualize module and mechanism depends both on the reference process (on whether we are interested in morphogenesis or epigenesis, or in the evolutionary process itself) and on whether we are thinking analytically or synthetically. The first arises because we are talking about disjunct types of process in which biological entities participate differently; the second selects our operational criteria. The practical consequence of both is merely that biologists must be careful not to assume any particular correspondence between units that emerge from each perspective. (p. 308)

A classic example of a trait that is not a unit of developmental processes is the human chin (Gould & Lewontin, 1979; Wagner, 2014), which is a by-product of adjacent modular growth fields. Yet the chin could be a unit relative to a process of sexual selection. This sort of process relativity vexes attempts to identify "the" module and place it on a level relative to other entities.

In order to make the notion of decomposition under a process workable, it will be necessary to get clearer on what processes are. If biological entities are individuated by the processes they participate in, how are the processes individuated? A natural worry should be addressed from the start, which is that processes will have to be individuated by their participants, rendering the individuation condition objectionably circular. This type of worry is based on an abstract metaphysical picture of individuation rather than the messy world of empirical science (see Bueno et al., 2018). Metaphysically, individuation is a determination relation: it is irreflexive, asymmetric, and possibly transitive (Lowe, 2003). The asymmetry condition means that if x individuates y, then y cannot also be what individuates x. It is not difficult to see how, given asymmetry and irreflexivity, individuation is subject to a potential infinite regress in which, for any individuator x, one must ask what individuates x. The situation is analogous to the problem of regress in epistemic justification.

According to foundationalist epistemologies, chains of epistemic justification must terminate in "self-justifying" or "self-evident" principles, reasons, or perceptual givens. A similar foundationalism about individuation can be found among metaphysicians such as Lowe (2003), who maintains that chains of dependency in individuation must terminate in "self-individuating" things, thus violating irreflexivity.

However, the most plausible epistemologies of science are antifoundationalist, and antifoundationalism about individuation is also the more appropriate stance for understanding the workings of science. Scientific investigation starts from objects, events, and processes that are reasonably thought to be suitably individualized and revises the relative boundaries in a relational and iterative fashion as knowledge about the world grows. Some elements are held fixed during the revision of others, but ultimately all are revisable. This situation does not violate the irreflexivity or asymmetry of individuation; instead, individuation can be reciprocal if it is iteratively operating between entities that are *roughly* the same but with progressively revised boundaries. On this picture, processes can have important individuating roles without being foundational or self-individuating. An initial characterization may be enough to get an operational handle on processes, which would then be subject to further revision and refinement.

What are processes, then? In *The Strategy of the Genes*, Waddington (1957) wrote that "to provide anything like an adequate picture of a living thing, one has to consider it as affected by at least three different types of temporal change, all going on simultaneously and continuously" (p. 6). The three processes he identified were evolution, development, and metabolism. These processes are distinct because they have different participants, instantiate different regularities or patterns, and operate at different time scales. Finer-grained subprocesses can be distinguished within these—for example, selection and speciation, morphogenesis and growth, catabolism and homeostasis, with still finer distinctions within these. The status of these categories as *processes* is reflected in the fact that they involve patterns of change not only in the properties of things but also in the identities of things and components through time. The primary biological processes identified by Waddington are also supposed to be causal rather than epiphenomenal. This feature of processes is explicated in Salmon's (1984) process theory of causation in terms of their capacity to transmit "marks" across spacetime. Drawing on Bertrand Russell's notion of a "causal line," Salmon maintained that causal processes are "self-determined": they transmit their own uniformities of qualitative and structural features, and they exhibit regularities that allow inferences from what is happening during one part of the process to what is happening during another part of the process (Salmon, 1984, pp. 144–145).[5] Beyond these general features, a useful kind of further precision on the individuation of processes will likely have to come from consideration of the specific *sort* of process one is theorizing about.

To get compositional levels from processes, one needs a working understanding of "vertical" relations between processes as well as a notion of components participating in processes. The vertical relation should be a form of constitution rather than identity, because processes that are identical are at the same level. The central processes I focus on are evolution and development. In evolutionary developmental biology, theorists will sometimes say that evolution should be viewed as being, constitutively, the evolution of development (e.g., Gilbert et al., 1996, p. 362; Amundson, 2005). This is supposed to contrast with narrower views of evolution, such as Dobzhansky's view that evolution is change in allele frequencies. But what does it mean to say that evolution is *constituted by* development? Three aspects can be highlighted.

First, it means that evolution cannot change without development changing. If we take development in an inclusive sense that includes genotypic and phenotypic changes, there is no evolutionary change without change in development.[6] Second, although processes of development and evolution occur at different time scales, they occur during the same time intervals. At least for organisms that are thought to undergo development (eukaryotes) (Minelli, 2011), there is no process of evolution that occurs at a time and place where development does not also occur. Third, the entities that participate in evolution (primarily populations) are composed of the entities that participate in development (primarily organisms). Note, however, that development may include parts that are not part of an evolutionary process. Anything that is not heritable or that does not impact future generations, such as certain acquired somatic defects or achievements, is not taken up into the evolution of populations.

Starting from this example, a relation of process constitution can be formulated as follows:

Process constitution: process x is constituted by process(es) y_n iff: (1) x cannot change without y_n changing, but y_n can change without x changing, (2) x and y_n occur in the same spatial region and during the same time interval, and (3) the entities that participate in x are either exhaustively composed of the entities that participate in y_n, or they are composed of a subset of the entities that participate in y_n.

In most cases, the constitution base of a biological process x will be several qualitatively distinct processes (y_n) rather than one. A process y_1 that is only part of the full constitution base y_n can be termed a "partial constituent" of x.

There are several things to note about the three conditions on process constitution. Condition (1) is close to the notion of supervenience except in two respects. First, the relata are causal processes rather than properties. Second, the constituting process must be able to change without the constituted process changing. If constituting and constituted processes each supervened on the other, then they would be identical under the principle of the identity of indiscernibles. Normally, property supervenience is formulated in a way that is compatible with the properties being identical, but here the aim is to build in a distinction between constitution and identity for an account of levels. Condition (2) for process constitution excludes "horizontal" or diachronic determination relations from counting as cases of constitution. Condition (3) ensures that the constitution relation is anchored in the right region of space. But note that it is consistent with the phenomenon of "part loss" across scales. Development can include parts and events that are not part of an evolutionary process. The same is true for molecular metabolic processes and developmental process and perhaps also for neurophysiological processes and mental processes. Conditions (1) and (3) reflect the same underlying relationship between lower and higher levels: the lower level typically includes more than the higher level. This can be a consequence of the fact that higher-level perspectives and processes select only certain parts from the detail of the lower level but also because the scalar differences between levels are such that lower-level effects are not strong enough to make a difference at the higher level.

The above notion of constitution yields a straightforward *interlevel* relation for processes:

Interlevel (processes): two processes are at different levels in the same hierarchy iff they stand in a constitution relationship. Constituting processes are at a lower level than constituted processes.

Note that this interlevel relation, combined with condition (1) on process constitution, generates the expectation that lower-level processes will have shorter time scales than higher-level processes (DiFrisco, 2017). If l_i can only change when l_{i-1} is changing, but l_{i-1} can change without l_i changing, then the regular changes in l_{i-1} should be occurring at a faster rate.

From here we can put forward the following interlevel relation for *entities*:

> **Interlevel (entities)**: two entities are at different levels in the same hierarchy iff there is a constitution (not identity) relation between the processes they participate in. Participants in constituting processes are at a lower level than participants in constituted processes.

An organism is at a different level than its population because they figure into distinct processes (e.g., development and evolution), and these processes are related by constitution. Since development is the constituting process, the organism is at a lower level than the population. Note that process constitution gives levels a restricted scope. There is no constitutive relation between population processes and climate processes in an ecosystem (although there could certainly be causal relations between them), and so populations and climate processes are neither on the same level nor on different levels within the same hierarchy.

To formulate an *intralevel* relation, we can start from the notion of partial constitution: two processes are on the same level if they are partial constituents of the same overarching process. This should not be viewed as a necessary condition, because in many cases, we want to put processes on the same level without there being a relevant overarching process—for example, two processes of evolution or selection in different populations. Likewise, we may want to put two processes on the same level when they are not direct constituents of the same process but only constituents of constituents of the same process. This would be the case for spatially separated cellular-level processes that are constituents of different tissue- or organ-level processes that are constituents of the same metabolic or developmental process. The intralevel relation for processes should therefore accommodate this sort of flexibility:

> **Intralevel (processes)**: two processes are at the same level in the same hierarchy iff (1) neither process constitutes the other and either (2) they are partial constituents (y_1 and y_2 of y_n) of the same process (x), or (3) they are sufficiently similar in some relevant feature among the cluster of characteristic intralevel features.

Characteristic features of intralevel processes include the broad Wimsattian variety: similarity in structure, function, dynamics, or scale or interacting more strongly or frequently with each other.

The same approach can be applied to the intralevel relation for entities. To start, two entities are at the same level in the same hierarchy if neither is part of the other and if they participate in processes that are partial constituents of the same overarching process. Hence, an organism is on the same level as another organism because they both participate in processes (development) that are partial constituents of the same overarching process (evolution), and neither is part of the other. Without further qualification, however, this intralevel relation will run into similar difficulties as Craver's (2007) levels of mechanisms whenever we consider multiple partitions of constituency (see DiFrisco, 2017; Eronen,

2013, 2015). Again, the most promising way forward will be to rely on a nonessentialistic "cluster" interpretation of the intralevel relation:

> **Intralevel (entities)**: two entities are at the same level in the same hierarchy iff (1) neither is part of the other, and either (2) they participate in processes that are partial constituents (y_1 and y_2 of y_n) of the same overarching process (x), or (3) they are sufficiently similar in some relevant feature among the cluster of characteristic intralevel features.

The "sufficiently similar" in conditions (3) no doubt introduces some indeterminacy into intralevel placements for entities and processes. This can be filled in for different contexts of inquiry that invoke levels. Qualitative similarity and belonging to the same type will be more important intralevel criteria for (R3) contexts—especially for generalization, comparison, and inference across systems that are not causally interacting. For example, it may be useful to invoke levels in comparative studies of major transitions in individuality between two populations that never interact with each other. In such cases, relevant similarities can ground statements, inferences, or models about a "group level" or "colony level" that includes both populations at once. By contrast, similarity in scale as well as interacting more strongly or frequently will be more important criteria for (R2) contexts, when the concern is to discover the causal structure of systems whose parts are potential interactants. While these kinds of qualifications may help reduce some of the ambiguity of same-level placements, some ambiguities will remain. There is likely is no other way to capture all of the interesting cases where we want to place entities at the same level than tolerating some degree of nondiscreteness.[7]

This framework for understanding levels leaves a number of issues open. Among these is what it means to "participate in a process." The notion of dispositional properties or conditional powers will probably be important here (see Gillett, 2013a). For now, I note some points in favor of this type of levels scheme.

• By connecting compositional levels directly to the category of processes, we make composition sensitive to causal relevance, which cannot be done with the standard object–property ontology of levels. Accordingly, unlike Oppenheim and Putnam's (1958) structuralist account, this levels scheme is consistent with "multiple type composition" in the sciences and thus gives us a better grasp of composition relations between types (R1).

• As a contextual specification of Wimsatt's (1976, 2007) general levels of organization, the restriction to a single overarching process reduces (without eliminating) the amount of ambiguity in levels placements and nondiscreteness of levels. As a result, one can expect that "levels break down" at higher scales than in the general scheme, and it should be easier to instantiate regularities and predictions about the focal process at higher levels (see section 6.5).

• Unlike Craver's (2007, 2015) levels of mechanisms, there is a stable intralevel relation, although it is based on a cluster of characteristics. In addition, the items at levels need not be tokens and need not figure into mechanistic explanations specifically (e.g., whole-organism or population-level processes).

• Including processes and their constituency relations in the account of compositional levels provides a categorially richer picture of levels of organization than is available from structural hierarchies. Of particular interest is that we can represent processes themselves as units of generalization, or as maxima of regularity and predictability, rather than attributing this role to entities only (R3). This is important when the same process has a shifting participant base, such that the regularity is to be found in the process rather than the entities that transiently participate in it. This phenomenon is common in comparative developmental biology.

In order to put these ideas to work, I want to now take a more detailed look at a specific scientific contextualization of levels in evolutionary developmental biology.

6.5   Levels of Developmental Evolution

"Developmental evolution" refers to the confluence of two causal processes, one constituting and one constituted. This confluence involves a two-way filter of each process from the perspective of the other. On the one hand, development is viewed from evolutionary perspective, which filters out constituents of development that are not heritable and that do not vary. On the other hand, evolution is viewed from a developmental perspective, which filters out evolutionary effects that are not caused by development. That evolution has nondevelopmental causes is compatible with the thesis that evolution is constituted by change in development. According to Van Valen's (1973) pithy formula, "evolution is the control of development by ecology." Ecology has controlling influences, but development is what is controlled—that is, ecological effects on evolution are manifested in development as the substrate of evolutionary change.[8]

In fact, the individuation of these processes is not precise and fixed but is a dynamic element in the ongoing growth of evolutionary and developmental biology. This is especially true of development, since developmental biology is not characterized by an overarching theory describing a unified process of "development" but rather by families of diverse approaches to *aspects* of development (Love, 2015; see Minelli & Pradeu, 2014, and essays therein). These aspects are subprocesses within development, such as transcriptional regulation, differentiation, cellular processes, growth, tissue mechanics, and so on. "Development" can be viewed as the complex embedding process, of which these subprocesses are partial aspects. Since this process of development has mobile boundaries and a yet-unresolved structure, what it means to "filter" evolution through a developmental perspective will change as developmental biology changes. It is important to point out, however, that this broad sense of development differs from what is more often referred to as development, which is *early* development, embryogenesis, and morphogenesis of major characters. The narrow process of development may include major causes of evolutionary change, but only the wider process can be what constitutes evolutionary change.

6.5.1   Levels and Variational Independence

Researchers in evo-devo sometimes refer to general levels of organization (i.e., molecules, cells, tissues, organs, organisms), but more often, what is at stake is levels of organization decomposed under the process of developmental evolution. These more restricted levels do not include everything that exists in the organism but just the main parts that have causal or constitutive relevance for the process of developmental evolution (see table 6.1).

Let us now map these items onto the levels scheme from the previous section. Starting with interlevel relations, note that, in the right column of table 6.1, the relation between

Table 6.1
A schematic catalogue of the levels of developmental evolution

Entity level	Process level
population	evolution
organism	development, ontogenesis
character	development
cells/tissues	*morphogenesis*
gene network	regulatory interactions, signaling, transcriptional state specification
gene	gene expression and regulation

Note: Categories emphasized in bold italics are those that possess a significant degree of variational independence (see text and table 6.2).

lower and higher processes is one of constitution (or partial constitution) without identity. As for intralevel relations, different processes, such as processes of gene expression, can be on the same level by being constituents of the same overarching process, such as a complex regulatory cascade or intercellular signaling. However, most processes of gene expression in a developing organism are not direct constituents of the same individual process. They can be placed on the same level, if one likes, on the basis of their close similarity in many of the characteristic intralevel features (structure, function, dynamics, scale, and interacting more strongly or frequently with each other than with other processes). The same intralevel relation can be derived for the other process levels.

In the left column, the entities are at different levels because they participate in processes that are related by (partial) constitution. They are also at different levels because they compose one another, given that one of the conditions on process constitution is that the processes have participants that are related compositionally. But the strategy here is not to *lead* with a notion of composition, instead relying on causal processes to *select* the functionally relevant composition relations from more inclusive levels of organization. This is why the compositional relations between entity levels are not exhaustive (e.g., cells are composed of more than just gene networks). Entities like genes are on the same level when they participate in processes (gene expression) that are partial constituents of the same overarching process (e.g., transcriptional regulation, signal transduction). Again, however, most activities of gene expression in a developing organism are not direct constituents of the same regulatory process but are separated by intermediary processes. Genes can then be placed on the same level on the basis of their similarity in the characteristic intralevel features. Again, the same analysis of inter- and intralevel relations can be repeated for the other entities in the left column.[9]

More entities and processes can be added to the hierarchy depending on the grain of analysis. In general contexts of discussion, however, workers in evo-devo often work with *fewer* levels: primarily, genes, gene networks, morphogenetic processes, and phenotypic characters (see table 6.2). The reason why these items are singled out at levels is that they are considered to have the greatest influence on the process of developmental evolution— they are the primary "levels" at which the process determines its course. To participate in an evolutionary process, it is necessary that some item possess heritable variation, so that it can reliably reappear over generations. To be a "unit" or "module" of heritable variation, the item must vary independently or "quasi-independently" of other items. The condition

Table 6.2
Description of variational independence across levels

Level	Variational independence
Character level	• Different processes can cause the same (homologous) character.
	• The same processes can cause different characters.
Process level	• Different networks can cause the same morphogenetic process.
	• The same networks can cause different morphogenetic processes.
Network level	• The regulatory topology of a network is not uniquely specified by its component genes, so the same genes can compose different networks.
	• Different individual genes can have the same functional roles in a gene regulatory network, so the same networks can have different genetic components.
Genetic level	• Individual genes are paradigmatic units of heritable variation, varying independently of variations at higher levels.

of *variational independence* captures what it means for something to be a functioning unit of developmental evolution. Brigandt (2007) proposes that variational independence is what makes a body part a distinct character in developmental evolution. The same condition of variational independence (or "quasi-independence") can be used for understanding levels of developmental evolution if it is combined with the conditions on levels from the previous section.[10]

Variational levels of developmental evolution: some class of items constitutes a level of developmental evolution iff: (1) that class varies (quasi-)independently of other classes of items, and (2) the classes are at different levels in the same process hierarchy (see "interlevel (processes)" and "interlevel (entities)" above).

The variational independence of phenotypic characters from the morphogenetic processes that produce them has been recorded observationally since the early days of embryology (see Scholtz, 2005, for examples). A well-studied example is the segmented anteroposterior axis of insects, which is a conserved character despite its being produced by different segmentation processes in different insect groups. In short-germ segmentation (found in more basal insect orders), segments are specified sequentially from anterior to posterior. In long-germ segmentation (found in more derived holometabolous insects), segments are specified simultaneously by subdivision of the blastoderm (Clark et al., 2019; Liu & Kaufman, 2005).

The variational independence of morphogenetic processes from the networks that cause them is increasingly coming to be appreciated as dynamical modeling of morphogenesis is incorporated into the research methods of evo-devo (DiFrisco & Jaeger, 2019; Jaeger, 2019; Jaeger & Monk, 2014; von Dassow & Munro, 1999). An example of the same network underlying different processes again comes from studies of insect segmentation. The same gap gene network produces different segmentation phenotypes in fruit flies and scuttle flies due to time-variable maternal inputs, differences in regulatory strengths within the network, and the fact that different subnetworks are in a critical state in the two groups (Jaeger & Monk, 2014; Verd et al., 2019). An example of different networks underlying the same morphogenetic process can be found in the vertebrate segmentation clock (Krol et al., 2011). Vertebrae develop out of segmented precursors in the mesoderm known as somites. Somitogenesis is described by the "clock and wavefront" model (Cooke & Zeeman, 1976): somites are produced by the interaction between oscillating waves of gene expression (clock) and a mechanism that halts the waves in a periodic fashion (wavefront). Comparative studies have revealed that, even though the segmentation process is conserved across all vertebrates, there are substantially different networks with different individual genes in distant vertebrate groups (mouse, chicken, and zebrafish) (Krol et al., 2011).

The variational independence of gene networks from individual genes can be expressed in terms of dissociation between gene functioning in networks (regulatory role) and gene structure (base pair sequence). Structurally different genes can have the same functions, and significant amounts of functional redundancy from the genetic level to the network level are well documented. In the vertebrate segmentation clock, for example, the presence of many genes capable of producing the same kind of oscillatory behavior may have allowed genetic divergences to accumulate over evolutionary time (Krol et al., 2011). Likewise, the same genes can have different functions and acquire new functions with apparent ease in evolution (True & Carroll, 2002). Many key regulatory genes, such as genes from the *Notch* and *Wnt* families, are used and reused in completely different networks and developmental processes. So even though gene networks are entirely composed of individual genes, their identity is not determined by the structural identity of the component genes. Instead, it is the *interactions* between genes in a network, as well as the network's spatiotemporal and cell-biological context, that gives them causal specificity in the process of morphogenesis. Their status as crucial *units* of variational independence derives from the fact that, as Wagner (2014) writes, genes organized in networks "form a functional unit in which developmental causality is realized at the level of the network rather than at the level of the single gene" (p. 117). The move toward explaining development in terms of complex networks of interacting genes rather than individual genes has been one of the major breakthroughs of developmental genetics in the past decades (Peter & Davidson, 2015; Salazar-Ciudad & Jernvall, 2013; see DiFrisco & Jaeger, 2019).

6.5.2   Theoretical Roles for Levels of Developmental Evolution

How do the levels of developmental evolution map onto the three major roles (R1) to (R3) for levels distinguished earlier? On (R1), these levels identify general composition relationships under a dynamical aspect in which composition is relativized to a focal process. Not all components are placed into levels, so following a "constructive monist" approach to levels (see section 6.2), (R1) is partly sacrificed for the other roles.

On (R2), there is some truth to the general idea of stronger interactions within levels than between levels, but the process of development centrally involves *interlevel* causation (see Baedke, this volume; Green, this volume; Woodward, this volume). More important for (R2) is the notion of causal independence (see Woodward, this volume), which appears in this context as variational independence. There is a tension here between independence and levels, since constitution is a type of dependence relationship. How can higher-level things be causally independent if they are composed of, or constituted by, lower-level things? In many cases, the causal independence of higher-level variables can take the form of a summation or screening-off. For example, conditional upon the temperature of a sample of gas at thermal equilibrium, variations in the kinetic energies of its component molecules that are consistent with this temperature do not make a difference to its pressure (see Woodward, this volume). We have seen a strong form of causal independence in which the higher-level variable is robust under perturbations to the components and also possesses greater generality. The segmentation clock in each vertebrate species is caused by a unique regulatory network and set of individual genes, so it is ultimately dependent on these factors. But these lower-level factors vary across taxonomic groups, whereas the morphogenetic process (somitogenesis) character (somites) and certain key signaling pathways (Wnt, FGF, Notch) do not. The development of somites is evolutionarily most stable at the "process level." Causal explanations of somites in terms of the morphogenetic process therefore have greater generality than those in terms of the participant genes.

This leads directly into the role of generalizing over levels (R3). Discovering comparative generalizations is a crucial task in evo-devo because there are too many species to study individually. Levels, as depicted in table 6.1 and especially table 6.2, contribute to this task by demarcating stable bases of comparative generalization. Importantly, these levels include more than just genetic and network levels, although those levels have been the major focus of comparative evo-devo so far. The generalizability of causes of development at different levels is expressed in Salazar-Ciudad and Jernvall's (2013) notion of a "causality horizon." Essentially, the causality horizon between compared species is the lowest level at which the causes of a morphological feature are the same. Vertebrate somitogenesis has a *high* causality horizon—namely, the level of the morphogenetic process (oscillating clock and wavefront). Below the causality horizon, identifying the more fundamental cause of a phenotype in one species will not successfully predict the cause of a corresponding phenotype in another species (Salazar-Ciudad & Jernvall, 2013, p. 286). Accordingly, the causality horizon represents "the boundary of successful, or appropriate, reductionism in development" (Salazar-Ciudad & Jernvall, 2013, p. 286).

The causality horizon demarcates "maxima of regularity and predictability" (Wimsatt, 2007, p. 209) in developmental evolution (R3). That these maxima can be *processes* in addition to the standard compositional entities is an insight that becomes accessible by shifting from the frame from structural levels and general levels to decomposition under a specific causal process.

6.5.3   Beyond Levels of Selection and Replicator–Interactor Levels

I close this section by noting a few of the ways in which levels of developmental evolution represent a richer conceptualization of levels than existing evolutionary frameworks. There are no such frameworks that take a purely structural perspective, probably because structural frameworks would have difficulty accommodating selection for function. More standard frameworks are based on effect functions, such as levels of selection and the replicator–interactor hierarchy (for a classic, in-depth criticism of these frameworks, see Griesemer, 2005).

Much theorizing about evolution in biology and philosophy takes place in the hierarchical frame of levels of selection (Lloyd, 2017; Okasha, 2006). One issue with levels of selection as an ontological framework is that it presupposes a background notion of levels of organization that remains unanalyzed (Griesemer, 2005), and *general* levels of organization are probably too blunt an instrument for this role (Eronen & Ramsey, 2021). A more basic limitation is that selection is only a subprocess of evolution. Levels decomposed under the process of selection do not capture the sources and structures of variation, which are developmental and genetic, but only the *outcomes* of development. Accordingly, levels of selection alone are too limited to serve as the hierarchical framework for explaining things like major transitions in individuality, which involve changes in development and inheritance.

As Griesemer (2006) writes, "Because the origin of new levels involves the evolution of new developmental processes and not only the evolution of adaptations, there can be no escaping the need for an account of how heredity and development intertwine to co-produce conditions for evolvability" (p. 344).

Hull's (1980) "replicator–interactor" framework was designed as an improvement over Lewontin's (1970) levels of selection framework that would distinguish the units of variation or heredity from units of selection. The replicator–interactor distinction is a hierarchical generalization of Dawkins's (1978) "replicator–vehicle" distinction, in which replicators and interactors can potentially be found at many levels of organization as long as entities at that level fulfill the requisite functional role. A replicator is any entity that passes on its structure through direct replication or copying. An interactor is any entity that "directly interacts as a cohesive whole with its environment in such a way that replication is differential" (Hull, 1980, p. 318). Although the replicator–interactor framework purports to provide a decomposition of levels under the evolutionary process (Hull, 1980, pp. 312–313), this decomposition is defective for two reasons.[11]

First, it abstracts only outcome-based definitions from causal processes (Griesemer, 2006). In the case of replication, the outcome is structural similarity between original and copy, while material processes of inheritance are backgrounded (Griesemer, 2006). In the case of interactors, the outcome is differential replication, while ecological processes are backgrounded. This first defect leads right back to the same limitations of levels of selection just described. The second defect is that the replicator–interactor framework includes only two narrow subprocesses out of the wider process of developmental evolution—namely, inheritance and selection. This leads to a severely impoverished picture of evolution. By excluding development and its subprocesses, we lose any chance of capturing the causal structures that link genetic inheritance and selection. Moreover, by excluding development, the framework *cannot* incorporate insights about the hierarchical structure of variation, including features like network-level causation, variational independence above the genetic level, and shifting causality horizons in evolution.

6.6   Conclusion

The notion of levels of organization has multiple cross-cutting roles in the life sciences. Even when we restrict the discussion to only the most major of these roles (R1) to (R3), it is often difficult to actually identify compositional types (R1) that track causal structures (R2) that hold with generality (R3). I attributed this difficulty to the dissociability of structure and function in complex evolved systems. A consequence of this dissociability is that, as long as we wish to satisfy all three roles at once, it will be necessary to find some way of bridging structural and functional perspectives on levels. One way is to combine perspectives while relying on a characterizational or "cluster" approach to level placements, as I interpret Wimsatt (1976, 2007). Another way is to distinguish perspectives in terms of the causal processes they track and decompose levels under a single unified causal process. In the process-relative approach, placement at levels is based on constitution relationships between causal processes. This has the effect of restricting the structures or components to those that participate in the focal process, thereby making level placements less ambiguous, while also including dynamical categories in the range of generalizations about levels (R3). These two approaches are complementary and opt for different trade-offs between generality and specificity. The general picture can be reached from the process-relative approach by aggregating multiple perspectives and identifying levels where there are robust points of convergence between them (see Wimsatt, this volume; Griesemer, this volume).

The roles for levels start to acquire more concrete significance when we descend from the thin air of metaphysics of science to an empirical context. The case study investigated in section 6.5—decomposition under the process of developmental evolution—yields a hierarchy concept with several distinctive features. Not all parts and processes of organisms are placed into levels (R1), and those that are possess the special property of variational independence (R2) in either the entity or process at a given level. Variational independence can ground the notion of a "causality horizon" in comparative evo-devo, which marks the limits of reduction as well as privileged levels of generalization (R3). The end result is an evolutionary picture of levels that is deeper than the effect-functional frameworks of levels of selection and replicator–interactor levels, while also including them as partial subprocesses of developmental evolution.

The approach proposed here is not a complete framework for thinking about levels in evolutionary and developmental biology, but it is hoped that this initial attempt can open up a more expansive role for levels in future theorizing. One future direction will be to integrate metabolic or physiological perspectives on development. A key insight of Waddington's three-process hierarchy of evolution–development–metabolism was that any theory that does not include all three is destined to be incomplete in some crucial respect (Waddington, 1957, p. 6). Work in evo-devo often operates with the narrow understanding of development as the acquisition of specific regulatory states in cell populations, which is explained in terms of modular gene networks. In the background are more systemic, nonmodularized metabolic processes such as growth, resource competition, tissue-level mechanics and regulation, and character–character interaction generally. Integration of these processes within a broader conception of development might show that modularity of gene networks does not translate to modularity at higher levels of development (Gawne et al., 2020; Nijhout & Emlen, 1998). A second future direction will be to move from descriptions of variational independence across levels to explanations of its occurrence (DiFrisco, forthcoming; DiFrisco et al., 2020). This might follow general explanations deriving from the tradition of Simon (1962) and Wimsatt (1976, 2007) of why there are levels of organization at all, but contextualized specifically to the process of evolution.

Acknowledgments

Thanks to Dan Brooks, Carl Gillett, Jim Griesemer, Davide Serpico, and Bill Wimsatt for helpful comments. Thanks also to all participants of the workshop on Hierarchy and Levels of Organization in the Biological Sciences for insightful discussion.

Notes

1. Note that Craver (2015, 19n22) neither endorses nor rejects token identity across levels, although he does recognize the descriptive feature of "part loss" when we ascend to higher levels. See section 4 for further details.

2. The same point has also been made by Alan Love and Bill Wimsatt.

3. Thanks to Jim Griesemer for prompting this idea.

4. See also Griesemer (2018) for a complementary exploration of the idea of "process-relativity of individuation" or "process realism with entity relativism" in the context of biological individuality.

5. The causal nature of processes could also be explicated in terms of our ability to *manipulate* it at one time and observe changes in the process at a later time.

6. Amundson (2005) calls this the "causal completeness principle" (p. 176).

7. At this point, the question arises whether descriptive complexity and perspectival pluralism re-enter the process-relative levels but at a finer grain than with general levels of organization. This seems to some extent unavoidable, but the goal is to reduce ambiguity rather than eliminate it entirely. The structure of perspectives-within-perspectives nonetheless does not seem to be genuinely fractal or iterated at all scales of conceptual resolution.

8. Those promoting *eco-evo-devo* can be interpreted as advocating for either the inclusion of more ecological variables into the study of developmental evolution or, more radically, the inclusion of ecological parts and processes within the boundaries of an "extended development" (Gilbert & Epel, 2009; see Griesemer, 2019, this volume). In this chapter, I wish to use the term "development" in a way that is compatible with both extended and restricted views of organism boundaries.

9. Note that the notion of a "character" is not used here in the wide sense from systematics of any "characteristic," which would apply to all lower levels as well. "Characters" in the narrow sense are body parts or gross phenotypic features, and there are often hierarchical relationships within a single character complex (e.g., digits as part of the autopod or vertebrae as parts of the vertebral column).

10. Note that variational independence across levels is neither randomly nor uniformly distributed, and the conserved correspondences possess special significance for a mechanistic theory of character identity. See DiFrisco et al. (2020).

11. The same criticism applies to more recent frameworks based on the distinction between "genealogical" and "economic" hierarchies, which is a hierarchical generalization of the replicator–interactor distinction.

References

Allen, T. F. H., & Starr, T. B. (1982). *Hierarchy: Perspectives for ecological complexity*. Chicago, IL: University of Chicago Press.

Amundson, R. (2005). *The changing role of the embryo in evolutionary thought: Roots of evo-devo*. Cambridge, UK: Cambridge University Press.

Baedke, J., & McManus, S. (2018). From seconds to eons: Time scales, hierarchies, and processes in evo-devo. *Studies in History and Philosophy of Science Part C: Studies in History and Philosophy of Biological and Biomedical Sciences, 72,* 38–48.

Bechtel, W. (2012). Identity, reduction, and conserved mechanisms: Perspectives from circadian rhythm research. In S. Gozzano & C. S. Hill (Eds.), *New perspectives on type identity* (pp. 43–65). Cambridge, UK: Cambridge University Press.

Brigandt, I. (2007). Typology now: Homology and developmental constraints explain evolvability. *Biology and Philosophy, 22,* 709–725.

Brigandt, I. (2017). Bodily parts in the structure-function dialectic. In S. Lidgard & N. K. Nyhart (Eds.), *Biological individuality: Integrating scientific, philosophical, and historical perspectives* (pp. 249–274). Chicago, IL: University of Chicago Press.

Brooks, D. S. (2017). Layer cakes and guilt by association. *Biological Theory, 12*(3), 142–156.

Bueno, O., Chen, R.-L., & Fagan, M. B. (Eds.). (2018). *Individuation, process, and scientific practices*. Oxford, UK: Oxford University Press.

Clark, E., Peel, A. D., & Akam, M. (2019). Arthropod segmentation. *Development, 146,* dev170480.

Clarke, E. (2016). Levels of selection in biofilms: Multispecies biofilms are *not* evolutionary individuals. *Biology and Philosophy, 31,* 191–212.

Cooke, J., & Zeeman, E. C. (1976). A clock and wavefront model for control of the number of repeated structures during animal morphogenesis. *Journal of Theoretical Biology, 58,* 455–476.

Craver, C. F. (2007). *Explaining the brain*. Oxford, UK: Oxford University Press.

Craver, C. F. (2015). Levels. In T. Metzinger & J. M. Windt (Eds.), *Open MIND*. Frankfurt am Main, Germany: MIND Group.

Craver, C. F., & Bechtel, W. (2007). Top-down causation without top-down causes. *Biology and Philosophy, 20,* 715–734.

Dawkins, R. (1978). Replicator selection and the extended phenotype. *Zeitschrift für Tierpsychologie, 47,* 61–76.

DiFrisco, J. (2017). Time scales and levels of organization. *Erkenntnis, 82*(4), 795–818.

DiFrisco, J. (2018). Token physicalism and functional individuation. *European Journal for Philosophy of Science, 8*(3), 309–329.

DiFrisco, J. (2019). Kinds of biological individuals: Sortals, projectibility, and selection. *British Journal for the Philosophy of Science, 70*(3), 845–875.

DiFrisco, J. (forthcoming). Toward a theory of homology: development and the de-coupling of morphological and molecular evolution. *British Journal for the Philosophy of Science*.

DiFrisco, J., & Jaeger, J. (2019). Beyond networks: Mechanism and process in evo-devo. *Biology & Philosophy, 34*, 54.

DiFrisco, J., Love, A. C., & Wagner, G. P. (2020). Character identity mechanisms: A conceptual model for comparative-mechanistic biology. *Biology & Philosophy, 35*, 44.

Eldredge, N., & Salthe, S. (1984). Hierarchy and evolution. *Oxford Surveys in Evolutionary Biology, 1*, 184–208.

Eronen, M. I. (2013). No levels, no problems: Downward causation in neuroscience. *Philosophy of Science, 80*(5), 1042–1052.

Eronen, M. I. (2015). Levels of organization: A deflationary account. *Biology and Philosophy, 30*(1), 39–58.

Eronen, M. I., & Ramsey, G. (2021). *What are the 'levels' in levels of selection?* Unpublished manuscript.

Gawne, R., McKenna, K. Z., & Levin, M. (2020). Competitive and coordinative interactions between body parts produce adaptive developmental outcomes. *BioEssays*, 1900245.

Gilbert, S. F., & Epel, D. (2009). *Ecological developmental biology: Integrating epigenetics, medicine, and evolution*. Cambridge, UK: Sinauer Associates.

Gilbert, S. F., Optiz, J. M., & Raff. R. A. (1996). Resynthesizing evolutionary and developmental biology. *Developmental Biology, 173*, 357–372.

Gillett, C. (2013a). Constitution and multiple constitution in the sciences: Using the neuron to construct a starting framework. *Minds & Machines, 23*, 309–337.

Gillett, C. (2013b). Understanding the sciences through the fog of functionalism(s). In P. Huneman (Ed.), *Functions: Selection and mechanisms* (pp. 159–181). Dordrecht, Netherlands: Springer.

Goodwin, B. C. (1963). *Temporal organization in cells*. London, UK: Academic Press.

Gould, S. J., & Lewontin, R. C. (1979). The spandrels of San Marco and the panglossian paradigm: A critique of the adaptationist programme. *Proceedings of the Royal Society of London B, 205*, 581–598.

Griesemer, J. R. (2000). Reproduction and the reduction of genetics. In P. Beurton, R. Falk, & H.-J. Rheinberger (Eds.), *The concept of the gene in development and evolution: Historical and epistemological perspectives* (pp. 240–285). New York, NY: Cambridge University Press.

Griesemer, J. R. (2005). The informational gene and the substantial body: On the generalization of evolutionary theory by abstraction. In M. R. Jones & N. Cartwright (Eds.), *Idealization XII: Correcting the model, idealization and abstraction in the sciences* (pp. 59–115). Amsterdam, Netherlands: Rodopi.

Griesemer, J. R. (2006). Genetics from an evolutionary process perspective. In E. M. Neumann-Held & C. Rehmann-Sutter (Eds.), *Genes in development* (pp. 199–237). Durham, NC: Duke University Press.

Griesemer, J. R. (2018). Individuation of developmental systems: A reproducer perspective. In O. Bueno, R.-L. Chen, & M. B. Fagan (Eds.), *Individuation, process, and scientific practices* (pp. 137–164). Oxford, UK: Oxford University Press.

Griesemer, J. R. (2019). Towards a theory of extended development. In G. Fusco (Ed.), *Perspectives on evolutionary and developmental biology: Essays for Alessandro Minelli* (pp. 319–334). Padova, Italy: Padova University Press.

Hou, C., Kaspari, M., Vander Zanden, H. B., & Gillooly, J. F. (2010). Energetic basis of colonial living in social insects. *Proceedings of the National Academy of Sciences, 107*, 3634–3638.

Hull, D. L. (1980). Individuality and selection. *Annual Review of Ecology, Evolution, and Systematics, 11*, 311–332.

Jaeger, J. (2019). Dynamic structures in evo-devo: From morphogenetic fields to evolving organisms. In G. Fusco (Ed.), *Perspectives on evolutionary and developmental biology: Essays for Alessandro Minelli* (pp. 335–355). Padova, Italy: Padova University Press.

Jaeger, J., & Monk, N. (2014). Bioattractors: Dynamical systems theory and the evolution of regulatory processes. *Journal of Physiology, 592*, 2267–2281.

Kauffman, S. A. (1971). Articulation of parts explanation in biology and the rational search for them. In R. C. Buck & R. S. Cohen (Eds.), *PSA 1970* (pp. 257–272). Dordrecht, Netherlands: Reidel.

Krol, A. J., Roellig, D., Dequéant, M.-L., Tassy, O., Glynn, E., Hattem, G., Mushegian, A., Oates, A. C., & Pourquié, O. (2011). Evolutionary plasticity of segmentation clock networks. *Development, 138*, 2783–2792.

Laubichler, M., & Wagner, G. P. (2000). Organism and character decomposition: Steps towards an integrative theory of biology. *Philosophy of Science, 67,* S289–300.

Lewontin, R. C. (1970). The units of selection. *Annual Review of Ecology and Systematics, 1,* 1–18.

Liu, P. Z., & Kaufman, T. C. (2005). Short and long germ segmentation: Unanswered questions in the evolution of a developmental mode. *Evolution & Development, 7*(6), 629–646.

Lloyd, E. (2017). Units and levels of selection. In E. N. Zalta (Ed.), *The Stanford encyclopedia of philosophy.* Retrieved from https://plato.stanford.edu/archives/sum2017/entries/selection-units/

Love, A. (2007). Functional homology and homology of function: Biological concepts and philosophical consequences. *Biology and Philosophy, 22,* 691–708.

Love, A. (2015). Developmental biology. In E. N. Zalta (Ed.), *The Stanford encyclopedia of philosophy.* Retrieved from https://plato.stanford.edu/archives/fall2015/entries/biology-developmental/

Lowe, E. J. (2003). Individuation. In M. J. Loux & D. W. Zimmerman (Eds.), *The Oxford handbook of metaphysics* (pp. 77–95). Oxford, UK: Oxford University Press.

Minelli, A. (2011). Animal development, an open-ended segment of life. *Biological Theory, 6,* 4–15.

Minelli, A., & Pradeu, T. (Eds.). (2014). *Towards a theory of development.* Oxford, UK: Oxford University Press.

Nijhout, H. F., & Emlen, D. J. (1998). Competition among body parts in the development and evolution of insect morphology. *PNAS, 95,* 3685–3689.

Okasha, S. (2006). *Evolution and the levels of selection.* Oxford, UK: Oxford University Press.

O'Neill, R. V., DeAngelis, D. L., Waide, J. B., & Allen, T. F. H. (1986). *A hierarchical concept of ecosystems.* Princeton, NJ: Princeton University Press.

Oppenheim, P., & Putnam, H. (1958). Unity of science as a working hypothesis. In H. Feigl, G. Maxwell, & M. Scriven (Eds.), *Minnesota studies in the philosophy of science* (pp. 3–36). Minneapolis: University of Minnesota Press.

Peter, I. S., & Davidson, E. H. (2015). *Genomic control process: Development and evolution.* Amsterdam, Netherlands: Elsevier.

Potochnik, A., & McGill, B. (2012). The limitations of hierarchical organization. *Philosophy of Science, 79,* 120–140.

Rueger, A., & McGivern, P. (2010). Hierarchies and levels of reality. *Synthese, 176,* 379–397.

Salazar-Ciudad, I., & Jernvall, J. (2013). The causality horizon and the developmental bases of morphological evolution. *Biological Theory, 8*(3), 286–282.

Salmon, W. C. (1984). *Scientific explanation and the causal structure of the world.* Princeton, NJ: Princeton University Press.

Salthe, S. (1985). *Evolving hierarchical systems: Their structure and representation.* New York, NY: Columbia University Press.

Scholtz, G. (2005). Homology and ontogeny: Pattern and process in comparative developmental biology. *Theory in Biosciences, 124,* 121–143.

Seibt, J. (2014). Non-transitive parthood, leveled mereology, and the representation of emergent parts of processes. *Grazer Philosophische Studien, 91,* 165–190.

Simon, H. A. (1962). The architecture of complexity. *Proceedings of the American Philosophical Society, 106,* 467–482.

True, J. R., & Carroll, S. B.. (2002). Gene co-option in physiological and morphological evolution. *Annual Review of Cell and Developmental Biology, 18,* 53–80.

Van Valen, L. (1973). Festschrift. *Science, 180,* 488.

Verd, B., Monk, N. A., & Jaeger, J. (2019). Modularity, criticality, and evolvability of a developmental gene regulatory network. *eLIFE, 8,* e42832.

von Dassow, G., & Munro, E. (1999). Modularity in animal development and evolution: elements of a conceptual framework for EvoDevo. *Journal of Experimental Zoology (Molecular Development and Evolution), 285,* 307–325.

Waddington, C. H. (1957). *The strategy of the genes.* London, UK: Routledge.

Wagner, G. P. (2014). *Homology, genes, and evolutionary innovation.* Princeton, NJ: Princeton University Press.

Wimsatt, W. C. (1972). Complexity and organization. In K. F. Schaffner & R. S. Cohen (Eds.), *PSA 1972* (pp. 67–86). Dordrecht, Netherlands: Reidel.

Wimsatt, W. C. (1976). Reductionism, levels of organization, and the mind-body problem. In G. G. Globus, G. Maxwell, & I. Savodnik (Eds.), *Consciousness and the brain: A scientific and philosophical inquiry* (pp. 205–267). New York, NY: Plenum Press.

Wimsatt, W. C. (2007). *Re-engineering philosophy for limited beings.* Cambridge, MA: Harvard University Press.

7 Manipulating Levels of Organization

Alan C. Love

Overview

Despite their widespread invocation in scientific practice and pedagogy, levels of organization as hierarchical representations have come under increasing philosophical scrutiny. However, another significant dimension of scientific practice is manipulation. Scientists not only purport to represent levels but also claim to manipulate them. In this chapter, I examine the manipulation of relatively stable configuration states in developmental biology with special attention to the origin of tissue-level organization from cell-level organization during embryogenesis. I use two forms of experimental practice for illustration: mixed cellular aggregates and tissue engineering. These successful experimental practices that rely on diverse forms of manipulation help to demonstrate that levels of organization cannot be reduced to principles of scale and do not support the claim that levels are only defined locally in terms of mechanisms. Additionally, they do not correspond to comprehensive, abstract "layer-cake" perspectives that are global in scope and map directly onto the disciplinary structure of the sciences. Overall, manipulation practices show that developmental biologists concentrate on *transitions between* particular levels, and the experimental establishment of these transitions contributes to interpreting cell and tissue configuration states as ontological levels of organization.

7.1 Levels of Organization: Scrutiny and Skepticism

Open an introductory biology textbook from any publisher and you will find levels of organization near the outset. "Biological organization is based on a hierarchy of structural levels, each level building on the levels below it" (Campbell & Reece, 2002, p. 2). As is made clear in the instructor's guide and main text, levels of organization are considered a *unifying* theme for the biological sciences: "Investigating biological organization at its many levels is fundamental to the study of life" (p. 3). Although it is not often remembered, this view of the role that levels play in biological reasoning has roots in the organicist movement of the early twentieth century; organicists attempted to steer a middle way between strict forms of physicochemical mechanism and various vitalist approaches appealing to nonphysical forces (Eronen & Brooks, 2018, §1). The role of levels was not only for peda-gogical formation but also for structuring investigation.

The picture of levels of organization found in textbook discussions has some resonance with a logical empiricist outlook where nature is organized into a hierarchy of levels such that every whole at one level can be decomposed into constituents at a lower level: social groups, multicellular organisms, cells, molecules, atoms, and elementary particles (Oppen-heim & Putnam, 1958). This encouraged a "layer-cake" view of the sciences as offering theories at corresponding levels of a strict compositional hierarchy: sociology, organismal biology, molecular biology, chemistry, and physics. However, the philosophical aim of logical empiricists was different from the rationale of biological practitioners. They were attempting to counterbalance an increasing specialization of the sciences with a meta-scientific study "promoting the integration of scientific knowledge" (Oppenheim & Putnam, 1958, p. 3). The goal was to reduce one branch of science with a theory of objects in its domain (e.g., biology) to another branch of science with a theory of the constituents of those objects in its domain (e.g., chemistry). These reductive aims contradict the textbook perspective: "Biology balances the reductionist strategy with the longer-range objective of understanding how the parts of the cells, organism, and higher levels of order, such as ecosystems, are functionally integrated" (Campbell & Reece, 2002, p. 4).

Despite their widespread invocation in scientific practice and pedagogy, levels of organ-ization have come under increasing philosophical scrutiny (see Eronen & Brooks, 2018, for an overview). In some cases, there is outright skepticism (Eronen, 2013, 2015a; Potoch-nik, 2017; Potochnik & McGill, 2012; Thalos, 2013). This is often expressed by showing that there are multiple, conflicting views of levels of organization and that confusion can be removed by shifting to principles of scale. More radical forms of skepticism also exist: "We will deny … that the world comes in 'levels'. Contemporary science … gives no inter-esting content to this metaphor" (Ladyman & Ross, 2007, p. 54). An alternative orientation is to restrict claims about levels to the context of mechanisms (Craver, 2007; Craver & Bechtel, 2007; Povich & Craver, 2018; for criticism, see Eronen, 2015a). More global accounts of levels of organization, whether from scientific textbooks or logical empiricist philosophy, are rejected, but levels are defined in terms of a mechanism that exhibits a behavior as a consequence of the activities of its component parts being organized in a particular fashion. "It is the set of working parts that are organized and whose operations are coordinated to realize the phenomenon of interest that constitute a level" (Bechtel, 2008, p. 146). The result is a localist view of levels of organization that is functional rather than structural and (purportedly) "accurately describes the multilevel explanatory struc-tures one finds in biology and other special sciences … [and] captures the common idea of 'levels of organization'" (Povich & Craver, 2018, p. 188).

A third family of perspectives seeks to identify an account of levels of organization that makes sense of scientific practice but is also broader. These are often associated with the insight that "levels of organization can be thought of as local maxima of regularity and predictability in the phase space of alternative modes of organization of matter" (Wimsatt, 2007, p. 209). In some cases, this involves attending to what underwrites regularity and predictability, such as time scales (DiFrisco, 2017). In other cases, allied concepts, such as biological parthood and kind membership, are utilized to generate an account of levels: "Object X is located on a lower level than object Y iff X is a biological part of Y or X

belongs to the same general biological kind as one or more biological parts of Y" (Kaiser, 2015, p. 183; see also Gillett, this volume). This permits the multifunctionality of components at levels because they are not defined solely as working parts of one mechanism. In still other cases, levels of organization are understood as local to particular domains, such as proteins, although not necessarily commensurable with hierarchical organization from other domains (even at similar scales) and distinct from the mechanisms that might contain members of these domains as components (Love, 2012).

The majority of these philosophical analyses focus on levels of organization in terms of hierarchical representation.[1] Levels of organization as a concept is understood to be an epistemic tool for representing ontological structure and function in nature. If the concept does this poorly or inconsistently, there are reasons for rejecting it in favor of other notions (e.g., scale), restricting the notion of levels of organization in a suitable fashion (e.g., only levels within a mechanism), or hesitating about whether levels have any ontological status (e.g., skepticism). The criterion of adequacy is (or at least is advertised as) correspondence to successful scientific practice. However, another significant dimension of scientific practice is manipulation. Scientists not only purport to represent levels of organization but also claim to manipulate them and appear to do so successfully. This suggests an alternative reading of the standard picture of levels of organization depicted in textbooks. If "investigating biological organization at its many levels is fundamental to the study of life," then analyzing the *experimental* investigation of biological organization at one or more of its putative levels could secure an account of (some) levels of organization (or relatively stable configuration states; see Woodward, this volume) that makes sense of scientific practices and also yields something broader. Levels of organization could be understood as an epistemic tool for *manipulating* ontological structure in nature.

In this chapter, I examine the manipulation of levels of organization in developmental biology with special attention to the origin of tissue-level organization from cell-level organization during embryogenesis. I use two forms of experimental practice for illustration: mixed cellular aggregates and tissue engineering. These experimental practices that rely on diverse forms of manipulation help to demonstrate that levels of organization cannot be reduced to principles of scale as some skeptics have argued. However, they also do not lend credence to abstract "layer-cake" perspectives about levels that are comprehensive in character, are global in scope, and map directly onto the disciplinary structure of the sciences. These diverse, successful practices sustain a modest realism about specific stable configuration states as levels of organization. Overall, manipulation practices show that developmental biologists concentrate on *transitions between* particular levels, and the experimental establishment of these transitions contributes to interpreting cell and tissue configuration states as ontological levels of organization—an inference from successful biological practice to scientific metaphysics.

7.2 Thinking about Manipulation

Just as representation is an abstract category that covers a diversity of representational forms and choices, so also manipulation is an abstract category that covers a diversity of experimental forms and choices. An awareness of this diversity is important for identifying successful practices of manipulation. Furthermore, given that conceptions of levels of organization are always embedded in a framework that includes the criteria relevant for their identification and use (Brooks, 2017), we must attend to differences in criteria of manipulation in the same way that we attend to differences in criteria of representation. Successful manipulation is relative to criteria that derive from a combination of different forms of manipulation (types, strategies, and means), the research question in view, and investigative aims (e.g., characterization, explanation, prediction, or control).

There are two broad *types* of manipulation: disruptive and constructive. A *disruptive* manipulation results in an intentional change to the system's normal operation, whether a precise modulation to achieve a quantitative difference or the wholesale disabling of a mechanism. That the manipulation results in a change is intentional, but what exactly that change is need not be intended. Disruptive manipulations are common in exploratory experimentation when biologists are characterizing phenomena. Simply "breaking things" can be illuminating for designing further experiments that control additional variables and make it possible to achieve specific outcomes. A *constructive* manipulation, in contrast, involves not only bringing about a change but also doing so in a way that yields an antici-pated result (i.e., a predicted outcome). This is often associated with testing hypotheses about the operation of a mechanism or teasing apart details in a causal process. The ability to constructively change the value of a variable precisely yields causal explanations that identify relationships of invariance between variables over particular ranges of values under controlled conditions (Woodward, 2003).

Strategies of manipulation can be divided into three categories: diminution, substitution, and augmentation. Diminution involves the removal of an entity or activity (i.e., the elimi-nation of a variable) or a decrease in concentration or rate (i.e., setting the value of the variable lower than is typically observed or expected). In developmental biology, tactics that implement this strategy include genetic knockdown techniques (Housden et al., 2017) and the ablation of cells with lasers (He & Masland, 1997). Augmentation involves the addition of an entity or activity (i.e., the inclusion of a variable) or an increase in concen-tration or rate (i.e., setting the value of the variable higher than is typically observed or expected), such as genetic overexpression (Prelich, 2012) or the transplantation of tissue sections from one location to another (Bergqvist et al., 1985). Substitution involves the replacement of an entity or activity (i.e., the substitution of a variable) in order to discover what results (if anything) from replacing modular features of processes or mechanisms. Common tactics include transgenic techniques, such as introducing orthologues or para-logues (Onuma et al., 2002) and xenografting (Alvarado-Mallart, 2005).

In addition to manipulation types and strategies, we also can distinguish three *means* of manipulation: genetic, chemical, or physical. Genetic means, such as transgenic tech-niques, involve the introduction of native or foreign ribonucleotide sequences or the modi-fication of nucleic acids using the genetic approach (e.g., RNA interference). Chemical means involve the introduction or modification of a chemical species, which can be native to the developing system (e.g., retinoic acid or hormones), although often they are intro-duced at novel times and places or in varying amounts (Butcher, 2017). Physical means involve the introduction or modification of physical forces (e.g., compression, tension, or electrical fields). These are inspired by dynamics pertinent to the phenomenon under scru-tiny, such as changes in the morphological organization of tissues in response to unbalanced stresses (Kim et al., 2017).

Although types, strategies, and means of manipulation have inherent criteria related to their use, whether they count as successful experimental manipulations also depends on criteria derived from a research context that has specific investigative aims and a particular research question. Investigative aims such as characterization—what a phenomenon consists of or does—and explanation—accounting for what comprises the phenomenon or how it is produced—are typically combined with prediction and control. Manipulations are successful when they meet both inherent criteria for types, strategies, and means, and standards for characterization, explanation, prediction, and control with respect to a research question of interest (e.g., how does neural crest tissue delaminate into migratory cells?).

Unsurprisingly, all of this strongly parallels what we find in the domain of representation where different investigative aims in the context of distinct research questions require distinct kinds of representational strategies, types, and means (van Fraassen, 2008). Recognizing the diversity of experimental manipulations and the heterogeneous basis for their criteria of success provides us with a framework for describing analytically how and why developmental biologists proceed when investigating the relatively stable configuration states of cells and tissues, especially the transitions between them.

7.3 Experimental Practices in Developmental Biology: Manipulating "Levels"

A central aim of developmental biology is the characterization, explanation, prediction, and control of transitions between individual cells and cellular aggregates, both normal and pathological (Baum et al., 2008). The transition from individual cells to structured cellular aggregates is a critical step in many processes of embryogenesis and involves cells differentiating into distinct identities (Kiecker et al., 2016). The reverse transition—from structured cellular aggregates to individual cells—is a deadly step in the etiology of cancers, making it possible for them to spread throughout the body (Hanahan & Weinberg, 2011), but also a part of normal ontogeny, such as in neural crest cell migration (Le Douarin & Kalcheim, 1999). Both transitions between these particular stable states of organization are of interest to developmental biologists, and a variety of experimental approaches aid in manipulating them.

Individual cells are often considered "at" the cellular level and cell aggregates are often considered "at" the tissue level. However, this does not immediately yield an argument in favor of these as autonomous "levels" of organization. Developmental biologists sometimes refer to these as "states" rather than levels. Also, individual, mobile cells can be labeled mesenchymal and sheets of cells aggregated into shaped structures are often identified as epithelial (Baum et al., 2008). In order to avoid prejudging the question of whether "levels" exist, consider these as relatively stable configuration states with a preferred status. Two examples of these experimental practices illustrate how developmental biologists manipulate these configurations.

7.3.1 Mixed Cell Aggregates

A classic manipulative approach of experimental embryology involved artificial systems called "mixed cell aggregates." As the name suggests, these are conglomerates of heterogeneous cells isolated from different times or locations in an embryo. The basic procedure was to mix dissociated cells of one histological type with dissociated cells of another type into a "heterotypic" cell aggregate and observe what happens. Variations on this procedure included mixing dissociated cells from the same histological type ("homotypic") or using dissociated cells from other species in either homotypic or heterotypic combinations. These manipulation procedures were aimed at specific research questions related to the tissue-configuration state: "One of the most striking and fascinating problems in biology is the emergence of characteristic organization in embryonic tissues, the orderly assemblage and arraying of cell groups, and intercellular constituents into a diversity of structural and functional entities" (Moscona, 1960, p. 45). The method of manipulation was tailored to addressing this problem: "to approach tissue formation as a problem in experimental synthesis; that is, to study, by compounding discrete cells into collective complexes, the manner of bonding and ordering of cells into histogenetic systems" (p. 46). Characterization of the phenomena was an explicit aim: "information might be gained on distinctive grouping tendencies and patterns of behavior in cells" (p. 46).

Understanding aspects of the characteristic organization of tissues that originate during embryogenesis is complicated because it is difficult to track individual cells as they differentiate, migrate, and assort. Additionally, ascertaining cellular potential requires exploring artificial conditions where the range of potency for cells can be expressed even if it is not manifested during normal ontogeny. Developmental biologists have long explanted groups of cells from their original context, placing them in conditions where the number of extraneous variables is reduced and observational discrimination is increased. To create mixed cell aggregates requires at least three steps. First, tissue-level organization must be disrupted so that there are individual cells of a type to combine. Second, these cells must be maintained stably in this configuration prior to combination with other cell types (i.e., they cannot die). Third, there must be a way to follow (track) different cells after aggregation in order to determine what cells have which potency and the conditions of their manifestation (in both time and space).

For example, researchers disaggregated the chick wing bud and mesonephros (a temporary excretory organ in birds) using the chemical trypsin to digest away cellular interconnections (Trinkaus & Groves, 1955). Previous studies in chick had found that small explants failed to differentiate in culture, whereas larger explants of the same cells always formed brain tissue. "It is evident from these observations that a description of the potencies of a given tissue mass composed of many cells does not necessarily reveal the capacities of the individual constituent cells" (p. 787). Three different types of aggregates were created: two homotypic explants (wing bud cells only; mesonephroi cells only) and one heterotypic explant (wing bud cells plus mesonephroi cells). After the explants were cultured in vitro for a number of days using different amounts and stages of disaggregated cells, characteristic features of tissue-level organization emerged in the explant: procartilage in wing bud explants, mesonephroi tubules in the mesonephros explants, and both of these features with regionalized spatial differences in the heterotypic explant.

Mixed cell aggregates are a disruptive type of manipulation (i.e., without a particular expected outcome) intended to characterize cell behavior that helps to answer questions about the "emergence of characteristic organization in embryonic tissues" (Moscona, 1960, p. 45). They involve a combination of augmentation and diminution strategies, including juxtaposing cell populations in a novel fashion (e.g., heterotypic aggregates) and removing the existing intercellular connections to see if they would re-form (e.g., homotypic aggregates). There also were elements of substitution in selecting disaggregated cell populations of different ages to include in the explants (e.g., 3.5-day versus 5-day wing bud cells). The experimental means were primarily chemical, both in terms of tissue disaggregation via trypsin and in vitro culture conditions for reaggregation. Researchers learned a number of things about the propensities of populations of cells to adopt relatively stable configuration states with characteristic tissue organization. Others have continued to do so (e.g., Aarum et al., 2003), typically with more sophisticated markers for tracking cells, as well as characterize phenomena related to different developmental questions, including cell differentiation (differential potency), morphogenesis (cell migration), and cellular adhesion in tissues (Steinberg, 2003).

7.3.2 Tissue Engineering

A descendant of mixed cell aggregates that is common in contemporary developmental biological research is tissue engineering (Bryant & Mostov, 2008). Instead of only disaggregating existing tissue architecture, undifferentiated cells can be grown under a variety of culture conditions to elicit distinct fates, patterns of movement, and tissue-level organization. Thus, instead of disrupting tissue-level organization, tissue engineering can begin with cellular components alone (typically stem cells) and manipulate their conditions of proliferation, such as by making them grow in a monolayer versus a three-dimensional scaffold (Ovsianikov et al., 2018). This approach reveals distinct patterns of differentiation and arrangement of various cell types into organized tissues. The variability of conditions facilitates the identification of potentiality and organization in a way inaccessible when using mixed cell aggregates, in part because other processes, such as morphogenesis and adhesion, can be controlled for (e.g., through preventing aggregation). A central motivation for tissue engineering is regenerative medicine, and a number of advances have been made in specific areas, such as for skin grafts (Berthiaume et al., 2011; Loebel & Burdick, 2018; MacNeil, 2007; Paschos et al., 2015).

The aim of regenerative medicine in tissue engineering marks a key difference in the manipulation types when compared with mixed cell aggregates. Instead of disruptive types where there is not an explicit expectation about the outcome of the manipulation, a constructive type is required because researchers need to bring about a change that yields an anticipated result. Tissue engineering is less about characterizing the relatively stable configuration states of cells and tissues and more about establishing relationships of invariance under intervention between causal variables (e.g., the type of cells to start with or conditions of culturing) and effect variables (e.g., the kind of tissue organization relevant for treating a severe injury). This is facilitated by sophisticated tools for tracking different cell types and a better understanding of the molecular signatures of these within specific tissues (Lamouille et al., 2014; Liu et al., 2008).

All strategies of manipulation are operative in tissue engineering. Increases or decreases in the density of cell populations have an effect on differentiation and tissue organization. Substitution with artificial materials is commonly used to mimic biological features, such as scaffolding (Ovsianikov et al., 2018), which facilitates types of proliferation or differentiation and, subsequently, tissue organization. Compared with mixed cell aggregates, there is a dramatic increase in the genetic means to undertake manipulations, including altering culture conditions through changed gene expression (e.g., increasing the rate of growth and proliferation of constituent cells) and generating increased efficiency in the process of producing cells for aggregation (Ikuno et al., 2017). Multicellular states of organization can be induced even with synthetic genetic circuits (Toda et al., 2018). However, chemical and mechanical means are also prominent and used more precisely than was possible previously (Fahy et al., 2018; Gliga et al., 2017). Manipulative capacities for types, strategies, and means have increased both quantitatively and qualitatively. Quantitative increases pertain to precision in using different means to introduce changes that exemplify all three kinds of strategies. Qualitative changes pertain to types of manipulation; it is now possible to induce experimental alterations constructively and not just disruptively (i.e., to achieve predicted outcomes and control their manifestation). These manipulative capacities have been adopted for different research questions (e.g., cell differentiation or regeneration) and yielded epistemic advances in how tissue configurations are formed out of cellular states and how transitions between these states of organizations occur in normal ontogeny and pathology (Bryant & Mostov, 2008).

7.4   Interpretation: From Biological Practice to Scientific Metaphysics

Developmental biologists routinely manipulate two stable configuration states: individual cells and organized aggregates of cells (i.e., tissues). These manipulations are successful with respect to criteria derived from different strategies, types, and means in combination with investigative aims and research questions about ontogeny. This success suggests there is something about the world that licenses the reliable response of these states to experiments. How might this success speak to the existence and nature of "levels" of organization?[2]

The central inferential maneuver in the context of questions about scientific realism is the move from the success of science to metaphysical conclusions about what the world is ultimately like. Although scientific success is a good place to begin when identifying structures of reality (i.e., metaphysics), a persistent concern is that we know from history that scientists can be wrong—the successes of today can be the failures of tomorrow. This is the kernel of what is now referred to as the "pessimistic meta-induction" objection (Chakravartty, 2017, §3.3). Empirically successful theories of the past have subsequently been falsified and rejected. Why think anything different will obtain for our current theories? An assumption pervading most of these analyses is that the relevant success is located in the performance of theories or their component features (i.e., representations), such as whether a theory has made novel predictions.[3] However, as was argued long ago by Ian Hacking, "one can believe in some entities without believing in any particular theory in which they are embedded" (Hacking, 1983, p. 29). Hacking's memorable slogan for this orientation was illustrated from manipulative practices in electron microscopy: "if you can spray them then they are real" (p. 23). An additional motivation for shifting from theories to (especially manipulative) practices is that many have been stable across changes in scientific theories. Genetic approaches have been in use for more than 100 years despite major changes in our understanding of the nature of heredity and the concept of the gene (Waters, 2017).

This argument to structures of reality from the success of practices can be enhanced by drawing attention to the fact that scientific inquiry is not always organized around explanatory reasoning or a central theory (Waters, 2004, 2017). Waters highlights "pools of practical, descriptive, and evaluative knowledge" (Waters, 2004, §6) that correspond to experimental methodologies and procedures for the generation and maintenance of material entities, descriptive knowledge of recurring causal patterns, and evaluative knowledge that measures and tracks entities and causal patterns for their usefulness in subsequent research endeavors. Although infused with theory in different ways, this practice-based knowledge doesn't require an organizing theoretical framework and is a different locus for scrutinizing the success of science. And it is this practical, descriptive, and evaluative knowledge that has been achieved by developmental biologists for cell and tissue configuration states despite the absence of a unifying theory of development (Love, 2014).

Two other premises play a role in interpreting these successful practices as an argument for cellular and tissue configuration states as levels per se. The first is the ever-increasing ability to manipulate states of cell and tissue organization. There has been progress over time in the precision and accuracy of manipulation that can be achieved, independently of the theoretical framing of these organizational features, especially constructive types of manipulation in tissue engineering. Thus, this progress is not tied to a theoretical account of these configuration states. The second premise concerns the diverse ways in which these manipulations can be accomplished. This diversity is in view for discussions of realism based on robustness (Wimsatt, 1981, 2007). Assuming these techniques are sufficiently independent, then it is less likely that every one of them is unexpectedly mistaken or in error. Robustness arguments are inherently complex, and discussing them in detail is beyond the scope of the present chapter (Kuorikoski & Marchionni, 2016; Stegenga, 2009; Wimsatt 1981, 2007). Here I rely on and extend robustness arguments that have been offered in the context of biological practices (Eronen, 2019) by requiring that robustness involve the actual manipulation of phenomena rather than only having multiple ways of measuring or detecting a phenomenon.

On Eronen's (2019) account (see also Eronen, 2015b), robustness provides epistemic justification for taking some entity as a real feature of the world. This justification comes in degrees, and the amount of justification is understood in terms of robust evidence for the entity in question. I narrow "robust evidence" to "ability to manipulate," which limits the reach of the justification because it is a more stringent condition. This ability to manipulate is relative to a community of scientific investigation at a historical time. As a consequence, the ability to manipulate can increase or decrease over time. The independence of the manipulation methods is explicit in that the different means of manipulation (chemical, genetic, and physical) are independent by virtue of involving different kinds of causal mechanisms and implicit because there is often independence in terms of relying on different causal mechanisms even within the same means. This is observable when researchers use distinct approaches within the same means to prosecute the same strategy, such as multiple genetic means of accomplishing diminution (Housden et al., 2017). The entities in question here (cell and tissue levels of organization) and their properties (features of their distinctive configuration states) are responsive to the manipulation practices that experimentally modify them and exhibit a stability across theoretical change, which has been the primary concern of the pessimistic meta-induction. However, it should be emphasized that the ontological implications drawn here are not intended to support a broader entity realism across the life sciences (cf. Eronen, 2019).

A few final elements need to be emphasized. As noted, developmental biologists experimentally manipulate transitions between stable cellular and tissue configuration states. These are preferred states of organization—not just experimental manipulations at a scale—that generalize (see below) and (largely) exhibit part–whole compositional relations indicative of nested levels. Hence, the successful manipulations do not merely concern two entity types, cell and tissue, but point to preferred configuration states that exhibit generalized part–whole, compositional relationships. The appellation of "levels" is a codification of the general, stable, nested-hierarchical, structural organizational states exhibited in cells and tissues that are privileged by virtue of successful experimental manipulations of transitions between them.

To reiterate, the argument from successful biological practice to the reality of cellular and tissue levels of organization is based on the (1) existence of successful manipulation practices, (2) progressive increase of success in these manipulation practices over time, and (3) diversity of successful manipulation practices operative in the experiments of developmental biologists. This argument is not impervious to objections that developmental biologists are potentially mistaken in thinking they have successful manipulation practices for cellular and tissue organization, or that these have increased over time, or that different manipulation practices pertain to the same configuration states. However, the burden of proof rests with those advancing one or more of these objections; plausible reasons in their favor must be provided.

7.4.1 Generalization

There remains an outstanding question about the degree to which this argument from successful manipulation practices in developmental biology to cellular and tissue levels of organization generalizes. In what ways can we expect these results to apply in different contexts, especially where levels are routinely invoked in biological reasoning? At least three different dimensions of generalization can be distinguished: (a) from successful manipulation practices to things not manipulated, (b) from successful manipulation practices in developmental biology to similar phenomena conceptualized and investigated in other biological disciplines, and (c) from successful manipulation practices at the cellular and tissue levels to different types of levels, either "above" (e.g., organisms or populations) or "below" (e.g., molecular).

The most plausible dimension of generalization pertains to extending the same categories for levels of organization (i.e., cellular and tissue) to similarly arrayed phenomena (i.e., "cells" and "tissues") that have not been manipulated. It would be overly restrictive to limit the interpretation of cell and tissue configuration states as levels of organization only to those entities that have been actually manipulated. Given that the same relatively stable configuration states obtain even in those cases where researchers do not intervene, there is no reason to withhold the category of levels until actual manipulation of both cellular and tissue organization has occurred. Within developmental biology, this form of generalization underwrites the possibility of regenerative medicine. In most cases, we have not manipulated the cellular and tissue organization of particular individuals but presume that what has been learned about transitions between migratory mesenchyme and aggregated epithelia can be marshaled for treatments of individuals exhibiting metastatic cancers.

A similar line of reasoning applies to a generalization of cell and tissue levels to other taxa from the successful manipulation of experimental organisms. The export of the conclusion is based on comparative embryological, botanical, and zoological observation. These investigations have demonstrated repeatedly that stable configuration states of cells and tissues, corresponding to what is manipulated in mixed cell aggregates and tissue engineering, are present across the tree of life. However, this generalization is not universal; there are taxa that only exhibit cellular aggregation temporarily, such as during a life history stage (e.g., slime molds) and without the same level of stability or organizational architecture (e.g., biofilms). Additionally, there are cases where stable tissue configuration comes at the expense of cellular organization, such as in the permanent tissues of plants (e.g., wood) or in mineralized tissues in animals (e.g., bone or teeth), where cells are no longer alive and thus a cellular level of organization no longer exists (even if there was one at an earlier time). Thus, these generalizations require calibration to recognize whether cellular and tissue levels obtain.

Our second dimension of generalization is whether successful manipulation practices in developmental biology export to similar phenomena conceptualized and investigated in other biological disciplines.[4] Here the answer is less clear. There can be key differences in what features of organization are tracked in different disciplines. For example, in physiological studies, specific functional properties such as hardness or stiffness for structural support do not require a structural cellular configuration state, as in cases of wood or bone. And what counts as a tissue configuration state can differ depending on the process in view, such as electrical conductance in nerves or chemical filtering in the parenchyma of kidneys. If we turn to evolutionary studies, the cellular "level" can be treated as an autonomous individual (i.e., an organism) that exhibits behavioral propensities not uniquely associated with all cells. Additionally, these behavioral propensities may discriminate between kinds of tissue-level organization, only some of which are deemed sufficiently autonomous, such as a snowflake cluster versus hyphae in yeast (Ratcliff et al., 2015; Wloch-Salamon et al. 2017). However, despite these divergences in the conceptualization of cellular and tissue organization, there is overlap. Even if physiological properties no longer have an underlying cellular level of organization, one often was required developmentally. Overall, generalizations across disciplinary domains require special scrutiny to ascertain when and whether stable cellular and tissue configuration states are actually "levels" across different disciplinary approaches.

The final dimension of generalization concerns whether the successful manipulation of cellular and tissue levels permits claims about other types of levels, either "above" (e.g., organisms) or "below" (e.g., molecular). The three-pronged basis of the argument in favor of cellular and tissue levels of organization—existence, progressive increase, and diversity of successful manipulation methods for these configuration states—suggests that these premises might not be fulfilled for different configuration states that are putatively levels, although organoid models are an example of increasingly successful manipulations "above" tissues that begin with stem cell organization (Sato & Clevers, 2013). Independent arguments for the existence, progressive increase, and diversity of successful manipulation methods for different configuration states and transitions between them could yield a similar conclusion. However, this alone would not yield a layering of levels unless there are successful methods for manipulating transitions between these independently established configuration states and those established through developmental biological manipulations (i.e., cellular and tissue levels). Levels could be combined into a single hierarchy as long as transitions between configuration states were demonstrable experimentally. One also might question the existence of a molecular "level" since there is not (to date) successful practices that facilitate a constructive transition to a cellular level of organization. Furthermore, the amount of manipulative control available for "higher" configuration states and transitions between them, especially constructive experimental strategies, is less abundant and not commensurate in precision with what is available for cellular and tissue configuration states. Therefore, a generalization of the argument for cellular and tissue levels to other kinds of levels based on successful manipulation methods is unclear because (1) the fulfillment of the premises might be different enough to prevent a layering of levels (i.e., no demonstrated transitions), and (2) the number and precision of manipulation practices available may be substantially less for other configuration states (e.g., putative "higher" levels).

7.5   Comparison: Situating Manipulation-Based Levels among Alternatives

We are now in position to compare the manipulation-based account of levels of organization for cells and tissues with other existing accounts. Recall that the account emerged from a criterion of adequacy—correspondence to scientific practice—that encouraged concentrating on the manipulation of stable configuration states in addition to hierarchical representation. I analyzed two examples where developmental biologists have successfully manipulated configuration states related to cells and tissues, and argued in terms of robustness that these successful practices point toward a scientific metaphysics of cell and tissue levels based on the existence, progressive advance over time, and diversity of successful manipulation practices related to general, stable, nested-hierarchical, structural states of cellular and tissue organization. Different generalizations of this argument, especially to other putative levels, are more or less plausible based on the availability and heterogeneity of manipulation practices for the relevant configuration states, including demonstrating transitions between them.

This modestly realist account of levels of organization is novel in two ways. It argues directly for only two levels (cell and tissue), rather than a multilayer stratification, and primarily in the context of developmental biology, rather than across all of biology. Yet this account is also general in that it finds application across taxa that exhibit cellular and tissue organization. These generalized levels are based in structure (material composition, arrangement, and shape) rather than function (activities performed or displayed) and defined mostly by part–whole relationships. However, this is not strict nesting because a critical element of tissue organization is extracellular matrix (ECM) that lies outside of cells (Mouw et al., 2014; Sheng et al., 2017). ECM "parts" of a tissue have configuration states but do not comprise a level from which one can transition to or from cellular or tissue organization. Additionally, there is no assumption of transitivity in compositional organization (e.g., that cellular components can be further decomposed into parts at another level), although the relevant constructive experimental strategies make feasible combining cell and tissue levels with others (e.g., organ). Violations of strict nesting and transitivity arise from the concentration on *transitions between* privileged, stable states of organization and the experimental establishment of transitions between cell and tissue configuration states (Baum et al., 2008). Both are crucial for an inference from successful biological practice to scientific metaphysics—genuine levels of organization.

The first family of approaches to compare involves skepticism about levels. If we only appeal to a principle of scale that is entirely continuous, then there are no resources for placing entities at distinct and discrete positions in a hierarchy of configuration states (Eronen, 2015a; Potochnik & McGill, 2012). This is what was demonstrated in the case of manipulation-based cell and tissue levels of organization. Arguably, the notion of scale avoids some problems associated with more global accounts of levels; arranging items on a scale only requires measuring some quantitative property, such as size. If we relax complete skepticism and acknowledge advantages that derive from using scales of different kinds (including time: DiFrisco, 2017), then a manipulation-based account of cellular and tissue organization is potentially compatible because it does not demand a global nested hierarchy and recognizes degrees of plausibility in how this account generalizes. For example, one might achieve a complementary notion of "quasi-discreteness" with scalar clustering.[5] A collection of entities could be sorted into types, and upon measurement of their size, clusters of distinctive values might be detectable for types relative to a threshold of significant separation.

At the other end of the spectrum, logical empiricist accounts of levels (the traditional "layer-cake" perspective) are in conflict with the manipulation-based account derived from developmental biological practices. The resulting cellular and tissue levels of organization are not comprehensive in character or global in scope. In fact, their generalization to other types of putative levels is one of the dimensions where we have reason to be concerned; the relevant existence, progress, and diversity of manipulation methods are not universally available. There is a further conflict because the relationship between cell and tissue configuration states is not strictly nested or "type" compositional (see DiFrisco, this volume): tissue-level organization exhibits structure that arises in part from ECM (Mouw et al., 2014; Sheng et al., 2017). An additional problem is that developmental biology, as a scientific discipline, simultaneously studies (at least) two levels of organization and therefore the disciplinary structures of science do not correlate in any straightforward fashion with compositional levels (see discussion in Potochnik, 2017). There is little that seems recoverable from this perspective since its theoretical commitments cannot be relaxed without a severe loss of its core claims and motivation (e.g., systematicity).

What about a levels-within-mechanisms perspective? If read strictly, then there is tension with the manipulation-based account because the experimental means pertain to different kinds of causal mechanisms and the questions addressed by manipulations reflect different processes composed of distinct behaviors, such as differentiation, adhesion, or migration. Working parts at the cellular level can be organized and have their operations coordinated to realize different phenomena of interest (i.e., as working parts of different mechanism types). It also falsifies the claim that a levels-within-mechanism perspective "accurately describes the multilevel explanatory structures one finds in biology and other special sciences … [and] captures the common idea of 'levels of organization'" (Povich & Craver, 2018, p. 188). However, if we relax the perspective so that levels can but need not have their meaning within the context of mechanisms (i.e., accurately capturing *some* multilevel explanatory structures in biology), then there are possibilities for affirming both levels-within-mechanisms and manipulation-based cell and tissue levels of organization. This might fit within a patchwork orientation that draws on different types of scale, mechanisms, or manipulation to yield a pluralist picture of levels of organization akin to what

DiFrisco suggests: "Rather than a universal and discrete division of causal interactions or types, to the contrary, it is both more plausible and closer to the scientific use of the concept to think that a hierarchy offers reliable generalizations about the structure of causal interactions in complex systems" (DiFrisco, 2017, p. 801; see also Gillett, this volume).

Further comparisons can be made with other perspectives on levels. The manipulation-based account of cell and tissue levels fits Wimsatt's insight that "levels of organization can be thought of as local maxima of regularity and predictability in the phase space of alternative modes of organization of matter" (Wimsatt, 2007, p. 209), although it puts an emphasis on how those alternative modes of organization are defined in terms of reliable and repeatable manipulations of transitions between these maxima. Additionally, manipulation-based levels are "constituted by families of entities usually of comparable size and dynamical properties, which characteristically interact primarily with one another" (Wimsatt, 2007, p. 204), but there is more caution in how general this viewpoint is given constraints on the ability to manipulate these families of entities and their dynamical properties similarly (if at all), especially transitions between them. Kaiser's (2015) view of biological parthood could mesh with manipulation-based levels since cells are on a lower level than tissues because cells are biological parts of tissues. However, features of tissue-level organization, such as ECM, are more difficult to situate in this perspective (e.g., should ECM be thought of as a "biological part" with internal organization of its own?). Since Kaiser's account of biological parthood is not anchored in practices of manipulation, discrepancies might trace back to this methodological divergence.

Finally, a manipulation-based account may or may not be in conflict with domain-specific accounts of levels (Love, 2012). This depends on whether the hierarchical representations in a domain map onto the stable configuration states that can be manipulated effectively. This also relates to which successful manipulation practices in developmental biology can be generalized to similar phenomena (i.e., "cells" and "tissues") conceptualized and investigated in other biological disciplines or different types of levels, either "above" (e.g., organisms) or "below" (e.g., molecular). Both situations required caution, and therefore further work is needed to ascertain any coordination or conflict between these types of accounts. That further work is needed constitutes a primary methodological lesson of our comparative analysis: once global hierarchies, complete skepticism, and monist accounts of levels (e.g., only within mechanisms) are jettisoned or relaxed, the remaining space of possibilities, whether in terms of representation or manipulation, is large, heterogeneous, and complex.

7.6 Concluding Remarks

Some will find the conclusion of this analysis unsatisfying. It does not yield a snappy slogan similar to what Hacking (1983) offered when he first advanced entity realism. It does not indicate which view among many is correct or incorrect, apart from rejecting systematic global levels, universal skepticism, and univalent local viewpoints. However, an advantage of the view of cell and tissue levels of organization argued for in terms of successful manipulation practices from developmental biology is that it neither overgeneralizes nor undergeneralizes. Both have been recurring mistakes in philosophical discussion. Overgeneralization can be seen in both the layer-cake view of reality derived from a global compositional hierarchy and a complete skepticism that nature exhibits levels of any kind. Undergeneralization can be seen in the movement toward more localized accounts, such as levels-within-mechanisms, which sacrifice significant elements of successful manipulation and representation practices to secure a unitary understanding of what it means to talk about levels of organization. Navigating a middle way between too much and too little generalization is the standard modus operandi of the sciences, especially in biology. It is what we should expect if the success of scientific practice (i.e., epistemic work) operates in an opportunistic fashion, representing and manipulating in whatever way is accurate and fruitful, thereby achieving piecewise approximations to reality, including levels of organization.

Acknowledgments

I am grateful to the organizers and participants of the 36th Altenberg Workshop in Theoretical Biology, "Hierarchy and Levels of Organization in the Biological Sciences," where the ideas in this chapter were first presented. Their feedback and input helped shape the argument and structure of the final paper. Thanks also to Dan Brooks, James DiFrisco, Sara Green, and Bill Wimsatt for providing substantial comments on an earlier version of the manuscript.

Notes

1. "References to levels of organization and *related hierarchical depictions* of nature are prominent in the life sciences and their philosophical study, and appear not only in introductory textbooks and lectures, but also in cutting-edge research articles and reviews" (Eronen & Brooks, 2018, emphasis added).

2. Importantly, levels of organization as a concept can be related to productive epistemic work in biology independent of the ontological status of levels. My argument concerns the relationship between epistemic work and ontological status, but it does not exhaust the possibilities for epistemic work per se. Also, manipulating one "level" to access processes at another "level" can be seen as a deliberate exploitation of "levels leakage," a methodological implication I ignore here (see Wimsatt, 2007, chap. 10).

3. "Scientific realism is a positive epistemic attitude toward the content of our best theories and models … a realism about whatever is described by our best scientific theories" (Chakravartty, 2017).

4. One might consider this a question about whether these stable configuration states should be promoted from "perspectives" to "levels" since it can be argued that a level of organization should have its constituent features and processes exhaustively characterized across theoretical or disciplinary orientations (see Wimsatt, 2007, chap. 10).

5. Thanks to James DiFrisco for this specific suggestion.

References

Aarum, J., Sandberg, K., Haeberlein, S. L. B., & Persson, M. A. A. (2003). Migration and differentiation of neural precursor cells can be directed by microglia. *Proceedings of the National Academy of Sciences USA, 100,* 15983–15988.

Alvarado-Mallart, R.-M. (2005). The chick/quail transplantation model: Discovery of the isthmic organizer center. *Brain Research Reviews, 49,* 109–113.

Baum, B., Settleman, J., & Quinlan, M. P. (2008). Transitions between epithelial and mesenchymal states in development and disease. *Seminars in Cell & Developmental Biology, 19,* 294–308.

Bechtel, W. (2008). *Mental mechanisms: Philosophical perspectives on cognitive neuroscience.* London, UK: Routledge.

Bergqvist, A., Jeppsson, S., Kullander, S., & Ljungberg, O. (1985). Human uterine endometrium and endometriotic tissue transplanted into nude mice. Morphologic effects of various steroid hormones. *The American Journal of Pathology, 121,* 337–341.

Berthiaume, F., Maguire, T. J., & Yarmush, M. L. (2011). Tissue engineering and regenerative medicine: history, progress, and challenges. *Annual Review of Chemical and Biomolecular Engineering, 2,* 403–430.

Brooks, D. S. (2017). In defense of levels: Layer cakes and guilt by association. *Biological Theory, 12,* 142–156.

Bryant, D. M., & Mostov, K. E. (2008). From cells to organs: Building polarized tissue. *Nature Reviews Molecular Cell Biology, 9,* 887–901.

Butcher, R. A. (2017). Small-molecule pheromones and hormones controlling nematode development. *Nature Chemical Biology, 13,* 577–586.

Campbell, N. A., & Reece, J. B. (2002). *Biology* (6th ed.). San Francisco, CA: Benjamin Cummings.

Chakravartty, A. (2017). Scientific realism. In E. N. Zalta (Ed.), *The Stanford encyclopedia of philosophy.* Retrieved from https://plato.stanford.edu/archives/sum2017/entries/scientific-realism/

Craver, C. F. (2007). *Explaining the brain: Mechanisms and the mosaic unity of neuroscience.* Oxford, UK: Oxford University Press.

Craver, C. F., & Bechtel, W. (2007). Top-down causation without top-down causes. *Biology and Philosophy, 22,* 547–563.

DiFrisco, J. (2017). Time scales and levels of organization. *Erkenntnis, 82,* 795–818.

Eronen, M. I. (2013). No levels, no problems: Downward causation in neuroscience. *Philosophy of Science, 80,* 1042–1052.

Eronen, M. I. (2015a). Levels of organization: A deflationary account. *Biology & Philosophy, 30,* 39–58.

Eronen, M. I. (2015b). Robustness and reality. *Synthese, 192,* 3961–3977.

Eronen, M. I. (2019). Robust realism for the life sciences. *Synthese, 196,* 2341–2354.

Eronen, M. I., & Brooks, D. S. (2018). Levels of organization in biology. In E. N. Zalta (Ed.), *The Stanford encyclopedia of philosophy.* Retrieved from https://plato.stanford.edu/archives/spr2018/entries/levels-org-biology/

Fahy, N., Alini, M., & Stoddart, M. J. (2018). Mechanical stimulation of mesenchymal stem cells: Implications for cartilage tissue engineering. *Journal of Orthopaedic Research, 36,* 52–63.

Gliga, A. R., Edoff, K., Caputo, F., Källman, T., Blom, H., Karlsson, H. L., Ghibelli, L., Traversa, E., Ceccatelli, S., & Fadeel, B. (2017). Cerium oxide nanoparticles inhibit differentiation of neural stem cells. *Scientific Reports, 7,* 9284.

Hacking, I. (1983). *Representing and intervening: Introductory topics in the philosophy of natural science.* Cambridge, UK: Cambridge University Press.

Hanahan, D., & Weinberg, R. A. (2011). Hallmarks of cancer: The next generation. *Cell, 144,* 646–674.

He, S., & Masland, R. H. (1997). Retinal direction selectivity after targeted laser ablation of starburst amacrine cells. *Nature, 389,* 378–382.

Heerboth, S., Housman, G., Leary, M., Longacre, M., Byler, S., Lapinska, K., Willbanks, A., & Sarkar, S. (2015). EMT and tumor metastasis. *Clinical and Translational Medicine, 4,* 6.

Housden, B. E., Muhar, M., Gemberling, M., Gersbach, C. A., Stainier, D. Y. R., Seydoux, G., Mohr, S. E., Zuber, J., & Perrimon, N. (2017). Loss-of-function genetic tools for animal models: Cross-species and cross-platform differences. *Nature Reviews Genetics, 18,* 24–40.

Ikuno, T., Masumoto, H., Yamamizu, K., Yoshioka, M., Minakata, K., Ikeda, T., Sakata, R., & Yamashita, J. K. (2017). Efficient and robust differentiation of endothelial cells from human induced pluripotent stem cells via lineage control with VEGF and cyclic AMP. *PLoS ONE, 12,* e0173271.

Kaiser, M. I. (2015). *Reductive explanation in the biological sciences.* Cham, Switzerland: Springer.

Kiecker, C., Bates, T., & Bell, E. (2016). Molecular specification of germ layers in vertebrate embryos. *Cellular and Molecular Life Sciences, 73,* 923–947.

Kim, H. Y., Jackson, T. R., & Davidson, L. A. (2017). On the role of mechanics in driving mesenchymal-to-epithelial transitions. *Seminars in Cell & Developmental Biology, 67,* 113–122.

Kuorikoski, J., & Marchionni, C. (2016). Evidential diversity and the triangulation of phenomena. *Philosophy of Science, 83,* 227–247.

Ladyman, J., & Ross, D. (2007). *Every thing must go: Metaphysics naturalized.* Oxford, UK: Oxford University Press.

Lamouille, S., Xu, J., & Derynck, R. (2014). Molecular mechanisms of epithelial–mesenchymal transition. *Nature Reviews Molecular Cell Biology, 15,* 178–196.

Le Douarin, N. M., & Kalcheim, C. (1999). *The neural crest* (2nd ed.). Cambridge, UK: Cambridge University Press.

Liu, X., Yu, X., Zack, D. J., Zhu, H., & Qian, J. (2008). TiGER: A database for tissue-specific gene expression and regulation. *BMC Bioinformatics, 9,* 271.

Loebel, C., & Burdick, J. A. (2018). Engineering stem and stromal cell therapies for musculoskeletal tissue repair. *Cell Stem Cell, 22,* 325–339.

Love, A. C. (2012). Hierarchy, causation and explanation: Ubiquity, locality and pluralism. *Interface Focus, 2*, 115–125.

Love, A. C. (2014). The erotetic organization of developmental biology. In A. Minelli & T. Pradeu (Eds.), *Towards a theory of development* (pp. 33–55). Oxford, UK: Oxford University Press.

MacNeil, S. (2007). Progress and opportunities for tissue-engineered skin. *Nature, 445*, 874–880.

Moscona, A. A. (1960). Patterns and mechanisms of tissue reconstruction from dissociated cells. In D. Rudnick (Ed.), *Developing cell systems and their control* (pp. 45–70). New York, NY: The Ronald Press Company.

Mouw, J. K., Ou, G., & Weaver, V. M. (2014). Extracellular matrix assembly: A multiscale deconstruction. *Nature Reviews Molecular Cell Biology, 15*, 771–785.

Onuma, Y., Takahashi, S., Asashima, M., Kurata, S., & Gehring, W. J. (2002). Conservation of *Pax 6* function and upstream activation by *Notch* signaling in eye development of frogs and flies. *Proceedings of the National Academy of Sciences USA, 99*, 2020–2025.

Oppenheim, P., & Putnam, H. (1958). The unity of science as a working hypothesis. In H. Feigl, M. Scriven, and G. Maxwell (Eds.), *Concepts, theories, and the mind-body problem* (pp. 3–36). Minneapolis: University of Minnesota Press.

Ovsianikov, A., Khademhosseini, A., & Mironov, V. (2018). The synergy of scaffold-based and scaffold-free tissue engineering strategies. *Trends in Biotechnology, 36*, 348–357.

Paschos, N. K., Brown, W. E., Eswaramoorthy, R., Hu, J. C., & Athanasiou, K. A. (2015). Advances in tissue engineering through stem cell-based co-culture. *Journal of Tissue Engineering and Regenerative Medicine, 9*, 488–503.

Prelich, G. (2012). Gene overexpression: Uses, mechanisms, and interpretation. *Genetics, 190*, 841–854.

Potochnik, A. (2017). *Idealization and the aims of science*. Chicago, IL: University of Chicago Press.

Potochnik, A., & McGill, B. (2012). The limitations of hierarchical organization. *Philosophy of Science, 79*, 120–140.

Povich, M., & Craver, C. F. (2018). Mechanistic levels, reduction, and emergence. In S. Glennan & P. Illari (Eds.), *The Routledge handbook of mechanisms and mechanical philosophy* (pp. 185–197). Abingdon, UK: Routledge.

Ratcliff, W. C., Fankhauser, J. D., Rogers, D. W., Greig, D., & Travisano, M. (2015). Origins of multicellular evolvability in snowflake yeast. *Nature Communications, 6*, 6102.

Sato, T., & Clevers, H. (2013). Growing self-organizing mini-guts from a single intestinal stem cell: mechanism and applications. *Science, 340*, 1190–1194.

Sheng, Y., Fei, D., Leiiei, G., & Xiaosong, G. (2017). Extracellular matrix scaffolds for tissue engineering and regenerative medicine. *Current Stem Cell Research & Therapy, 12*, 233–246.

Stegenga, J. (2009). Robustness, discordance, and relevance. *Philosophy of Science, 76*, 650–661.

Steinberg, M. (2003). Cell adhesive interactions and tissue self-organization. In G. B. Müller & S. A. Newman (Eds.), *Origination of organismal form: Beyond the gene in developmental and evolutionary biology* (pp. 137–163). Cambridge, MA: MIT Press.

Thalos, M. (2013). *Without hierarchy: The scale freedom of the universe*. Oxford, UK: Oxford University Press.

Toda, S., Blauch, L. R., Tang, S. K. Y., Morsut, L., & Lim, W. A. (2018). Programming self-organizing multicellular structures with synthetic cell-cell signaling. *Science, 361*, 156–162.

Trinkaus, J. P., & Groves, P. W. (1955). Differentiation in culture of mixed aggregates of dissociated tissue cells. *Proceedings of the National Academy of Sciences USA, 41*, 787–795.

van Fraassen, B. C. (2008). *Scientific representation*. New York, NY: Oxford University Press.

Waters, C. K. (2004). What was classical genetics? *Studies in the History and Philosophy of Science, 35*, 783–809.

Waters, C. K. (2017). No general structure. In M. H. Slater & Z. Yudell (Eds.), *Metaphysics of science: New essays* (pp. 81–107). New York, NY: Oxford University Press.

Wloch-Salamon, D. M., Fisher, R. M., & Regenberg, B. (2017). Division of labour in the yeast: *Saccharomyces cerevisiae*. *Yeast, 34*, 399–406.

Wimsatt, W. C. (1981). Robustness, reliability, and overdetermination. In M. B. Collins (Ed.), *Scientific inquiry and the social sciences* (pp. 124–163). San Francisco, CA: Jossey-Bass.

Wimsatt, W. C. (2007). *Re-engineering philosophy for limited beings: Piecewise approximations to reality*. Cambridge, MA: Harvard University Press.

Woodward, J. (2003). *Making things happen: A theory of causal explanation*. New York, NY: Oxford University Press.

8 The Origin of New Levels of Organization

Jan Baedke

Overview

Hierarchies including levels of genes, cells, tissues, organs, organisms, and ecosystems are ubiquitous in the biosciences. Describing such hierarchies gives order to the complexity of nature. Recently, some philosophers of science involved in the "new mechanism philosophy" have argued for conceptualizing biological hierarchies strictly in terms of composition and constitution. This approach is the latest prominent version of theories of hierarchies that share the assumption that one does not need to address how the levels of organization described in hierarchical models originally came into existence. In fact, it is usually presupposed that levels of organization already exist and do not change in kind. This perspective does not consider that levels in multicellular organisms are built up over developmental time and have emerged as qualitative novelties in the evolutionary past. Against this background, I first discuss the shortcomings and biases of this prominent view of biological hierarchies and levels. Second, by drawing on both historical debates in early twentieth-century organicism and contemporary cases from developmental biology and evolutionary developmental biology (evo-devo), I illustrate the different ways how changes in the compositional and temporal organization of levels can occur. This broader perspective on changes in organization (rather than merely on changes in part–whole relations) introduces a complementary view of biological hierarchies, so-called dynamic hierarchies, that can more easily address the creation of new levels, such as multicellularity, in developmental evolution.

The universe consists of a series of levels of complexity and organisation, hierarchical in thought and successive in time, for the simpler preceded the more complicated.
—Joseph Needham (1943b)

8.1 Introduction

Philosophers of science and scientists usually describe the hierarchical organization of living systems as a hierarchy of parts and wholes. Specifically, parts on one level are linked to a whole, usually located on a different higher level. Common part–whole examples are genes and phenotypes, cells and tissues, organisms and ecosystems, and so on. Compositional levels are usually related "vertically" (i.e., from one, say lower, level to another, say higher, one) in a synchronic manner through things (genes, cells, tissues, organisms, etc.) that constitute one another.

In recent years, this general perspective on hierarchy has been incorporated into a theory of mechanistic levels by the so-called new mechanistic philosophy (NMP), especially by Carl Craver (Craver, 2007, 2015; Povich & Craver, 2018). While being rather critical about Oppenheim's and Putnam's (1958) classical layer-cake account (especially about its argument for generality), NMP adopts its basic compositional view on hierarchy. The aim is to reconstruct how scientists—usually molecular and cell biologists—mechanistically explain when they use hierarchical models. Despite disagreeing on a number of other points, there is a consensus within NMP that hierarchical mechanistic models link activities of parts of a system with activities of the whole system (Bechtel & Richardson, 1993; Craver, 2007, 2015; Craver & Bechtel, 2007; Machamer et al., 2000; Povich & Craver, 2018).

A number of authors have argued that this framework of mechanistic levels is insufficient to grasp what levels are in all scientific investigations. Specifically, many philosophers have noted that multiple criteria are needed to capture what levels are, including hierarchical distinctions involving time scales (the speed of a process), size scales (differences between the size of entities), and force scales (the strength of interactions) (see Baedke & Mc Manus, 2018; DiFrisco, 2017; Eronen, 2015; McGivern & Rueger, 2010; Potochnik & McGill, 2012; Wimsatt, 1976). In addition, it has been argued that Craver's theory—the most comprehensive theory of hierarchies within NMP to date—lacks an adequate criterion for the "sameness of levels" (Eronen, 2013, 2015). Other criticisms of levels note that in many biological hierarchies, there are not well-defined sets of lower-level properties that higher-level properties of a system supervene on (see, e.g., Potochnik & McGill, 2012, pp. 127–128).[1]

In this chapter, I develop another critique of the theory of hierarchy, as presented by NMP, focusing on the vertical dynamics between levels and the question of where levels of organization come from. In living systems, different levels of biological organization are vertically *built up* over time, from single cells to multicellular organisms. Thus, levels of organization and their associated processes (genetic, cellular, organismic, and ecological) should not be treated as a given or unchangeable in kind but as evolvable and as biological *explananda* (Mc Manus, 2012). This view is different from, and complementary to, NMP's compositional view of hierarchy, which presupposes that levels are simply there. This is maybe because of the fact that NMP is focused primarily on levels in molecular mechanisms, rather than on the mechanisms' temporality (either developmental or evolutionary). As a consequence, advocates of compositional hierarchies never address how the levels depicted in multilevel diagrams and schemata of mechanistic explanations come into existence. They only discuss how causal processes located at each level gradually change (i.e., how a change on the level of the whole is brought about by a gradual change in the causal properties of its parts on another level).

In the following analysis, I first describe the idea that levels of organization can be understood as being built up over developmental and evolutionary time (sections 8.2 and 8.3). Next, I discuss different ways in which the organization of individual levels can change by means of examples from early twentieth-century organicism and contemporary evo-devo (sections 8.4 and 8.5). This includes changes in the compositional organization of living systems, such as changes in relations and number of parts, as well as changes in their temporal organization, such as in the rate of developmental processes. In some of these cases, like in the evolution of multicellularity, organizational changes can give birth to new levels of organization (section 8.6). Finally, a first attempt is made to describe the epistemic and explanatory roles this view of "dynamic hierarchies" (focusing on the change and origin of organizational patterns) plays and how it structures scientific problems in developmental evolution (section 8.7).

8.2 The Construction of Levels

NMP offers various examples of hierarchically organized biological mechanisms, including mechanisms of protein synthesis (Machamer et al., 2000), photosynthesis (Tabery, 2004, pp. 4–8), synapse transmission (Andersen, 2012, pp. 423–425; Tabery, 2009, pp. 655–657), synaptic plasticity and action potential (Craver, 2002, pp. S85–S88), Mendelian genetics (Darden, 2005; Glennan, 2005, p. 446), and cellular metabolism (Bechtel & Abrahamsen, 2005, pp. 428–429). Somewhat surprisingly, in all of these cases, the way in which the various levels of organization in biological mechanisms, which are depicted in multilevel diagrams, actually come into existence is never addressed (see also, e.g., Craver, 2007, 2015). In other words, this view of hierarchies presupposes that levels are always present or, implicitly, that addressing the problem of their origin is of minor importance.

This perspective underlies the idea that the states of a biological system may change over time, but these changes occur within an unchanging hierarchical architecture. A gene, protein, synapse, or cell always plays the same compositional role—each remains the part for the same whole, or the whole for the same part(s)—whatever changes may occur in a mechanism.[2] This view may be illustrated by an analogy. The components of a mousetrap usually do have different capacities. Once assembled, they may change (e.g., they may show signs of wear), as might the mousetrap along with them (e.g., the wear allows mice to survive and escape), but the mousetrap's components, no matter how they are put together, usually do not allow one to build a car engine, for instance. In other words, in the first case the changed components still play the same role. They remain parts constituting the same whole (the mousetrap), without changing it in kind. I suggest that this specific perspective on how levels (i.e., parts and wholes) change stems from a bias in the choice of examples of mechanistic hierarchies that have been investigated. In fact, the cases discussed by NMP advocates leave aside more dynamic hierarchies in developmental biology and evolutionary developmental biology (evo-devo) where the reorganization of parts typically leads to qualitatively different wholes.

As a consequence of its narrower view on changes in levels, NMP's view of hierarchies is not interested in developing a narrative of how levels of organization are constructed over developmental and evolutionary time. By excluding such cases, levels of mechanisms fall short of capturing the concept of levels of organization in its full significance (i.e., in a broader sense, as, e.g., described by Wimsatt, 1976, 2000, 2007). For instance, NMP's limited view of hierarchies does not address questions such as "How do processes of single cells turn into a multicellular adult organism during development?" and "How do new body plans arise in evolution?" However, these very questions are central to hierarchical approaches in developmental biology and evo-devo. For example, Vrba and Eldredge (1984) highlight this different perspective on biological hierarchies:

Hierarchy is a central phenomenon of life. Yet it does not feature as such in traditional biological theory.... We urge that interlevel causation should feature centrally in explanatory hypotheses of evolution.... A general theory of biology is a theory of hierarchical levels—how they *arise* and interact. (p. 146, emphasis added)

In a similar vein, in his 1993 book *Development and Evolution*, Stanley W. Salthe confesses that his earlier work on hierarchy theory (Salthe, 1985) was missing this very dynamic or developmental perspective. He admits that "there was no attempt in it to deal with change" (p. ix). Instead, one needs to remember that "hierarchies take a long time to develop" (p. 5). With questions like this in view, evo-devo researchers seek to explain the *change* and *origin* of organization, like the origin of body plans (i.e., sets of characters common to a group of phylogenetically related animals at a certain stage in development).

The vertical buildup of (what are usually considered as) levels of organization is a highly plastic procedure and sensitive to environmental influences. Take, for instance, the epigenetic mechanisms of cellular differentiation. These mechanisms do not construct tissues and organisms the way one puts together the components of a mousetrap. Instead, in dynamic living systems, the entities and their properties (as well as the levels of organization at which those entities and properties are located) arise in a stochastic and not merely aggregative manner (Baedke, 2018; Burggren, 2015; Wagner, 1996). For example, during ontogenesis, the properties of cells emerge in a highly contingent and spontaneous (Huang, 2012) as well as stochastic (Kupiec, 2014) way from collective and nonlinear interactions in gene networks. Random and environmentally sensitive molecular interactions between DNA and chromatin proteins cause stochastic gene expression, which codetermine the outcome of cell differentiation. In more general terms, lower-level entities typically have the collective capacity to develop into several qualitatively different higher-level entities (see also Oyama, 2000). This view is supported by findings of developmental plasticity and environmental responsiveness on various levels of organization, which have recently broadened the range of multiple stable phenotypes shown to coexist within one genotype—a range of variation that can allows for "sudden, broad evolutionary changes" (Huang, 2012, p. 149). In addition, this possibility of broad evolutionary changes is supported by the fact that variation occurs in living systems not only in gradual and continuous manners but also in discontinuous ways (see Bergman & Siegal, 2003; Jaeger et al., 2012; Jaeger & Sharpe, 2014; Lange et al., 2014). Examples of such plastic "either/or-development" are predator-induced polyphenism in water fleas or environmental sex determination in reptiles.

This view of contingent development from single cells to multicellular organisms more easily leads to questions about how different (and possibly new) levels of organization (with distinct cellular, tissue-level, and organismic processes) arise over time. These questions are usually not asked within NMP's framework of mechanistic levels, as it does not consider the temporality of vertical interlevel relations. However, as has been pointed out repeatedly (Baedke 2020; Mc Manus, 2012; Parkkinen, 2014; Ylikoski, 2013), developmental mechanisms, in fact, build up levels diachronically. Thus, in developmentally oriented fields, levels are not conceptualized as already existing and not changing in kind (i.e., only having the underlying parts changes their properties). Instead, in multicellular organisms, levels are built up plastically during contingent development. In addition, through discontinuous changes, they can evolve to qualitatively novel levels in evolution.[3]

Looking at developmental biology and evo-devo helps us shift the focus in theories of biological hierarchies away from changes in (the relations between) parts and wholes and toward changes in (the relations between) the organization of parts and wholes.[4] To understand how such (changes in) organization work(s) in explanations, one needs to answer, first, what actually organization is and what it means for the organization of a particular level to change. For example, Craver (2007, p. 135) suggests that Wimsatt's (2000) account of emergence as nonaggregativity can serve as a fruitful starting point for clarifying the concept of organization. According to this view, organization occurs in those situations (e.g., nonlinearity or nondecomposability) in which the whole is more than the sum of its parts.[5] In these cases, the whole depends in some way on the organization of its parts. While this account offers criteria to identify when and in which situations organization exerts influence on wholes, it offers little information on *why* (i.e., due to which exact changes of organization) wholes fail to be the sum of their parts (see Kuorikoski & Ylikoski, 2013).

Another problem with the way organization is introduced in many accounts of levels is that organization, if addressed at all, is usually treated as an *explanans*, not as an *explanandum*. Organization may explain why wholes on one level are more than their parts located on another, but the reason why the parts (or the wholes) are organized in the way they are is usually not addressed. As a consequence, the question of why cells, organisms, or ecosystems have a specific kind of organization (that allows locating them on different levels) is usually neglected. Any epistemic program that seeks to answer this question has to understand biological organization as an *explanandum* that needs to be addressed.

These two blind spots of NMP's view of biological hierarchies—organizational changes of levels during the developmental construction of hierarchical order and organization as an *explanandum*—will be addressed below. Unfortunately, due to its specific focus on mechanistic levels and molecular case studies, NMP offers a limited body of literature and examples of how organization is supposed to change and how this change can lead to novel kinds of organization. Luckily, we find various conceptual analyses that address these questions throughout the history of developmental biology and evo-devo.

8.3 Organicism and Genetic Hierarchies

The idea that hierarchies should be understood as dynamic hierarchies in which levels arise over time is rooted in theoretical discussions in early twentieth-century organicism.[6] In 1930 and 1931, theoretical biologist Joseph Henry Woodger investigated the ideas of level and organization in three papers on the "Concept of Organism" (Woodger, 1930a, 1930b, 1931). He stressed that "hierarchical order may be generated in nature through a temporal process" (1930a, p. 9). Thus, one cannot understand the hierarchical organization of a "living organism in abstraction from time" (p. 10). Only a developmental and evolutionary perspective allows understanding how biological hierarchies become formed *over time*, such as how cells become members of a particular tissue level at which they perform a particular function in the overall hierarchy of the organism. Woodger calls such hierarchies "*genetic hierarchies*":

We see that if we begin with a single organism (or pair of organisms) a hierarchy is generated in which the original starting point constitutes the highest level, the F_1 generation constitutes the next level and so on.... Such a hierarchy would be a *genetic hierarchy*. Cell-division... also generates [such] a hierarchy. (Woodger, 1930a, pp. 9–10)

In other words, the buildup of genetic hierarchies can occur over developmental and transgenerational (or evolutionary) time. Figure 8.1 shows two such hierarchies.

In both of these hierarchies, it is not presupposed that levels are simply there. A level endures over a specific period of time before it gives birth to another level. Thus, according to Woodger, the terms "higher" and "lower" in genetic biological hierarchies should in fact be understood as "earlier" and "later" during developmental or evolutionary time. This idea was picked up by Joseph Needham, especially in his Herbert Spencer lecture, "Integrative Levels: A Revaluation of the Idea of Progress," in 1937. By reflecting on biological evolution (and how it relates to sociocultural "progression"), he states,

In the development of the individual organism, as in that of organisms in general, progression took place from low to high complexity, from inferior to superior organisation. There had been a time when a certain level of organisation *had not existed*, there would come a time when far higher levels would appear. (Needham, 1943a, pp. 245–246, emphasis added)

For Needham, it is clear that in nature, we have to deal with a succession of levels, not with static and preexisting levels. The main idea of his concept of "integrative levels" is that what is a whole at a lower and older level becomes a part at the next higher and newer level. For example, cells are integrated into multicellular organisms and organisms into social systems.[7]

In order to understand this procedure of integration, one has to understand how wholes give birth to "later" wholes and thus how they are, over time, continuously recruited into the working of these new wholes as parts. Organicists agreed that one has to understand how organization in living systems changes over developmental and evolutionary time to answer this question. Needham summarizes this view nicely: "Biological organisation is the basic problem of biology; it is not an axiom from which biology must start" (Needham, 1943c, p. 18). Organization is construed as the most central *explanandum* to understand how the hierarchical order of living beings is created. It is not the starting point of a biological explanation, but its goal.

But what exactly is organization, and how does it change over developmental and evolutionary time? Organicists in the early twentieth century introduced a number of theoretical viewpoints in order to address these questions. Subsequently, these ideas had a strong impact on the way developmental biologists and evo-devo researchers came to understand how levels of organization evolve. Two theoretical viewpoints can be adopted that describe how the organization of levels can change: one focuses on changes in *compositional organization* and the other on changes in *temporal organization*.

8.4   Changes in Compositional Organization

Joseph Woodger distinguishes three ways in which organization can change over time in order to answer "how we are to conceive change in an organized entity so that it may become specifically different from another organized entity" (Woodger, 1930b, p. 450). These organizational changes are depicted in figure 8.2. Motif *A* is the initial organized

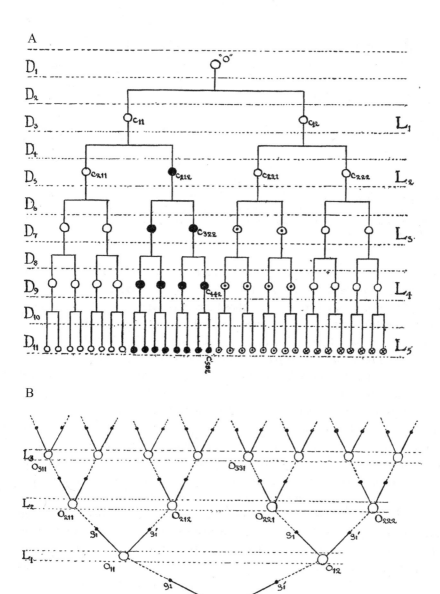

Figure 8.1

Two genetic hierarchies. (A) Cell division hierarchy with respect to zygote *o*. (B) Transgenerational hierarchy with respect to a zygote o_1. A: During cellular differentiation, cells endure over a certain period of time. Intervals between the broken lines represent durations ($D_{1...11}$). A given cell endures throughout the duration represented by the interval between two adjacent lines. *o* endures throughout the duration D_1, which is earlier than any other duration containing members of the system (e.g., cell c_{11} in duration D_3). When cells share the same temporal order (i.e., *D*), they are located on the same level ($L_{1...5}$). This idealized hierarchy presupposes that division occurs simultaneously in all cells located on one level. Marked cells constitute the descendants of different cells on different levels (black: descendants of c_{212} on L_3, L_4, L_5; dot: descendants of c_{221} on L_3, L_4, L_5; cross: descendants of c_{222} on L_5). B: Hierarchy of relations between two zygotes (e.g., o_1 and o_{11}) across generations through gametes (e.g., g_1). Zygotes that share the same history (o_{211} and o_{212}) as descendants of the same zygote (o_{11}) are located on the same level (L_2). Gametes are not members of levels but intermediaries in the production of zygotes. Durations are not indicated. (Woodger, 1930b, p. 455 (A); 1930b, p. 460 (B); reproduced with permission from The University of Chicago Press)

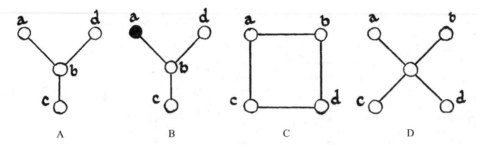

Figure 8.2
Different changes in organizational motifs. Graphic symbols represent a single organized entity and the three types of change it may undergo. Circles depict parts (structural features) that have specific causal relations (see lines) to other parts within the whole organized entity. For further description, see text. (Woodger, 1930b, p. 450; reproduced with permission from The University of Chicago Press)

entity, B is a change in the properties in one of the parts (e.g., a mutation in a gene), C is a change in the relationships between the parts, and D is a change in the number of parts. B is the kind of change recent theories of compositional hierarchies have discussed most often. Woodger identifies it as a kind of change studied, especially in genetics.

The other organizational changes C and D are more complex. They include changes in *relations between parts* and *number of parts*. These changes are largely ignored in today's NMP. Woodger holds these two kinds of change C and D to be central for embryological investigations that study the formation and change of organization within the whole organism. This research focus on C and D remains important today, both in developmental biology as well as in the evo-devo research program.[8] As we will see later (section 8.6), these fields support Woodger's idea that that C and D can lead to qualitatively novel levels of organization patterns in a discontinuous manner. But before turning to the question of how levels arise, for now, let us focus on how the organizational patterns typical of certain levels can change.

Changes in relations between parts (C) and number of parts (D) during organismic development often lead to so-called *evolutionary novelties*. Such a novelty is usually assigned if a feature in a group of organisms (taxon) has no homologous precursor in any taxon in the ancestral lineage (see Peterson & Müller, 2016). Homologies can only be assigned relative to a particular level of organization (Abouheif, 1997; Hall & Kerney, 2012). This means that for a novelty to appear on one (say organ) level of organization, this does not also require a novelty at another (say genetic) level to appear. Besides considerations of morphological novelties (and their nonhomology), evo-devo has developed a number of ways to identify the novel character of organizational patterns brought about, among others, by the changes C and D (see Brigandt & Love, 2012). This includes behavioral novelties (Palmer, 2012), novel ecologically roles (Erwin, 2012), and novel evolutionary scenarios and adaptive benefit (Hallgrímsson et al., 2012). One may also add scaffolding to this list (see Griesemer, this volume).[9] These concepts describe a wide field of game-changing events that range from qualitatively novel population dynamics to the opening of previously unexplored niches (e.g., in the evolution of wing morphology).

Woodger's changes in relationships between parts (C) are today discussed as *heterotopy* in evo-devo. This concept refers to an evolutionary change in the spatial location of a developmental event. The "ontogenetic location" of the parts a, b, c, and d relative to one another is altered in such a way that these parts are placed into novel relational contexts (see figure 8.2C). Heterotopic change can go along with qualitatively novel and nonhomologous phenotypic patterns, as well as to completely new ecological and evolutionary scenarios. For example, the recruitment and recontextualization of limb patterning genes that were ancestrally used to develop insect legs, mouthparts, or antennae are crucial for the origin of beetle horns (Moczek et al., 2006; see also Love & Urban, 2016). These horns are nonhomologous to other structures of the insects.[10] In addition, they have novel functional significance and fitness consequences as male beetles use them in combat over access to females.

Also, Woodger's organizational change in the number of parts (D) is central to the understanding of the origin of qualitative novel morphological patterns. For example, Striedter (2005) has shown that it can be the case that some parts actually evolved *ex nihilo* in evolution, but they did not develop from nothing. This is so because, as Striedter (2005, p. 205) exemplifies with respect to the brain, new interactions between morphogenetic fields can lead to the emergence of new areas of the brain, such as particular areas of the neocortex, that simply have no homologues—at least not in terms of special homology.[11] In such cases of so-called *phylogenetic proliferation*, there are no (obvious) previous lineages to which we could ascribe these new parts of the brain.

These old and new investigations share the idea that during developmental and evolutionary time, the compositional organization—not only the causal properties of single parts—of biological hierarchies can change in a number of ways. However, this view of the variability of organization does not only concern compositional hierarchies.

8.5 Changes in Temporal Organization

Besides these views on how changes in the compositional organization of tissues and organs occur in developmental evolution, one can find other narratives in the history of embryological thought and in contemporary evo-devo that are based on a quite different understanding of biological hierarchies. They draw not on the concept of composition but on scale, especially *time scale*.

Craver (2015, p. 19) has argued that scale relations between levels should always be understood as "accidental consequence" of more fundamental componency relations. According to this view, the cell is located on a different level than the organism (and is explanatory relevant for the latter) not because it is smaller than the organism or because it has a faster "life cycle" than the organism, but because it plays a particular role as a (occasionally smaller and faster replicating) part in the whole organism. In contrast to this "composition first" view of biological hierarchy, which understands scalar characterizations as always being secondary to a compositional characterization, I show below that time scales provide a perspective on biological hierarchies, especially in developmental biology and evo-devo, that can do without descriptions of componency relations.

According to this view—different from the NMP perspective—the "distance" between two "time-scalar levels" depends on how much the rates of the processes located on these levels differ (Baedke & Mc Manus, 2018; DiFrisco, 2017; Salthe, 1985, 1993; Simon, 1962, 1973; Tëmkin & Eldredge, 2015). The rate of a process can refer to its frequency, rhythm, or the time the process takes to overcome perturbations (relaxation time). In other words, given their relation to other (faster, slower, or equally fast) process rates in a local context, processes are located on the same or different time-scalar levels. Then, depending

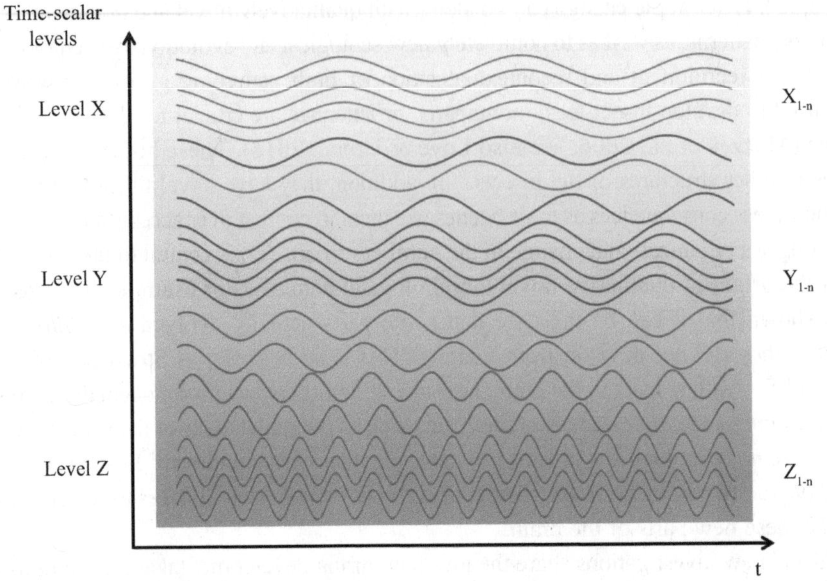

Figure 8.3
A time-scalar hierarchy. A hierarchy of three vertically continuous time-scalar levels (*X, Y, Z*). Processes on one level occur with similar rates. Processes X_{1-n} occur with a slower rate than processes Y_{1-n}, which are slower than processes Z_{1-n}. Processes on each level show a constant characteristic rate over time (*t*). (Baedke & Mc Manus, 2018, p. 42; reproduced with permission from Elsevier)

on explanatory interest, processes that on average occur at the same or similar time scale are grouped together. For example, cellular processes that show on average a similar frequency of recurring cell cycles will appear on the same level. We end up with clusters of processes—scalar levels—that blend into each other (see figure 8.3).

The underlying idea of ordering different causal processes from one another depending on the rate with which they occur has a long history. For example, in early nineteenth-century embryology, Christian Heinrich Pander and Karl Ernst von Baer defended the idea that organization was produced during ontogenesis through a series of rhythmic foldings (see Wellmann, 2015). They argued that a spatiotemporally specific increase or decrease in growth rate of individual germ layers leads to morphological differentiation. In other words, different forms at higher scalar levels result from underlying processes occurring at different rates. This idea of ordering nature by means of the criterion of time scale remained influential in developmental biology in the twentieth century, especially in organicism. For example, Waddington (1957) developed a hierarchical model that placed different biological processes at distinct levels based on their rates:

Time and change is part of the essence of life. Not only so; to provide anything like an adequate picture of a living thing, one has to consider it as affected by at least *three different types of temporal change*. … These three time-elements in the biological picture differ in *scale*. On the *largest scale* is evolution; any living thing must be thought of as the product of a long line of ancestors and itself the potential ancestor of a line of descendants. On the *medium scale*, an animal or plant must be thought of as something which has a life history. … Finally, on the *shortest time-scale*, a living thing keeps itself going only by a rapid turnover of energy or chemical change; it takes in and digests food, it breathes, and so on. (p. 6, emphasis added)

Similarly, Brian Goodwin (1963) placed different cellular processes—genetic, epigenetic, and metabolic ones—on the same or different level(s), depending on whether they show the same (or a different) relaxation time. From this ordering, he derived assumptions about dynamic interlevel interactions: the more two cellular processes differ with respect to their rates, the less changes in the rate of one process, located on one level, will affect the other process on another level (see also Salthe, 1993).

This view on time-scalar hierarchies argues that, during development and evolution, novel animal forms and body plans can be created through changes in the rates of developmental processes. This link between rate changes and form changes was described by D'Arcy Thompson in 1917: "The form of an animal is determined by its specific rate of growth in various directions; accordingly, the phenomenon of rate of growth deserves to be studied as a necessary preliminary to the theoretical study of form, and mathematically speaking, organic form itself appears to us as a *function of time*" (quoted in Needham 1934, p. 79, emphasis added). Somewhat anticipating the later evo-devo perspective, D'Arcy Thompson's view was shared by Huxley et al. (1941), who stressed that an investigation of developmental relative growth rates "is one of the most fruitful ways of advancing our knowledge of phylogenetic development" (p. 225). Guided by this idea, for instance, Needham (1934) tried to understand how differences in brain morphology occur in evolution by comparing growth rates of factors (e.g., proteins, cholesterol) involved in brain development of rabbits, rats, and humans. Such studies should help to illuminate how changes in the temporal organization of developing organisms occur that are, based on one of the above concepts of novelty, different in kind. Today, such "scalar reorganizations" are central targets of investigations in evo-devo that focus on the phenomenon of *heterochrony*.

Heterochrony refers to an evolutionary change in the temporal organization of organisms' life cycles (Nicoglou, 2017; Smith, 2003). This includes changes in the onset and offset of developmental processes or the rate with which they occur. Often such changes involve rate changes in gene expression, which have morphologically and ecologically game-changing effects on higher levels of organization. For example, in marsupials, especially in kangaroos and wallabies, the forelimbs develop much faster than the hindlimbs as compared to placental mammals (see figure 8.4). This is due to a change of timing in lower-level expression patterns of key pattering genes, like *HOXD13*, in forelimbs. In contrast, in mammals like the mouse, fore- and hindlimb expression of these genes is similar (Albrecht et al., 2002). This change in temporal organization of limb development allows marsupials to develop a radically different development compared to placental mammals, as the prenatally faster developing and thus larger forelimbs enable the marsupial neonate (after a short pregnancy) to climb from the urogenital opening into the pouch and suckle (Chew et al., 2012, 2014). This opens up a completely different possibility space of reproduction and supports the movement of marsupials into niches with diverse environments as well as uncertain and often adverse conditions that "are not supported by free-hanging highly altricial young" (Edwards & Deakin, 2013, p. 45).[12]

8.6 The Level of Multicellularity

The above two views of change in the organization of living systems—change in compositional organization and temporal organization—are closely linked to research on how

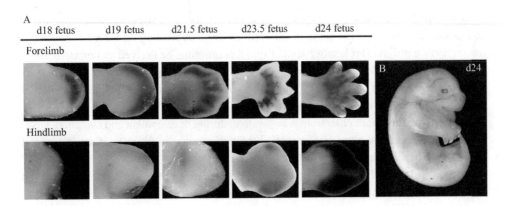

Figure 8.4
Limb development in tammars at selected stages before birth. (A) Forelimb and hindlimb development from days 18 to 24 of pregnancy. Dark gray coloring indicates mRNA of gene *HOXD13*. Expression of *HOXD13* in the forelimb starts distally from day 18 and becomes interdigital in the forelimb by day 21.5. In the hindlimb, *HOXD13* starts to be weakly expressed in the distal region by day 21.5. (B) Whole embryo at day 24, two days before birth. (Chew et al., 2012; reproduced with permission from BMC Developmental Biology)

qualitatively novel levels are built up in developmental evolution, especially during major evolutionary transitions (Maynard Smith & Szathmáry, 1995). Evo-devo researchers locate on newly arising and distinct levels nonhomologous body plans or behavioral, ecological, and evolutionary processes of novel kinds. Novel organizational patterns can be described as constituting a new level in two situations: (1) when they form a new whole that recruits previous wholes into its working as new parts (in compositional hierarchies) and (2) when they lead to the emergence of new phenomena that cluster around certain time scales not previously explored by living systems (in time-scalar hierarchies).

To exemplify this process of how organizational change may lead to the origin of novel levels, let us consider the case of the transition from uni- to multicellularity (see Niklas & Newman, 2016). In biology textbooks, this transition is usually treated as a prototypical example for the evolution of a new level of organization. Recent research in evo-devo suggests that in a "genetic hierarchy," the new level of multicellularity likely arose as a combination of two changes in compositional and temporal organizational at the dawn of Metazoa. To be more precise, this includes relocation of genes and developmental modules and, thus, a change in their relations to one another (heterotopy), as well as changes in timing and rates of developmental processes (heterochrony).

First, a heterotopic organizational change in location and context of developmental units (see figure 8.2C) has been documented in a number of model organisms to promote the transition to multicellularity. For example, in the Volvocine algae *Gonium*, existing genes involved in cell cycle regulation (e.g., CycD1) have been co-opted and recruited to promote the transition toward undifferentiated multicellularity (Olson & Nedelcu, 2016). Their relocation in *Gonium* has the effect that after mitosis, cells fail to break apart from division clusters. They stay attached to each other, which leads to the production of multicellular daughter colonies. Similarly, in the development of siphonocladean algae, like the green alga *Boergesenia forbesii*, the co-option and reuse of a preexisting developmental module responsible for wound healing led to a change toward so-called segregative cell division

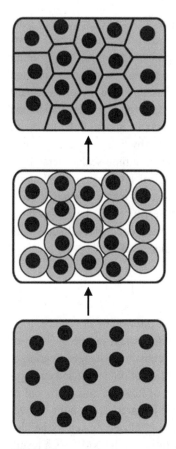

Figure 8.5
Segregative cell division. The protoplasm of a multinucleate organism (bottom) simultaneously separates into spherical portions (middle). These portions develop new cell walls, which leads to the formation of a multicellular (siphonocladous) body plan (upper). (After Niklas et al., 2013, p. 469)

(see figure 8.5) (Graham et al., 2009; Niklas et al., 2013). The original wound-healing response in Siphonocladales includes the retraction of the cell contents from the wound site, leading to a breakup of the protoplasm in numerous spherical protoplasts.[13] In more general terms, during this transition, changes in the compositional organization of previous wholes (single cells) allowed the recruitment and integration of these "old" wholes into the working a newly formed whole (multicellular cell) on a new level.

Second, besides this heterotopic reorganization, the transition to multicellularity also likely involved the temporal reorganization of developmental processes. This has been described, for example, in the development of Myxobacteria. Species in this group of prokaryotes, when starving, develop into an aggregative multicellular organism. This development then finally leads to the formation of multicellular structures. These so-called fruiting bodies have quite complex structures and different cell types. It has been shown that in the bacterium *Myxococcus xanthus*, changes in developmental timing do not disrupt the developmental process but likely induce higher-level morphological changes relevant for the origin of multicellularity (Escalante et al., 2012; see also Arias Del Angel, et al. 2017;

Bonner, 1982). An experimental change in the expression pattern of a key regulatory gene (*fruA*) that controls cell aggregation and fruiting body maturation leads to dramatic shortening of the time that is required to complete the development of multicellular fruiting bodies. Thus, development into a multicellular organism becomes more easily initiated and faster. This suggests the importance of temporal organization in the transition toward multicellularity. In more general words, the temporal reorganization of developmental processes in single cells promoted the origin of novel processes (the multicellular life span) that cluster around a specific time scale not previously explored by living systems (located between faster cell cycles and slower evolutionary processes in a time-scalar hierarchy).

As these studies exemplify, narratives of how the biological level of multicellularity arose in evolution conceptualize the buildup of this novel level as a combination of compositional and temporal reorganizations of processes on a different "older" level over time. In other words, the "genetic hierarchy" leading toward the multicellular level starts with organizational changes on the—eventually lower—unicellular level.

8.7   Toward a Theory of Dynamic Hierarchies

In contrast to common views of compositional hierarchies in philosophy of science, hierarchical research perspectives on developmental evolution focus less on the question of how changes in the activities of parts located on one level constitute changes in a larger whole located on another level. Instead, they continue a tradition of organicist and embryological thought that seeks to address the distinct questions of how the compositional or temporal organization of levels changes and how biological hierarchies are built up over developmental and evolutionary time. Their hierarchies are not closely linked to the idea of synchronic interlevel relations, such as interlevel constitution in the NMP. As a result, these hierarchies are not vertically static. In addition, organicists and evo-devo researchers describe hierarchies in which levels of organization not only change gradually. Rather, levels can change in kind and thus give birth to qualitatively different phenomena— ranging from nonhomologous traits to radically different ecological and evolutionary scenarios. Some of these phenomena qualify as forming a new level of organization.

Intuitively, we are more inclined to locate some of the discussed examples on a novel level compared to others. For example, while it has become widely accepted to represent the transition toward multicellularity as the emergence of a new level of organization, this usually does not hold for the origin of beetle horns, despite the fact that these novel structures go along with completely new ecological and evolutionary scenarios. This situation becomes more puzzling due to the fact that both examples share a similar pattern of organizational change (i.e., changes in relationships between parts or heterotopy). This tells us that identifying the way in which qualitatively novel organizational patterns are introduced during developmental evolution is not enough to establish a theory of "genetic hierarchies" or, as I have called it elsewhere, "*dynamic hierarchies*" (Baedke & Mc Manus, 2018).

We come to see that in developmental evolution, a toolkit of criteria has been developed that point to specific organizational changes that can occur on various levels. These criteria include morphological novelty, behavioral novelty, novel ecologically roles, novel evolutionary scenarios and adaptive benefit, as well as irreversible scaffolding. Each of these criteria can be considered a necessary but not sufficient criterion for the emergence of a novel level in evolution. In other words, not every of these novel organizational (compositional or time-scalar) patterns necessitate introducing a new level of organization in a hierarchical model. However, these criteria allow identifying phylogenetic discontinuities. What is more, they help to clarify the idea of emergence as nonaggregativity (Wimsatt, 2000), because they identify why (i.e., due to which exact set of changes in compositional and/or scalar organization) the organization of one level exerts influence on another, say higher, level. In a further step, one needs to decide on a case-by-case basis if this organizational influence leads to a discontinuous change in the other level so significant that it qualifies as a novel level. How can this decision be motivated?

Following the organicist view of genetic hierarchies, in hierarchical models of developmental evolution, a (structurally, behaviorally, or ecologically) novel organizational pattern can be described as a new level of organization, given four explanatory situations: (1) when organization is not (only) considered as an *explanans* but *explanandum*, (2) when one seeks to emphasize in a biological hierarchy an explanatorily relevant "earlier" and "later" (not just "higher" and "lower") between two organizational patterns over developmental or evolutionary time, and, as described before, (3) when one addresses the successful recruitment of an older compositional organizational into the working of a new whole, and (4) when one seeks to explain the emergence of new phenomena operating on previously unexplored time scales. When situations 1 and 2 coincide with 3 or 4, a description of developmental or evolutionary systems as dynamic hierarchies is epistemically fruitful, as it improves our grasp of these systems' temporal unfolding and increasing organizational complexity.

Evo-devo studies on the evolution of multicellularity are a good example of this. Narratives of the development and evolution of multicellularity usually show all of the above explanatory features. As a consequence, dynamic hierarchies of these cases are able to shed light on the developmental and evolutionary temporality of organization (e.g., the exact sequence of the transition toward multicellular cells or the time-scalar specificity of the new life cycle), as well as on how organizational complexity increases through the compositional integration of an older whole (single cell) into a new one (multicellular organism). It needs to be investigated which other major evolutionary transitions, like those from independent replicators (RNA) to chromosomes, from asexual clones to sexual populations, from solitary individuals to supraorganismic units (holobionts, colonies, etc.), and from primate societies to human societies with language (see Maynard Smith & Szathmáry, 1995), meet the above criteria in the same way as the transition toward multicellularity does.

One may understand these rather general remarks as cautious first attempts to develop a theory of dynamic hierarchies. One has to clarify in more detail the heuristic roles these hierarchies play in biological investigations as well as how they incorporate the local circumstances of their usage and structure scientific problems (*sensu* Brooks & Eronen, 2018; Brooks, this volume). To start with this undertaking, we can summarize, for now, that a view of dynamic hierarchies allows grasping different kinds of organizational change by treating levels of organization and the processes located on them not as pregiven. In addition, it understands levels to be much less consistent and stable than assumed, for example, by the NMP framework. This means, for instance, that within one hierarchical model, cellular processes do not necessarily remain cellular processes. Neither do organismic ones. In other words, while levels of organization seem to be robust and stable most of the time, there also exist discontinuous forms of variation (at genetic, cellular, and organismic levels) that may put them at risk of changing in kind. Organicism and evo-devo take this risk into consideration.

Against this background, studies on developmental evolution have developed comparative approaches that introduce a number of criteria and narratives of sameness and difference. These narratives provide context-dependent answers within a problem space (Brigandt & Love, 2012) of dynamic hierarchies. In this problem space, scientists ask, for example, "Is a new organizational pattern arising during development of a different kind compared to other (i.e., earlier) patterns, and should it thus be located on a separate level?" "What are the criteria according to which the novelty of a level's organization can be assessed?" and "When did a particular level arise in evolution that introduced novel organizational patterns into a hierarchy of ancestors?" Answers on these questions also depend on whether one seeks to develop a dynamic hierarchy from a compositional or time-scalar angle. Both perspectives may identify different levels (and the phenomena located on them) as novel. Thus, one needs to better understand how both can be fruitfully interlinked.

Finally, one should not forget that the importance of explanations of where, actually, levels of organization come from does not render unnecessary the standard view, according to which levels are simply there and gradually changing. These kinds of "fixed" or "static" hierarchies should be seen as complementary to dynamic hierarchies. Both conceptual frameworks serve different epistemic and explanatory roles (see Baedke & Mc Manus, 2018). The former is often used for identifying a particular system and its components, and the latter can convey a deeper understanding of how the different processes relevant in these systems arise and change. Thus, we also have to develop a deeper understanding of how different narratives on "static" hierarchies are integrated with narratives on dynamic hierarchies in biological investigations.

Acknowledgments

I thank Daniel Brooks, James DiFrisco, and William C. Wimsatt for the opportunity to participate in this project. I am grateful for comments received on earlier versions of this chapter by Ilya Tĕmkin, Alan C. Love, and James DiFrisco as well as by the participants of the workshop "Hierarchy and Levels of Organization" (KLI, March 2018). In addition, I gratefully acknowledge financial support from the German Research Foundation (DFG; project numbers BA 5808/1–1 and BA 5808/2-1).

Notes

1. In addition, NMP advocates have been criticized for a conceptual confusion about whether interlevel relations in compositional hierarchies are causal or constitutive relations. On this problem, see, for example, Ylikoski (2013), Romero (2015), and Gillet (this volume). Love and Hüttemann (2011) have emphasized that in biology, there exist causal or temporal part–whole relations that differ from atemporal relations traced in physics. They suggest that the concept of parthood is not necessarily linked to constitution, as usually assumed in the NMP. Below, I adopt this view of the temporal nature of parthood when discussing changes in compositional hierarchies.

2. For example, none of the above-listed hierarchical mechanisms in NMP (e.g., of protein synthesis or photosynthesis) provide an example in which this rule is violated.

3. Note, however, that the above-described features, like plastic responsiveness and discontinuous variation, do not necessarily entail levels of organization arising over time (see also section 7 below).

4. For the central role of hierarchies and interlevel explanations in evo-devo, see, for example, Vrba and Eldredge (1984), Love (2006), Hall and Kerney (2012), Moczek et al. (2015), and Wake (2015); for discussion, see Baedke & Gilbert (2020). Calcott (2009) suggests that in evo-devo, the question of how individuals change over evolutionary time is addressed by a series of hierarchical mechanistic models. While he argues that the reorganization of parts in developmental mechanisms of individuals leads to qualitatively different morphological structures over evolutionary time (e.g., a change from fins to tetrapod limbs), he offers only limited insights into the various ways in which organization can change in development and evolution. For a critical discussion of Calcott's view, see Baedke (2020).

5. Along the lines of this observation, Brigandt (2013, 2015) states that besides a structural analysis of a system by decomposing it into its lower-level parts and their causal properties, in order to explain and predict the dynamics of the whole, one has to focus on the system's organization: after decomposing a system, dynamical models have to put the parts back together again—they recompose the parts in different organizational motifs to model the behaviors of the whole (see also Bechtel & Abrahamsen, 2010).

6. On organicism, see Peterson (2016) and Baedke (2019). On the concept of level of organization and hierarchy in organicism, see Eronen and Brooks (2018).

7. Note that newer levels are not always higher levels, as assumed by Needham. For example, depending on the concept of organism applied, one may argue that tissue and organ levels arise later than the multicellular organism levels.

8. What is not discussed here, but of relevance for evo-devo, are changes in organizational motifs that include variations in promoting or inhibitory relations between the parts (see Alon, 2007). As Jaeger and Sharpe (2014) show, such changes in gene regulatory networks can lead to sudden, discontinuous, and drastic transitions in developmental dynamics.

9. Scaffolding describes hybrids between organisms and a living or nonliving scaffold for the use of reproduction or development. Examples are nonliving by-products or leftovers of metabolism (e.g., wood in trees) and hybrids between organisms of the same species (e.g., in viviparous animals, mothers temporarily scaffolding the development of the offspring) or different species (e.g., in symbionts and hosts of parasites).

10. But they may be called "deep homologies" due to their shared limb pattering genes on a lower level.

11. Recent studies suggest that neocortical circuit organization may, however, be described as a result of serial homology at the level of cell types and circuit patterns (Harris & Shepherd, 2015). Also, it is, of course, possible to talk about "field homologues" (i.e., developmental homologies at the level of morphogenetic fields) that may lead to nonhomologous adult structures (Striedter, 2005, chap. 6). Thus, we see how even those traits that evolved from nothing can develop from something.

12. Often developmental changes in rate, rhythm, and frequency are measured through their associated changes in concentration of gene products or in hormone titers. Such changes in amount are discussed in evo-devo as heterometry. For example, the evolution of the different beak patterns in Darwin's finches was made possible through heterometric changes in particular genes that encode proteins, which promote cell division and thus tissue growth (Abzhanov et al., 2004).

13. On other examples of relocation and co-option of developmental parts that promoted the evolution of multicellularity, see, for example, Arias Del Angel et al. (2017).

References

Abouheif, E. (1997). Developmental genetics and homology: A hierarchical approach. *Trends in Ecology and Evolution, 12*, 405–408.

Abzhanov, A., Protas, M., Grant, B. R., Grant, P. R., & Tabin, C. J. (2004). Bmp4 and morphological variation of beaks in Darwin's finches. *Science, 305*, 1462–1465.

Albrecht, A. N., Schwabe, G. C., Stricker, S., Böddrich, A., Wanker, E. E., & Mundlos, S. (2002). The synpolydactyly homolog (spdh) mutation in the mouse—a defect in patterning and growth of limb cartilage elements. *Mechanisms of Development, 112*, 53–67.

Alon, U. (2007). Network motifs: Theory and experimental approaches. *Nature Reviews Genetics, 8*, 450–461.

Andersen, H. (2012). The case for regularity in mechanistic causal explanation. *Synthese, 189*, 415–432.

Arias Del Angel, J. A., Escalante, A. E., Martínez-Castilla, L. P., & Beniítez, M. (2017). An evo-devo perspective on multicellular development of myxobacteria. *Journal of Experimental Zoology Part B, 328*, 165–178.

Baedke, J. (2018). *Above the gene, beyond biology: Toward a philosophy of epigenetics*. Pittsburgh, PA: University of Pittsburgh Press.

Baedke, J. (2019). O organism, where art thou? Old and new challenges for organism-centered biology. *Journal of the History of Biology, 52*(2), 293–324.

Baedke, J. (2020). Mechanisms in evo-devo. In G. B. Müller & L. Nuño de la Rosa (Eds.), *Philosophy of evo-devo*. Dordrecht, Netherlands: Springer.

Baedke, J., & Gilbert, S. F. (2020). Evolution and development. In E. N. Zalta (Ed.), *The Stanford encyclopedia of philosophy*. Stanford, CA: Metaphysics Research Lab, Stanford University.

Baedke, J., & Mc Manus, S. F. (2018). From seconds to eons: Time scales, hierarchies, and processes in evo-devo. *Studies in History and Philosophy of Biological and Biomedical Sciences, 72*, 38–48.

Bechtel, W., & Abrahamsen, A. (2005). Explanation: A mechanist alternative. *Studies in History and Philosophy of Biological and Biomedical Sciences, 36*, 421–441.

Bechtel, W., & Abrahamsen, A. (2010). Dynamic mechanistic explanation: Computational modeling of circadian rhythms as an exemplar for cognitive science. *Studies in History and Philosophy of Biological and Biomedical Sciences, 41*, 321–333.

Bechtel, W., & Richardson, R. C. (1993). *Discovering complexity*. Princeton, NJ: Princeton University Press.

Bergman, A., & Siegal, M. L. (2003). Evolutionary capacitance as a general feature of complex gene networks. *Nature, 424*, 549–552.

Bonner, J. T. (Ed.). (1982). *Evolution and development*. Berlin, Germany: Springer.

Brigandt, I. (2013). Systems biology and the integration of mechanistic explanation and mathematical explanation. *Studies in History and Philosophy of Biological and Biomedical Sciences, 44*, 477–492.

Brigandt, I. (2015). Evolutionary developmental biology and the limits of philosophical accounts of mechanistic explanation. In P. L. Braillard & C. Malaterre (Eds.), *Explanation in biology* (pp. 135–173). New York, NY: Springer.

Brigandt, I., & Love, A. C. (2012). Conceptualizing evolutionary novelty: Moving beyond definitional debates. *Journal of Experimental Zoology, 318B*, 417–427.

Brooks, D. S., & Eronen, M. I. (2018). The significance of levels of organization for scientific research: A heuristic approach. *Studies in History and Philosophy of Biological and Biomedical Sciences, 68–69*, 34–41.

Burggren, W. W. (2015). Dynamics of epigenetic phenomena: Intergenerational and intragenerational phenotype "washout." *The Journal of Experimental Biology, 218*, 80–87.

Calcott, B. (2009). Lineage explanations: Explaining how biological mechanisms change. *The British Journal for the Philosophy of Science, 60*, 51–78.

Chew, K. Y., Shaw, G., Yu, H., Pask, A. J., & Renfree, M. B. (2014). Heterochrony in the regulation of the developing marsupial limb. *Developmental Dynamics, 243*, 324–338.

Chew, K. Y., Yu, H., Pask, A. J., Shaw, G., & Renfree, M. B. (2012). HOXA13 and HOXD13 expression during development of the syndactylous digits in the marsupial *Macropus eugenii*. *BMC Developmental Biology, 12*, 2.

Craver, C. F. (2002). Interlevel experiments and multilevel mechanisms in the neuroscience of memory. *Philosophy of Science, 69*, S83–S97.

Craver, C. F. (2007). *Explaining the brain*. Oxford, UK: Oxford University Press.

Craver, C. F. (2015). Levels. In T. Metzinger & J. M. Windt (Eds.), *Open MIND* (pp. 1–26). Frankfurt, Germany: MIND Group.

Craver, C. F., & Bechtel, W. (2007). Top-down causation without top-down causes. *Biology and Philosophy, 22*, 547–563.

Darden, L. (2005). Relations among fields: Mendelian, cytological and molecular mechanisms. *Studies in History and Philosophy of Biological and Biomedical Sciences, 36*, 349–371.

DiFrisco, J. (2017). Time scales and levels of organization. *Erkenntnis, 82*, 795–818.

Edwards, M. J., & Deakin, J. E. (2012). The marsupial pouch: Implications for reproductive success and mammalian evolution. *Australian Journal of Zoology, 61*, 41–47.

Eronen, M. I. (2013). No levels, no problems: Downward causation in neuroscience. *Philosophy of Science, 80*, 1042–1052.

Eronen, M. I. (2015). Levels of organization: A deflationary account. *Biology and Philosophy, 30*, 39–58.

Eronen, M. I., & Brooks, D. S. (2018). Levels of organization in biology. In E. N. Zalta (Ed.), *The Stanford encyclopedia of philosophy*. Stanford, CA: Metaphysics Research Lab, Stanford University.

Erwin, D. H. (2012). Novelties that change carrying capacity. *Journal of Experimental Zoology, 318B*, 460–465.

Escalante, A. E., Inouye, S., & Travisano, M. (2012). A spectrum of pleiotropic consequences in development due to changes in a regulatory pathway. *PLoS ONE, 7*, e43413.

Glennan, S. (2005). Modeling mechanisms. *Studies in History and Philosophy of Biological and Biomedical Sciences, 36,* 443–464.

Goodwin, B. C. (1963). *Temporal organization in cells.* London, UK: Academic Press.

Graham, L. E., Graham, J. M., & Wilcox, L. W. (2009). *Algae* (2nd ed.). San Francisco, CA: Cummings.

Hall, B. K., & Kerney, R. (2012). Levels of biological organization and the origin of novelty. *Journal of Experimental Zoology B, 318,* 428–437.

Hallgrímsson, B., Jamniczky, H. A., Young, N. M., Rolian, C., Schmidt-Ott, U., & Marcucio, R. S. (2012). The generation of variation and the developmental basis for evolutionary novelty. *Journal of Experimental Zoology, 318B,* 501–517.

Harris, K. D., & Shepherd, G. M. G. (2015). The neocortical circuit: Themes and variations. *Nature Neuroscience, 18*(2), 170–181.

Huang, S. (2012). The molecular and mathematical basis of Waddington's epigenetic landscape: A framework for post-Darwinian biology? *BioEssays, 34*(2), 149–157.

Huxley, J. S., Needham, J., & Lerner, I. M. (1941). Terminology of relative growth–rates. *Nature, 148*(3747), 225–225.

Jaeger, J., Irons, D., & Monk, N. (2012). The inheritance of process: A dynamical systems approach. *Journal of Experimental Zoology B, 318,* 591–612.

Jaeger, J., & Sharpe, J. (2014). On the concept of mechanism in development. In A. Minelli & T. Pradeu (Eds.), *Towards a theory of development* (pp. 56–78). Oxford, UK: Oxford University Press.

Kuorikoski, J., & Ylikoski, P. (2013). How organization explains. In V. Karakostas & D. Dieks (Eds.), *EPSA11: Perspectives and foundational problems in philosophy of science* (pp. 69–80). Dordrecht, Netherlands: Springer.

Kupiec, J.-J. (2014). Cell differentiation is a stochastic process subjected to natural selection. In A. Minelli & T. Pradeu (Eds.), *Towards a theory of development* (pp. 155–173). Oxford, UK: Oxford University Press.

Lange, A., Nemeschkal, H. L., & Müller, G. B. (2014). Biased polyphenism in polydactylous pats carrying a single point mutation: The Hemingway model for digit novelty. *Evolutionary Biology, 41,* 262–275.

Love, A. C. (2006). Evolutionary morphology and evo-devo: Hierarchy and novelty. *Theory in Biosciences, 124,* 317–333.

Love, A. C., & Hüttemann, A. (2011). Comparing part-whole reductive explanations in biology and physics." In D. Dieks, W. J. Gonzalez, S. Hartmann, T. Uebel, & M. Weber, (Eds.), *Explanation, prediction, and confirmation* (The philosophy of science in a European perspective, Vol. 2) (pp. 183–202). Dordrecht, Netherlands: Springer.

Love, A. C., & Urban, D. (2016). Developmental evolution of novel structures—animals. In R. Kliman (Ed.), *Encyclopedia of evolutionary biology* (Vol. 3, pp. 136–145). Oxford, UK: Academic Press.

Machamer, P., Darden, L., & Craver, C. F. (2000). Thinking about mechanisms. *Philosophy of Science, 67,* 1–25.

Maynard Smith, J., & Szathmáry, E. (1995). *Major transitions in evolution.* Oxford, UK: Oxford University Press.

McGivern, P., & Rueger, A. (2010). Hierarchies and levels of reality. *Synthese, 176,* 379–397.

Mc Manus, F. G. (2012). Development and mechanistic explanation. *Studies in History and Philosophy of Biological and Biomedical Sciences, 43,* 532–541.

Moczek, A. P., Rose, D., Sewell, W., & Kesselring, B. R. (2006). Conservation, innovation, and the evolution of horned beetle diversity. *Development, Genes and Evolution, 216,* 655–665.

Moczek, A. P., Sears, K. E., Stollewerk, A., Wittkopp, P. J., Diggle, P., Dworkin, I., Ledon-Rettig, C., Matus, D. Q., Roth, S., Abouheif, E., Brown, F. D., Chiu, C.-H., Cohen, C. S., Tomaso, A. W. D., Gilbert, S. F., Hall, B., Love, A. C., Lyons, D. C., Sanger, T. J., Smith, J., Specht, C., Vallejo-Marin, M., & Extavour, C. G. (2015). The significance and scope of evolutionary developmental biology: A vision for the 21st century. *Evolution and Development, 17,* 198–219. Needham, J. (1934). Chemical heterogony and the groundplan of animal growth. *Biological Reviews, 9,* 79–109.

Needham, J. (1943a). Integrative levels: A revaluation of the idea of progress. In J. Needham (Ed.), *Time the refreshing river (essays and addresses, 1932–1942)* (pp. 233–272). London, UK: G. Allen and Unwin. (Original work published 1937)

Needham, J. (1943b). Metamorphoses of scepticism. In J. Needham (Ed.), *Time the refreshing river (essays and addresses, 1932–1942)* (pp. 7–27). London, UK: G. Allen and Unwin. (Original work published 1941)

Needham, J. (1943c). The naturalness of the spiritual world: A reappraisement of Henry Drummond. In J. Needham (Ed.), *Time the refreshing river (essays and addresses, 1932–1942)* (pp. 28–41). London, UK: G. Allen and Unwin. (Original work published 1939)

Nicoglou, A. (2017). The timing of development. In C. Bouton & P. Huneman (Eds.), *Time of nature and the nature of time* (pp. 359–390). Dordrecht, Netherlands: Springer.

Niklas, K. J., Cobb, E. D., & Crawford, D. R. (2013). The evo-devo of multinucleate cells, tissues, and organisms, and an alternative route to multicellularity. *Evolution and Development, 15,* 466–474.

Niklas, K. J., & Newman, S. A. (2016). *Multicellularity: Origins and evolution.* Cambridge, MA: MIT Press.

Olson, B. J. S. C., & Nedelcu, A. M. (2016). Co-option during the evolution of multicellular and developmental complexity in the Volvocine green algae. *Current Opinion in Genetics and Development, 39,* 107–115.

Oppenheim, P., & Putnam, H. (1958). Unity of science as a working hypothesis. In H. Feigl, M. Scriven, & G. Maxwell (Eds.), *Concepts, theories, and the mind–body problem* (pp. 23–36). Minnesota: University of Minnesota Press.

Oyama, S. (2000). *The ontogeny of information: Developmental systems and evolution* (2nd ed.). Durham, NC: Duke University Press.

Palmer, A. R. (2012). Developmental plasticity and the origin of novel forms: Unveiling cryptic genetic variation Via 'Use and Disuse.' *Journal of Experimental Zoology, 318B,* 466–479.

Parkkinen, V.-P. (2014). Developmental explanation. In M. C. Galavotti, D. Dieks, W. J. Gonzalez, S. Hartmann, T. Uebel, & M. Weber (Eds.), *New directions in the philosophy of science* (pp. 157–172). Dordrecht, Netherlands: Springer.

Peterson, E. L. (2016). *The life organic: The theoretical biology club and the roots of epigenetics.* Pittsburgh, PA: University of Pittsburgh Press.

Peterson, T., & Müller, G. B. (2016). Phenotypic novelty in EvoDevo: The distinction between continuous and discontinuous variation and its importance in evolutionary theory. *Evolutionary Biology, 43,* 314–335.

Potochnik, A., & McGill, B. (2012). The limitations of hierarchical organization. *Philosophy of Science, 79,* 120–140.

Povich, M., & Craver, C. F. (2018). Mechanistic levels, reduction, and emergence. In S. Glennan & P. Illari (Eds.), *The Routledge handbook of mechanisms and mechanical philosophy* (pp. 185–197). London, UK: Routledge.

Romero, F. (2015). Why there is not inter-level causation in mechanisms. *Synthese, 192,* 3731–3755.

Salthe, S. N. (1985). *Evolving hierarchical systems.* New York, NY: Columbia University Press.

Salthe, S. N. (1993). *Development and evolution: Complexity and change in biology.* Cambridge, MA: MIT Press.

Simon, H. (1962). The architecture of complexity. *Proceedings of the American Philosophical Society, 106,* 467–482.

Simon, H. (1973). The organization of complex systems. In H. H. Pattee (Ed.), *Hierarchy theory* (pp. 3–27). New York, NY: Braziller.

Smith, K. K. (2003). Time's arrow: Heterochrony and the evolution of development. *The International Journal of Developmental Biology, 47,* 613–621.

Striedter, G. F. (2005). *Principles of brain evolution.* Sunderland, UK: Sinauer Associates.

Tabery, J. (2004). Synthesizing activities and interactions in the concept of a mechanism. *Philosophy of Science, 71,* 1–15.

Tabery, J. (2009). Difference mechanisms: Explaining variation with mechanisms. *Biology and Philosophy, 24,* 645–664.

Tëmkin, I., & Eldredge, N. (2015). Networks and hierarchies: Approaching complexity in evolutionary theory. In E. Serrelli & N. Gantier (Eds.), *Macroevolution: Explanation, interpretation and evidence* (pp. 183–226). Heidelberg, Germany: Springer.

Vrba, E. S., & Eldredge, N. (1984). Individuals, hierarchies and processes: Towards a more complete evolutionary theory. *Paleobiology, 10,* 146–171.

Waddington, C. H. (1957). *The strategy of the genes.* London, UK: Allen and Unwin.

Wagner, A. (1996). Can nonlinear epigenetic interactions obscure causal relations between genotype and phenotype? *Nonlinearity, 9,* 607–629.

Wake, M. H. (2015). Hierarchies and integration in evolution and development. In A. C. Love (Ed.), *Conceptual change in biology* (pp. 405–420). Berlin, Germany: Springer.

Wellmann, J. (2015). Folding into being: Early embryology and the epistemology of rhythm. *History and Philosophy of the Life Sciences, 37,* 17–33.

Wimsatt, W. C. (1976). Reductionism, levels of organization, and the mind-body problem. In G. G. Globus, I. Savodnik, & G. Maxwell (Eds.), *Consciousness and the brain* (pp. 205–267). New York, NY: Plenum Press.

Wimsatt, W. C. (2000). Emergence as non-aggregativity and the biases of reductionisms. *Foundations of Science, 5,* 269–297.

Wimsatt, W. C. (2007). *Re-engineering philosophy for limited beings: Piecewise approximations to reality.* Cambridge, MA: Harvard University Press.

Woodger, J. H. (1930a). The "concept of organism" and the relation between embryology and genetics. Part I. *Quarterly Review of Biology, 5*, 1–22.

Woodger, J. H. (1930b). The "concept of organism" and the relation between embryology and genetics. Part II. *Quarterly Review of Biology, 5*, 438–463.

Woodger, J. H. (1931). The "concept of organism" and the relation between embryology and genetics. Part III. *Quarterly Review of Biology, 6*, 178–207.

Ylikoski, P. (2013). Causal and constitutive explanation compared. *Erkenntnis, 78*, 277–297.

9 Downward Causation and Levels

James Woodward

Overview

This chapter discusses a number of different notions of level, including levels as involving compositional relations and levels as relatively explanatorily autonomous. The notion of downward causation, understood as causation from upper to lower levels, is defended as legitimate. Downward causation is elucidated in terms of a relation called conditional causal independence: X is causally independent of Z conditional on Y when X is causally relevant (in the standard interventionist sense) to Z, Y is causally relevant to Z but conditional on the values of Y, changes in the value of X make no further difference to Z. When X is a lower-level realizer of Y and conditional independence holds, we can use Y rather than X to explain Z, and the Y to Z relation is conditionally autonomous from X-level facts. In such cases Y is an upper-level cause of Z and a downward cause of Z when Z is at a lower level than Y.

9.1 Introduction

This chapter is a defense of downward (or top-down) causation and, along with this, a discussion of levels—why we sometimes find it useful to think in terms of this notion and what its limitations might be. The connection between these topics arises in part because words like "downward" and "top-down" suggest a picture according to which the world is organized into "levels," with downward causation involving causes that are at a higher level than their effects. One possible view (see, e.g., Eronen, 2013) is that talk of levels makes no clear sense; if so, whatever might be involved in (what we call) top-down causation can't literally be causation from an upper to lower level. Put differently, it might seem that a prior challenge facing anyone who talks of downward causation is to provide an account of levels according to which talk of upper and lower levels makes sense and then evaluate whether there is causation from the former to the latter.

For a number of reasons, I'm not going to proceed in this way. Although I think that level notions do legitimate work and thus that we should not try to dispense with them, I also doubt that there is any single, consistent account that captures everything that people have had in mind in talking of levels. My view is that levels talk reflects a number of different considerations that are sometimes mutually reinforcing but also can push us to make very different—indeed inconsistent—judgments in assignments of levels. Privileging just one of these notions is likely to seem arbitrary and in any case will fail to do justice to the variety of motivations that underlie levels talk. It is also true, however, that these different notions are interrelated in various complex ways.[1] My focus in this chapter will be on three (of many possible) ways of thinking about levels, which I believe illustrate these claims:

(**1.1**) A notion rooted in compositional or part–whole relationships

(**1.2**) A notion tied to ideas about independence (including what I will call conditional independence) and, along with this, strategies for coarse graining and dimension reduction

(**1.3**) Closely related to this, a notion based on considerations of computational and epistemic tractability

As I will try to illustrate, lack of clarity about the relation between these different level notions is one reason why the notion of downward causation has seemed problematic. Conversely, thinking about downward causation provides a very useful point of entry into various ways of thinking about levels.

My discussion is organized as follows: I begin (section 9.2) with some brief remarks about the notions of level that will concern me. Section 9.3 explores what might be meant by downward causation. Sections 9.4 and 9.5 describe some examples that scientists have found it natural to describe in terms of downward causation. Sections 9.6 and 9.7 consider several objections that philosophers have advanced against the possibility of downward causation. I will argue that these objections are either misguided or do not apply to the examples in question. A crucial part of my argument will be that the putative examples of downward causation on which critics have focused are not, for the most part, what scientists have had in mind in talking of downward causation—the critics' objections do not apply to the examples that I give of downward causation. In particular, one idea to which I will be objecting is a picture according to which top-down causation involves a whole causally affecting its parts. I agree that, at least in many of the cases the critics have discussed, this is incoherent, but I also don't think it is what one should understand by downward causation.

Once I have sorted out these issues about downward causation, I will then introduce the notion of conditional independence (section 9.8) and use it to motivate some general remarks about levels. Talk of levels is ubiquitous in science, and this raises a number of questions: most obviously, there is the question of what might be meant by such talk. A related question concerns the legitimate function (if any) of such talk. Why do scientists apparently find such talk useful? What work does it do? Does it frequently mislead us, as some critics claim? I address such questions in sections 9.9 and 9.10.

9.2   Levels

9.2.1   Levels as Compositional

One familiar notion of "level" is compositional or mereological: objects or entities at a higher level are "composed" or "constituted by" objects at lower levels in a way that generates a hierarchy.[2] Here, "composed" means that the lower-level objects are (or at least are thought of as) "parts" of higher-level objects—or at least this is the paradigmatic notion of constitution.[3] Textbooks provide familiar illustrations of this idea. Atoms are composed of protons, neutrons, and electrons; molecules are composed of atoms; cells are composed of molecules; multicellular organisms are composed of cells; and so on. This is sometimes described as a "wedding cake" model of levels, since reality is regarded as divided into distinct "layers" based on part–whole relationships. We find this idea (among others) in Oppenheim and Putnam's (1958) classic paper, and it often seems to be the preferred conception of levels among metaphysically inclined philosophers. It is this notion of level that (I believe) underlies many philosophical objections to downward causation, since it encourages the idea that this involves causation from a whole to its parts.

9.2.2 Levels and Independence

Another notion of level is tied to claims about independence where (as I will understand this) it is a matter of relations among *variables*. (More pedantically, it is a relation among what in the world corresponds to variables—e.g., magnitudes such as mass and charge—but for brevity, I will write "variables" in what follows.) According to this conception, variables X and Z are at different levels when the behavior of Z is in some sense "independent" of the values taken by X, so that we can ignore (or largely ignore or ignore in many cases) X in constructing a causal explanation of Z, appealing instead to other variables at some different level. This notion of level is often tied to considerations having to do with the role of "scales"—spatial, temporal, and energetic—in constructing theories and models: sometimes when nature is kind, we have "separation" or near separation of scales, so that what happens at one length or energy scale can be understood largely independently of what happens at other scales, and this in turn leads us to think of interactions at one scale as at a different level than interactions at other scales.[4]

As an illustration, consider that, for the purposes of understanding what is going on within the nucleus and phenomena such as radioactive decay, two of the four fundamental forces—the strong and weak nuclear forces—are crucial. These forces are very strong at very short spatial scales. Gravity, another fundamental force, is effectively irrelevant for most purposes in modeling nuclear behavior. On the other hand, if we are interested in explaining/understanding chemical behavior—how atoms combine and form molecules and compounds—the strong and weak nuclear forces are effectively irrelevant, and yet another force—the electromagnetic force—plays a central explanatory role. In many cases, this separation of levels or scales—the fact that nature permits us to construct theories that explain aspects of nuclear behavior that appeal to factors that are different from those that are required to explain chemical behavior, so that we can do nuclear physics without doing chemistry and vice versa—is crucial for successful science. For one thing, without such separation, constructing models of many phenomena would be computationally intractable. This yields one reason why a notion of level is sometimes important in science.

What do we mean when we say for most explanations of chemical behavior, we don't need to invoke information of nuclear forces? Of course if those forces were sufficiently different, stable nuclei (and atoms) would not exist. So we don't mean that facts about those forces are completely irrelevant to chemical behavior. Rather (I suggest), what we have in mind is something like this: for purposes of chemistry, whatever is relevant about such forces can be represented by the values of a small number of variables, having to do, for example, with the mass and charge of the nucleus. Conditional on the values of such variables, additional more detailed information about the inner goings on of the nucleus is irrelevant to chemical behavior. So the notion of independence in play here is really a notion of *conditional independence*.

More generally, suppose that some of values of variable X are causally relevant to (not causally independent of) the values of variable Z but that it is also the case that there is some variable or set of variables Y (constructable from X by some coarse-graining operation that we can think of as representing the aggregate impact of the Xs) of much smaller dimensionality than X such that given the values of Y, additional variation in the values of X makes no further difference to the values of Z.[5] In such cases, I will say that X is *conditionally independent* of Z, given the values of Y. (As will become clear in section 9.8, which spells out this notion in more detail, conditional independence here is not conditional probabilistic independence but should be understood in terms of interventionist counterfactuals.) Such conditional independence allows us to appeal just to Y in explaining Z. For example, conditional on the value of the temperature of a gas, further variations in the kinetic energy of the individual molecules that are consistent with this temperature make, to a first approximation, no further difference to the pressure of the gas or to the values taken by certain other thermodynamic variables. When, as for these thermodynamic variables, this sort of conditional independence holds, we often find it natural to say that those variables are at the "same level" and, moreover, at a level that is "different" from the variables used to characterize the individual molecules that make up the gas.

It is important to understand that this second basis for level talk (which interactions are important and which either can be entirely ignored or subsumed into other, lower dimensional variables) is conceptually quite different from the composition-based notion of level or from notions tied directly to size considerations. Whether one object A is part of another B is obviously a distinct question from whether features of A are irrelevant, unconditionally or conditionally, to the behavior of B. Nuclei are "parts" of molecules and nucleons are parts of nuclei, but as noted above, detailed information about nuclear forces between nucleons can be safely ignored in understanding the chemical behavior of molecules. Electrons are also parts of molecules, but those aspects of the behavior of electrons that have to do with electromagnetic forces are crucial to understanding chemical behavior.

Of course, composition/size-based considerations and independence considerations are frequently related—the various components of a cell commonly interact more strongly with one another (are not independent with respect to one another, even given further information about some relatively small number of variables) than they do with the more distant components of other cells (the effects of which may be usefully represented by means of some small number of variables). Nonetheless, composition relations and relations of independence/dependence only imperfectly track one another: entities that are small in size relative to larger entities (or that are roughly the same size as components of those larger entities) can affect those larger entities: viruses and bacteria can contribute to the defeat of armies. Conversely (I shall argue), properties possessed by larger entities can causally affect properties of smaller entities, including properties of smaller entities that are components of those larger entities, as when the potential difference across a neuronal membrane affects the behavior of the ion channels that are part of the membrane— see section 9.5.

9.2.3 Levels and Tractability

A third notion (often tied, however, to the independence-based notion 9.2.2 above) ties "level" to considerations having to do with computational and epistemic tractability. Roughly speaking, two sets of variables (or phenomena characterized in terms of those variables) S_1 and S_2 will be at different levels in this sense when there are computational and other sorts of epistemic barriers to modeling or explaining systems that can be characterized by one set of variables (e.g., S_1) and relations among them in terms of variables and relationships from the other set (e.g., S_2). For example, because of such barriers, multicompartment models of fine-grained aspects of neuronal behavior cannot be aggregated to produce tractable models of whole neurons. In this respect, the two kinds of models and the relationships to which they appeal are at different "levels"—see section 9.9 for additional discussion. A closely related point is that some variables are only well defined or measurable at certain levels.[6]

9.3 Interventionism and Downward Causation

To talk about causal relationships between "levels," we require an account of causation. I will assume an interventionist account: Y causes Z when there is some possible intervention that changes the value of Y and, along with this, there is an associated change in the value of Z that occurs in a "regular" or "uniform" way. Here, "regular" means that in some range of background circumstances, the intervention setting the value of $Y=y$ is either followed by the same value of Z or the same stable probability distribution for the values of Z. When Y is at a "higher" level than Z, and this pattern of dependency obtains between Y and Z, Y downward causes Z. (Obviously, in characterizing downward causation in this way, we are not assuming a notion of level according to which variables that are causally related are automatically assumed to be at the same level—instead, we are assuming some other notion of level such as a composition-based notion.) This is how many scientists who make use of the notion of downward causation understand this notion. For example, George Ellis (2016) writes, "One demonstrates the existence of top-down causation whenever manipulating a higher-level variable can be shown to reliably change lower level variables" (p. 16)

We can further flesh out the idea of a "reliable" or regular change in Z under an intervention on Y in the following way.[7] Assume that the upper-level variable Y has a number of different lower-level "realizers." For example, if T is the temperature of a gas, many different possible arrangements of gas molecules, characterized in terms of the values of position and momentum variables for each of the molecules composing the gas, will realize the same value of this temperature.[8] Thus, a manipulation of an upper-level variable such as T, which sets it to some value t, can have lots of different possible realizations, corresponding to many different possible arrangements of gas molecules. However, for this manipulation to have a reliable (or uniform) effect on some second variable Z, we require that all of the different realizations of t (or almost all of them—see sections 9.8 and 9.9) should have the same uniform effect on Z, where again this means that they lead to the same value for Z or the same probability distribution for Z. In other words, given that T is set to the value t, it should not matter how that value is realized by the values of the associated lower-level variable as far as the effect of $T=t$ on Z is concerned—the effect of $T=t$ on Y should be in this sense "realization independent." If this condition is not met, T will not count as a cause (top-down or otherwise) of Z. This condition thus excludes "ambiguous" manipulations of candidate cause variables that have different effects depending on how the cause variable is realized (cf. Spirtes & Scheines, 2004). As we shall see in section 9.8, this nonambiguity requirement is a particular instance of the more general conditional independence requirement mentioned previously.

There are additional conditions that also must be met for downward causation to be present. One particularly important condition is that the putative cause-and-effect variables must be distinct in the right way—this condition is discussed in section 9.6. Within an interventionist framework, it is also important, as Ellis specifies, that the causal relata Y and Z are *variables*[9] where the mark of a variable is that it can assume several different values.[10] Variables include quantities like mass, position, voltage, and current, but they can also be binary or two-valued. According to interventionism, only variables or values of variables can stand in causal relationships: we haven't clearly specified what causal relationships we are talking about until we have specified the variables they involve. It is crucial to distinguish variables from things or entities. To anticipate an example discussed below, ion channels and cell membranes are things, not variables, and thus cannot literally stand in causal relationships. However, it is typically things or entities that stand in part–whole or compositional relationships. This is one reason why, within an interventionist framework and quite apart from further subtleties about causation, downward or otherwise, talk of wholes causing their parts seems incoherent—wholes and parts are things and hence cannot stand in causal relationships. Variables associated with wholes and parts can sometimes stand in causal relationships, but as we shall see, this needn't involve an objectionable kind of whole–part causation.[11]

9.4 Downward Causation Exemplified

Here are some putative examples of top-down causation—some drawn from recent books[12] and some from other sources.

(**4.1**) The use of mean field theories in which the combined action of many atoms on a single atom is represented by means of an effective potential V rather than by means of a representation of each individual atom and their interaction. Intuitively, V is at a higher level than the atom on which it acts (Clark & Lancaster, 2017).

(**4.2**) The influence of environmental variables, including social relations involving whole animals on gene expression within those animals as when manipulating the position of a monkey within a status hierarchy, changes gene expression controlling serotonin levels within individual monkeys. Here position within a social hierarchy is thought of (perhaps on the basis of compositional considerations) at a higher level than gene expression affecting serotonin levels.

(**4.3**) A red hot sword is plunged into cold water, and this alters the meso-level structure of the steel in the sword—cracks, dislocations, and grains in the sword. The treatment of the sword—heating and cooling—is at a higher level than these mesoscopic changes,[13] and the former downward causes the latter. (Example due to Bob Batterman.)

(**4.4**) Energy cascades. When a fluid is stirred in such a way that it exhibits large-scale turbulent motion, this motion is gradually transferred to motion at smaller scales—from large-scale eddies to much smaller-scale eddies. The large-scale motion may be on the scale of many meters, the small-scale motions on the scale of a millimeter where they are eventually dissipated as heat. Viscosity-related effects dominate at this smaller scale but are less important at larger scales. (Example due to Mark Wilson.)

9.5 The Hodgkin-Huxley Model as an Example of Downward Causation

Each of the previous examples is worth extended discussion, but to keep things tractable, I will largely focus on just one additional example—the Hodgkin–Huxley (HH) model of the action potential. This describes the factors causally affecting the overall shape of the action potential within an individual neuron. For reasons of space, I will not describe the model in detail, but the basic idea is that the neuron can be understood as a parallel circuit consisting of a capacitor that stores charge (the potential V across the neuronal membrane functions as a capacitor); a channel that conducts the sodium current I_{Na}, with an associated time- and voltage-dependent conductance g_{Na}; a channel that conducts a potassium current I_K with time- and voltage-dependent conductance g_K; and a leakage current I_l, which is assumed to be time and voltage independent. Since the channels are "part" of the cell membrane (they are embedded in it, so that the membrane is, at least on a compositional understanding of levels, at a "higher" level than the channels) and the behavior of the channels, including their conductances, is influenced by (among other factors) the potential difference V across the entire membrane, this looks like a plausible case of top-down causation, and indeed it is described as such by, for example, Denis Noble (2006). For future reference, we should also note that according to this model, the potential difference V across the cell membrane is itself causally changed by the various currents that occur in the ionic channels—as these change over time (and with different time courses), the total current I changes (in an apparent case of bottom-up causation) and V also changes, with these changes in V again changing the ion currents. It is this temporally extended pattern of mutual influence that accounts for the action potential. However, despite this apparent causal cycle, the fact that the ion channels are part of the cell membrane, and what is arguably the presence of downward causation, the HH model looks an intelligible causal representation—indeed one that is generally taken to be correct. How (if at all) can we make sense of this?

9.6 Downward Causation and Distinctness of the Causal Relata

As noted earlier, one objection to downward causation is that this involves wholes acting on their parts. This is thought to be objectionable because causes and effects must be "distinct," and wholes and parts are not sufficiently distinct to stand in causal relationships. (Objections of this sort can be found in Bechtel & Craver, 2007; Heil, 2017; and many others.)

What is meant by distinctness in this context? David Lewis's (2000) views are representative of a common understanding of this notion:

[For C to cause E] C and E must be distinct events—and distinct not only in the sense of nonidentity but also in the sense of nonoverlap and nonimplication. It won't do to say that my speaking this sentence causes my speaking this sentence or that my speaking the whole of it causes my speaking the first half of it; or that my speaking causes my speaking it loudly, or vice versa. (p. 78)

As this quotation suggests, the tendency in the philosophical literature has been to try to understand the relevant notion of distinctness in terms of the absence of logical relationships (nonimplication) or the absence of part–whole relations (spatial or temporal) and similar considerations, which leads immediately to the conclusion that wholes cannot cause their spatial or temporal parts because of a failure of distinctness. For example, the individual H_2O molecules making up a body of water are parts or constituents of that body, and one might object, as Heil (2017) does, to the claim that the position or motion of the whole body causes the position or motion of one of its molecular constituents on the grounds that these relata are not sufficiently distinct to stand in a causal relationship.[14] Similarly, Bechtel and Craver (2007) consider cases in which some temporally extended process is present that has a subprocess as temporal proper part or constituent and object to the claim the former can exert a causal influence on the former. To use their example, because "the change in the conformation of rhodopsin is a stage in the signal transduction pathway in visual perception, the change in conformation cannot be a cause of signal transduction" (p. 552). For similar reasons, they object to the claim that a mechanism considered as whole (i.e., as a collection of parts or constituents standing in ordinary causal relations with each other) can exert downward causation on the parts or constituents of that mechanism.

I agree with these claims (about the absence of downward causation in these examples) but do not think that the complaint of failure of distinctness among causal relata applies to the top-down relationship in the HH model or the other examples described in section 9.4. In other words, the relationship between V in the HH model and the behavior of the ion channels is *not* like the relationship between a body of water and its constituent molecules or like the relationship between a whole mechanism and its parts. To spell this out, suppose that P is a spatial or temporal part of W and let X be some variable that characterizes some feature of W and Y some variable that characterizes some feature of P. Then I claim that it is entirely possible for X and Y to be distinct in a way that allows for X to cause Y despite the parthood relationship between P and W. Indeed, in some such cases, it may make no sense to think of the variable Y as a part of X or as logically or semantically related to it in a way that precludes causation. For example, the ionic conductances g don't seem in any intuitive sense to be "part" of the variable V—it is hard to understand what this could possibly mean.[15] More important, even if one thinks of this relationship in terms of parts and wholes, these variables don't seem to exhibit the kind of failure of distinctness that variables like "saying hello" and "saying hello loudly" do. Similarly, variables describing the mesoscopic structure of the sword in (**4.3**) not are in any obvious sense "part" of the variable that describes how the sword has been heated and cooled.

In saying this, I don't mean to deny that variables can fail to be distinct in ways that preclude their standing in causal relationships—the concern about failures of distinctness is a legitimate worry. My point is rather that the usual ways of trying to characterize the kinds of failures of distinctness that matter for causal relatedness in terms of logical or mereological relations don't work very well. Here is a first pass at a proposal about distinctness that seems natural within an interventionist framework and which I have defended elsewhere (Woodward, 2015). The proposal is that variables are appropriately distinct (and thus suitable candidates for standing in a causal relationship as far as distinctness considerations go) when they satisfy a condition of independent fixability (**IF**):

(**IF**) A set of variables **V** satisfies independent fixability of values if and only if for each value it is possible for a variable to take individually, it is "possible" to set the variable to that value via an intervention, concurrently with each of the other variables in **V** also being set to any of its individually possible values by independent interventions. Here "possible" includes settings of values of variables that are possible in terms of the assumed, logical, mathematical, or semantic relations among the variables as well as certain structural or space–state relationships.

Thus, "possible" in **IF** should not be understood as restricted to combinations of settings that are causally possible, although of course if settings are causally co-possible, they are possible *tout court*. For example, in the usual state–space formulation of mechanics, the three-dimensional position and momentum components of each collection of particles at a time as well as the components for the same particle at different times are regarded as independently fixable, even though, once dynamical considerations are introduced, certain combinations of these may be causally excluded. Obviously, **IF** draws upon (and does not explicate) some antecedently understood notion of possibility that is broader than causal possibility. What **IF** adds is a focus on whether *operations* involving setting of variables to values are co-possible—this turns out to be crucial.

To further illustrate (**IF**), consider Lewis's example of saying "hello" and saying "hello" loudly. Expressed in terms of variables, suppose that X has two values (x_1 = saying hello loudly, x_2 = doing something other than saying hello loudly) and Y has values (y_1 = saying hello, y_2 = not saying hello). Then the values x_1 and y_2 are not co-possible, and there is a failure of distinctness. Notice that using (**IF**) to reach this conclusion does not require the assumption that saying hello is a "part" or a "constituent" of saying hello loudly (whatever that might mean), although it does require a judgment that it is in the relevant sense not possible (presumably because of logical or semantic relationships) to say hello loudly without saying hello.[16] As another illustration, position and momentum for an individual particle satisfy independent fixability (and hence are distinct), even though some may find it tempting to argue that there is a logical or part–whole relation between these variables (since momentum is the product of mass and the time derivative of position). As this last example illustrates, (**IF**) does not depend on our being able to make sense of constitutive or part–whole relations among variables and does not always yield the same conclusions as this last notion.

If we apply (**IF**) to the HH model, the question we should ask is whether V, the putative top-down cause, and the channel conductances and ionic currents, the putative effects of V, are distinct in a way that satisfies (**IF**). The answer to this question is "yes." First V is clearly manipulable in a way that is independent of the values taken by the conductances or ionic currents. This does not require questionable judgments about noncausal forms of possibility: it is shown by some of the experiments that were used by Hodgkin and Huxley to establish their model. These involved the use of a newly invented device called a "voltage clamp." This enabled the experimenters to impose a stable potential difference (at various levels they were able to choose) across the cell membrane in a way that depended only on the value set by the clamp. The clamp thus functioned as an (arrow-breaking) intervention device, with the membrane potential difference fixed by the device rather than by such endogenous causes as the operation of the ion channels. This allowed the experimenters to see and investigate (isolate) the effect of V on the ionic currents and conductances in a way that confirmed the predictions of the HH model. Similarly, various molecular interventions are possible that alter the individual ionic channel currents and conductances independently of V when the clamp device is used, and these again show behavior in accord with the HH model.

I claim that this independent manipulability, as captured by (**IF**), suffices to show that it is legitimate to think of V and the channel currents and conductances as sufficiently distinct to stand in causal relationships. A similar analysis applies to the other examples in section 9.4. More generally, the fact that claims of top-down causation often involve claims that variables that are predicated of wholes causally affect variables that are predicted of parts of those wholes is consistent with those variables being sufficiently distinct in the sense of **IF** to stand in causal relationships.

9.7 Causal Cycles

I noted above that the HH model appears to involve a causal cycle, at least if we confine ourselves to the variables employed in the model. A similar observation holds for many other putative examples of downward causation; often (not always) when an upper-level variable U is claimed to act on lower-level variable L, the value of U (perhaps at some later time) will result (causally) from the action of lower-level variables. For example, the position of a monkey in a dominance hierarchy causally affects the animal's serotonin level, but that level in turn affects position in the hierarchy—something that can be demonstrated by exogenously increasing an animal's serotonin level pharmaceutically with the result that the animal rises in the hierarchy.

One possible response to worries about cyclicity is to say that "underlying" any cyclic graph is a model with time-indexed variables with temporal lags that is acyclic. If these temporal lags (or the difference between the values of X_t and X_{t+1}) do not matter to the effects we are trying to capture, then the use of a cyclical representation may be unproblematic. For example, in the HH model, the response of the ion channel conductances to a change in voltage across the cell membrane (or, more accurately, to a change in the membrane at some distance from the channels) is not instantaneous, although this fact is not represented in the HH model. It is plausible that this temporal delay makes no difference to the generic shape of the action potential, which is why the model is successful despite omitting such information.

The issues around how to interpret graphs with cycles are complex, and I don't claim that time-indexing is always a satisfactory treatment. (Rather, different cases require different treatments, and there are a number of subtleties that I lack space to discuss.) It is worth noting, however, that graphs with cycles sometimes have a straightforward interventionist interpretation along the "usual arrow-breaking" lines. That is, one way of interpreting a bidirectional graph

$$U \leftrightarrow L \tag{1}$$

is as follows: (1) if we were to intervene on U, this would break the arrow directed into U from L while preserving the arrow directed out of U into L, thus replacing (1) with the following structure:

$$U \rightarrow L \tag{2}$$

If this interpretation correctly describes the causal facts, one would expect L to change as indicated under this intervention on U. Moreover, if one were to intervene on L, this would break the arrow from U directed into L, while preserving the arrow from L into U so that under this intervention, (2) is replaced with

$$L \rightarrow U \tag{3}$$

Again, if (3) is correct, U should change under this intervention on L.

In fact, as already noted, this is essentially what was done as part of the experimental confirmation of the HH model. The use of the voltage clamp constitutes an arrow-breaking intervention on V, and one looks to see whether under such an intervention on V, the channel conductances and currents respond in the way described by the equations, which they in fact do. Similarly, interventions on the channel conductances and currents followed by measurement of V can establish the existence of a causal relationship running upward from these to V. Thinking of these results as implied by (1) thus provides a coherent interpretation of that graph. In the case of the relationship between status and serotonin levels, we can take a monkey with currently low status and move him to another less competitive troop where, because of his abilities, he will rise to a higher status (suppose he is bigger than all of the monkeys in the second troop and smaller than many in the first troop) and observe the predicted increase in serotonin levels, which provides evidence for top down $U \rightarrow L$ causation. As noted above, we can also change his serotonin levels by an exogenous pharmaceutical intervention and observe the resulting change in his status. For any given monkey, his serotonin level and his status in the absence of such interventions or after the effects of the interventions have been allowed to equilibrate will presumably reflect the joint operation of processes operating in both causal directions, in which case the bidirectional graph may be particularly appropriate.

It is sometimes claimed that the cyclic graphs are inconsistent with the directionality or asymmetry of causal relationships. It seems to me that this conflates two issues. It is plausible that causal claims have a kind of directionality built into them, and this mandates the use of directed (rather than undirected) graphs to represent such relationships. However, a graph can be directed while still containing cycles.

9.8   Conditional Independence

In section 9.2, I briefly introduced a notion of conditional independence. In this section, I spell this notion out in more detail and relate it to my previous discussion, explaining how it bears on notions of level and downward causation.

I begin with an "ideal" case.[17] Suppose that we have a set of variables L_i with very high dimensionality that are causally relevant (by the standard interventionist criterion of relevance) to some explanandum E (or set of explananda E_i), which may be either upper or lower level. Suppose also there is a set of upper-level variables U_k of much smaller dimensionality with the following property: interventions on the values of U_k are also causally relevant to the E and, furthermore, conditional on the values of the U_k when these are fixed by interventions, further variations in the values of the L_i, produced by independent interventions, make no difference to (are irrelevant to) the values of E. In a bit more detail, let us say that the variables L_i are *unconditionally relevant* (alternatively, irrelevant or independent) to E if there are some (no) changes in the values of each L_i when produced by interventions that are associated with changes in E. Variables can be unconditionally relevant but *conditionally irrelevant*. A set of variables L_i is irrelevant to variable E *conditional* on additional variables U_k (conditionally irrelevant to E) if the L_i are unconditionally relevant to E, the U_k are unconditionally relevant to E, *and* conditional on the values of U_k when these are fixed by interventions, changes in the value of L_i produced by interventions and consistent with these values for U_k are irrelevant to E. Less precisely but perhaps more intuitively, the U_k "screen off" the L_i from E, when screening off is interpreted in terms of interventionist counterfactuals rather than in terms of conditional probabilistic independence.[18] If it is possible to find such a set of variables U_k (and perhaps also if they meet certain additional conditions of the sort gestured at in footnote 7), we can replace the L_i with them insofar as we are just interested in describing difference-making relationships bearing on E—that is, in identifying those variables' variations in which make a difference for E. The U_k do just as good a job as the L_i in this respect. And of course, identifying such difference-making relationships is what explanation and causal analysis is all about according to the interventionist.

Examples that look roughly like this are quite common. To return to an example mentioned briefly in section 9.2, suppose we are interested in explaining the macroscopic behavior of a gas as characterized by such variables as temperature, pressure, and volume A given temperature t for the gas will correspond to or can be realized by any one of a very large number of collections of molecules with different positions and momenta—six such variables for each molecule in the gas, so that this variable has over 10^{24} independent dimensions. But (except for a measure zero set of cases) the impact of any of these profiles on the macroscopic variables depends entirely on their aggregate or average behavior, which is summarized by the values of the thermodynamic variables. Given the values of the macroscopic variables, further details having to do with the exact positions and momenta of the individual molecules are conditionally irrelevant to many aspects of the behavior of the gas.

Similarly, suppose, as the HH model in effect claims, that as an empirical matter, given the value of the overall membrane potential V, further lower-level detail captured by lower-level variables (e.g., variables describing the fields associated with the individual atoms and molecules making up the membrane) is conditionally irrelevant to the shape of the action potential, the gating behavior of the ion channels, and so on.[19] To the extent this is true, we may legitimately appeal just to V to explain these explananda. Under these conditions, V is a legitimate downward cause.

9.9   Levels and Conditional Independence

How does all this relate to assignments of levels? In the case of the gas, we have a rationale for treating the thermodynamic variables as at the "same" level, since the right sort of conditional independence relation holds among them and separates them from "lower-level" information about the position and momenta of individual molecules. In addition, compositional considerations reinforce this assignment of levels. In contrast, in the case of the HH model, compositional and perhaps other considerations suggest that the causal relata (V and the channel conductances) are at different "levels," even though there is interaction between these levels. But the underlying logic is the same: it is legitimate to treat (in the right sort of setup) temperature as a cause of pressure because a given value of temperature has a uniform effect on other thermodynamic variables like pressure (uniform in the sense that given that value, further variations in molecular arrangements realizing the temperature make no difference to those thermodynamic variables), and it is legitimate to treat V rather than some more detailed description of the potential differences resulting from the exact arrangement of electrons along the cell membrane as a cause of aspects of the behavior of the ion channels because the different realizations of V have a similarly uniform effect on that behavior.

Several additional remarks may help to clarify how the conditional independence idea is to be understood. Note first that it is relativized to a particular target explanandum (or perhaps a set of these). Variables L may be independent of explanandum E conditional on the values of variables U but L may not be independent of some other explanandum E^* conditional on U, so that if we wish to account for E^*, we do need to take the values of L into account. For example, if we wish to account for facts about the specific heats of gas, we must advert to quantum mechanical considerations rather than to macroscopic variables like pressure and temperature. Similarly, if our target explanandum is the overall shape of action potential, then conditional on the variables employed in the HH model, further information about molecular details may be irrelevant, but if we wish to explain other features of the system such as the behavior of dendritic trees, this particular conditional independence relation will no longer hold (cf. Herz et al., 2006).

In practice, as some of the examples already discussed suggest, there are often natural groupings of variables for which the same conditional independence relations hold—for example, conditional independence of various explananda characterized in terms of thermodynamic variables from lower-level molecular detail holds conditional on other thermodynamic variables, so that these variables form a natural grouping. This is one basis on which we group whole sets of variables into "levels."

Second, note the form taken by the conditional independence justification of the use of upper-level variables. It is common in the philosophical literature for defenses of upper-level causal claims or explanations to attempt to show that such explanations are *superior* to explanations in terms of lower-level variables and, moreover, superior in a way that is completely independent of "pragmatic" considerations having to do with human epistemic and calculational limitations.[20] The conditional independence justification does *not* claim this. Rather, what it attempts to do is to identify conditions under which it is *permissible* or *legitimate* to employ upper-level variables—permissible in the sense that this can be done without explanatory loss. There is no claim that an explanation in terms of lower-level variables (if we could produce one, which is frequently not the case[21]) would be inferior to an explanation in terms of upper-level variables.

Third, the condition described above, involving complete irrelevance of the lower-level variables to certain explananda conditional on the values of the upper-level variables, is obviously a kind of limiting case, although it is arguably not as rare as some philosophers suppose. The requirement of complete irrelevance may be relaxed in various ways.[22] We might require instead that the L_i be irrelevant to E conditional on U_k for most or "almost all" values of these variables or for values of those variables that are most likely to occur (perhaps around here right now). We might require that for those values of L_i for which exact conditional irrelevance fails, near-conditional irrelevance or independence holds—most, even if not all, of the variance in E explained by L_i is explained by U_k. We might think in terms of conditional irrelevance holding on some appropriate time or spatial or energy scale, even if not on others—for example, perhaps there are very fast variations in L_i occurring on a very fine-grained temporal scale that can make a difference to E even conditional on the values of the upper-level variables but the L_i very quickly settle down to constant equilibrium values that have an upper-level representation for which conditional irrelevance holds.[23]

My argument so far has been that considerations about conditional independence can be invoked to explain why it is permissible or legitimate to formulate causal claims in terms of upper-level variables, including causal claims that involve lower-level variables as effects—we may lose little or nothing by doing so in terms of the identification of difference-making factors for the effects in question. Moreover, such considerations provide one important basis for grouping variables into levels and for understanding when it is legitimate to collapse lower-level variables into more coarse-grained upper-level variables with fewer degrees of freedom. However, of course there is more to the story about why we actually employ such upper-level variables. It is at this point that various sorts of limitations of us humans (and perhaps all bounded agents) come into the picture. Some of these are calculational—we can't solve the 10^{23} body problem of calculating bottom up from the behavior of individual molecules to the aggregate behavior of the gas. In addition, we face the epistemic problem that we are unable to make the kinds of fine-grained measurements that would be required for such calculations to reach reliable results.[24]

To take another example, although there are fine-grained neural models employing large numbers (up to 1,000) of individual "compartments" (each of which represents a distinct circuit structure for a small portion of the neuron) that can be used to account for aspects of dendritic behavior and the role of neuronal spatial structure, these multicompartment models cannot, for reasons of computational tractability, be "aggregated up" to produce a model of the whole neuron. For that we require a different model like the HH model, which is a "single-compartment" model that neglects much of the spatial structure of the neuron but nonetheless is adequate to explain the overall shape of the action potential.[25] We thus find that not only is it permissible to formulate theories in terms of upper-level variables if we wish to explain certain explananda but that we often have no choice but to do this if we want models that are tractable or that we can calculate with. Put differently, we are sometimes in the fortunate situation that nature presents us with relations of conditional irrelevance/independence that we can then exploit to construct tractable models that would not otherwise be possible. When we build models and theories that exploit these opportunities, they will be structures in which upper-level causation appears. Note that although such computational considerations may reflect, at least in part, facts about us, the facts about conditional independence or near approximations to it which they exploit have to do with what nature is like—the latter are not just reflections of our computational limitations.

Considerations having to do with calculational and epistemic constraints of the sort just described thus represent another set of considerations (briefly described under 9.2.3 above) that influences judgments about levels. Among other considerations, models at different levels of detail may employ very different varieties of mathematical description that are difficult if not impossible to stitch together smoothly with the consequence that we cannot straightforwardly extend models and theories that are successful in accounting for the behavior of systems at certain scales (or levels of detail) to behavior at other scales or levels of detail. For example, models at one level may employ partial differential equations (which may be used to capture the role of spatial structure), at another level ordinary differential equations (which may abstract away from spatial structure), at another level Boolean or structural equations (which will neglect the underlying dynamics described by the differential equations), and at still another level Bayesian representations in terms of probability theory.[26] Thus, we often end up with situations in which each of a variety of different kinds of models have their own distinctive explananda, which they account for, and other possible explananda, which they cannot explain, either because the right sorts of conditional independence relations do not hold or for computational reasons or both. Again, this encourages us to think of such situations in terms of a separation into levels. The problem we then face is getting these levels to "talk to one another" when conditional independence partially fails.[27]

To relate these ideas to my earlier discussion of downward causation, consider the question of why we have a notion of downward causation at all and regard claims of downward causation as sometimes legitimate rather than (as some skeptics claim we should) insisting that the only true literally causal claims are those that relate variables that are all at the same lower level. My answer is that (1) when the candidate top-down cause has a uniform effect on some other variable regardless of how it is realized, we lose nothing by describing the situation in terms of upper-level rather than lower-level variables, and (2) for computational reasons, we may not be able to formulate an account of the effect in terms of lower-level variables in any case.

Finally, a brief remark about "autonomy." This word is used in many ways, but one natural meaning is that a framework or theory is autonomous or relatively so to the extent that one doesn't need information coming from some other theory or level to adequately model some range of phenomena. For example, as Batterman observes in his chapter for this volume, the Navier–Stokes equations are autonomous with respect to many explananda concerning fluid behavior in the sense that they account for those explananda without requiring information about the molecular details of the fluid. Obviously, autonomy, when so understood, is a relative or conditional notion in several senses. A theory may be autonomous in its ability to account for one set of explananda but not others, as Batterman also observes. A theory may need a whole lot of information from some other source to be adequate, or it may need relatively little information, perhaps of a very nondetailed generic sort. To pick up on the quotation at the beginning of Sara Green's chapter, certain generic facts about cars and the behavior of their drivers may be relevant to modeling traffic flow but not the details of the working of the internal combustion engine. I think of my remarks about conditional independence as one possible way of capturing these ideas.[28]

9.10 Conclusion

My remarks in the previous section attempt to provide a partial answer to the questions about the function of level talk in science posed in section 9.1. In addition to the role played by compositional considerations, such talk can be motivated by empirical facts about conditional independence and by considerations having to do with what it is possible to represent and calculate using various sorts of mathematical models. I see these factors as working together and interacting—the conditional independence facts create niches or opportunities for computationally tractable models, which succeed in explaining certain effects in virtue of abstracting away from certain conditionally irrelevant factors at more fine-grained levels of analysis.[29] In addition, ideas about levels can also play the useful heuristic role of providing plausibility arguments to theorists or modelers about which factors they may be able to ignore, prior to the construction of detailed models, although of course such arguments always need to be checked empirically. These are all considerations that help to explain why it is sometimes justifiable and indeed salutary and advantageous to make use of level-based arguments and reasoning strategies.

That said, we should also recognize the following complicating (and sometimes countervailing) considerations, which suggest caution about too much reliance on level-based considerations: first, finding cases in which conditional independence holds even approximately is (at least typically) not easy—it requires finding the "right" variables and the right strategies for representing the impact of high-dimensional variables in lower-dimensional ways. In some domains of inquiry, it may not be possible to find such variables at all (or at least variables that are well behaved in the sense of being cognizable and measurable)—instead, many different variables that we think of as at very different levels (where this is assessed in terms of compositional or other considerations) may all matter to the effects we are trying to explain, so that there is extensive "causal leakage"[30] across levels.[31] This does not make explanation impossible, but it certainly makes it more difficult. In such cases, heuristics based in ideas about sharp separations of levels can mislead.

We should also bear in mind that, as urged earlier, conditional independence facts are explanandum-relative—from the fact we can legitimately neglect certain factors in accounting for certain explananda E, it does not follow that we can legitimately neglect those factors in accounting for some other explanandum E^*, even if E^* seems intuitively similar to E or at the same level according to some notion of level such as one based on compositional considerations. Whether we can legitimately neglect such factors in accounting for E^* is always an empirical issue, which cannot be settled a priori. We should be particularly sensitive to the possibility that conflations among different notions of level can lead us to assume conditional independence in cases in which it is not warranted, as when we assume that composition-based differences in level automatically warrant assumptions about conditional independence.

Acknowledgments

Thanks to Sara Green, Bob Batterman, and Bill Wimsatt for helpful comments on an earlier version. Green's chapter in the present volume as well as Green and Batterman (2017) and Green (2018) provide many additional examples of downward causation and of modeling across levels (or scales) in biology. Batterman's chapter in this volume provides a number of illustrations of how talk of levels is tied to claims about scale separation and relative informational autonomy (closely connected to what I call conditional independence) and how, at the same time, in multiscale modeling, it is important to understand how information can be passed across scales. I see this work as complementing my own discussion.

Notes

1. The extent to which different criteria for level assignment lead to largely the same results (or not) is an important question on which I touch only in passing. There is a range of possible positions. One might think that, properly understood, different notions of level or criteria for level assignment produce judgments about levels that largely coincide. I believe this may be Bill Wimsatt's view. At the other extreme, one might think that the different criteria lead to results that diverge so much that they render talk of levels useless and misleading. My view is somewhere in the middle between these two possibilities but closer to Wimsatt's views than those of the complete level skeptics. Thanks to Wimsatt for pushing me on this point.

2. Space precludes detailed discussion of Wimsatt's rich and hugely influential early discussion of levels (e.g., Wimsatt, 1994), which includes all of the possibilities I discuss and much more. As will be apparent from his chapter in this volume, I share with Wimsatt the view that levels are (sometimes) real features of nature as well as his emphasis on the roles of independence and causal interaction in delimiting levels. I also agree that notions of level can be analytically very useful both in scientific theorizing itself and in philosophical reflection on science.

3. Some writers, such as Craver and Bechtel, also think of properties or "activities" as related by "compositional relations." I regard this as problematic, for reasons described below.

4. This notion of level as tied to independence (or near or relative independence) is also, I think, the primary notion motivating Simon's notion of near-decomposability, discussed in Wimsatt's chapter in this volume.

5. One of the simplest possibilities for such aggregation is some form of averaging as in the thermodynamic example mentioned immediately below. It is important to understand, however, that there are much more complicated possibilities, including in particular forms of aggregation that take into account or represent information about spatial or temporal correlations. These are discussed in Batterman's chapter in this volume.

6. A common illustration: the usual thermodynamic notion of temperature of a gas is only applicable to a collection of molecules at equilibrium—it is not well defined for an individual molecule. Similarly, in connection with the Hodgkin–Huxley model in section 9.5, the membrane potential is only defined (and measurable) as a feature of the whole membrane.

7. Ellis (2016) imposes a similar condition.

8. Realization is thus a relation between the *values* of an upper-level variable and various *values* of a lower-level variable, with many different values of the latter mapped into a single value of the former. It is *not* a relation between an upper-level variable and many lower-level variables. Realization is present when values of lower-level variables are averaged to yield a value for an upper-level variable, but as noted above, averaging is not the only form that realization can take.

9. Recall that this is shorthand for whatever in the world corresponds to variables or values of variables.

10. For more on variables and values of variables, see Woodward (2015, 2016, in press).

11. I believe the conditions described are necessary for downward causation, but I doubt that they are jointly sufficient. Although I lack space for detailed discussion, there is a natural candidate for an additional condition that is motivated by the fact that it seems possible for a variable to satisfy the conditions above and yet to be highly distributed, noncompact, or not simply connected and to not correspond to anything that we could measure or manipulate by upper-level measurement and manipulation procedures. The additional condition would require that the upper-level variable not have this character—intuitively, that it exhibit a certain kind of coherence, as when placing a gas in a heat bath has a coherent, coordinated effect on its component molecules. I won't try to explore this idea further, since I don't know how to state it precisely, and in any case, this additional condition seems to be satisfied in all of the examples of downward causation discussed below.

12. Valuable recent discussions of downward causation with many additional examples include Ellis (2016) and Noble (2006).

13. The heating and cooling affect the whole sword, not just components of it.

14. How does Heil's claim fit with the existence of energy cascades described in **4.4**? Heil objects to whole–part causation, but, to anticipate my discussion below, this is irrelevant to **4.4** since the transfer of energy from larger to smaller eddies takes time, so that the relation between the latter and the former is not a synchronic part–whole relationship.

15. Craver does attempt to elucidate what it is for one "activity" to be "constitutively relevant" to another by appealing to a "mutual manipulability" criterion. I lack space for discussion, but notice that the relationship between V and the channel conductances appears to satisfy Craver's criterion, which implies (in my view mistakenly) that the relationship between them is constitutive rather than causal. Moreover, systems with causal cycles appear to satisfy Craver's criterion, even though they involve causal relationships.

16. Older readers may remember "logical connection" arguments that claimed to show that desires and beliefs were "logically connected" to associated actions and hence not sufficiently distinct to serve as causes of them—for example, the desire D to drink beer could not cause drinking beer B because of a "logical connection" between

the two. A consensus eventually emerged that this was a flawed argument—despite the alleged logical connection between D and B, D can cause B. I see this as illustrating the dangers of relying on unclear ideas about logical connection and overlap in trying to elucidate distinctness among variables. Note that **IF** yields the correct judgment about this case—all of the different values of D and B are compossible: one can have the desire to drink beer without drinking, one can drink without the desire (e.g., out of a feeling of social obligation), and so on. Of course, if it is claimed that whenever one drinks, it follows a priori one has the desire (i.e., that drinking without the desire is excluded on conceptual grounds), **(IF)** will yield the conclusion that the relation is not causal, but this seems the correct assessment.

17. For a closely related set of ideas, see Chalupka et al. (2017). What follows has been substantially influenced by this paper and by discussion with Frederick Eberhardt.

18. In other words, we are to imagine that the value of $U_k = u$ is fixed by an intervention, while the value of L_i is set via interventions to any value that is consistent with u. If conditional independence holds, these further variations in L_i should have no influence on E.

19. Note what this claim says. It does *not* say that there are no local variations in the potential; it says that they do not matter for the explananda of interest.

20. For recent claims of this sort, see Weslake (2010) and Franklin-Hall (2016).

21. See my discussion below.

22. Of course, to the extent that we do this, we allow for manipulations that are in some respects ambiguous, so there is a trade-off around these considerations.

23. It is an important general fact about independence and conditional independence relations that they can be scale or grain relative in the sense that switching to different temporal or length scales or adopting certain procedures for aggregating lower-level variables can replace situations in which variables are dependent with situations in which related upper-level variables are independent or conditionally independent. For example, X might be correlated with Y on a very long time scale, but if relatively short time scales are relevant to the behavior of interest, it may be appropriate to treat X as constant, in which case it will be independent of Y.

24. Wimsatt makes the important point in correspondence that in the biological realm, such computational and epistemic limitations influence what are the correct causal relations and not just how we model these. If an organism can only perceptually detect relatively coarse-grained differences in, say, a prey or predator because of such limitations, then it is these coarse-grained features that causally affect behavior, rather than some finer-grained variable. My flight behavior is causally sensitive just to whether the animal before me is a tiger, rather than to fine-grained details of molecular realization.

25. Cf. Herz et al. (2006): "Single-compartment models such as the classic Hodgkin-Huxley model neglect the neuron's spatial structure and focus entirely on how its various ionic currents contribute to subthreshold behavior and spike generation. These models have led to a quantitative understanding of many dynamical phenomena including phasic spiking, bursting, and spike-frequency adaptation."

This short paper is very interesting in its discussion of neuronal models at different "levels" and the way in which models at each level are able to capture some aspects of neuronal behavior and not others and how "abstraction" (which basically amounts to neglect of certain features of the neuron which are irrelevant to the behavior one is trying to understand) can lead to models that can account for aspects of higher level neuronal behavior that, for computational reasons, cannot be captured by lower level models.

26. Thus, in Herz et al.'s (2006) catalog of different kinds of models, their level 1 to 3 models employ various kinds of differential equations while their level 5 models black box entire neurons and treat them as computing via Bayesian updating.

27. Some strategies for dealing with this problem are discussed in detail in Batterman's chapter in this volume.

28. I cannot resist two further remarks. First, ideas related to conditional independence might be used to capture part of what may be meant by talk of "emergence" and in a way that renders that notion unmysterious. That certain lower-level information is conditionally irrelevant to certain explananda should not be metaphysically puzzling and does not by itself imply the explananda are inexplicable in terms of the lower-level information. Second, traditionally, discussions of autonomy have been closely bound up with issues about reduction. Fodor (e.g., 1974), for example, says that psychology is autonomous to the extent that it is not reducible to neurobiology, where by "reduction," he has in mind something like Nagelian reduction. Understanding autonomy in terms of conditional independence does not map onto Fodor's picture in any simple way. If a psychological theory is fully type reducible to a true neurobiological theory, this would presumably mean that the psychological theory by itself was fully adequate in accounting for the psychological phenomena it was meant to explain since all the generalizations of the psychological theory follow from the true neurobiological theory. In this case, the psychological theory would be autonomous in the sense I propose, despite being reducible. Similarly, although multiple realizability is often taken to undermine the possibility of Nagelian reduction, it is compatible with autonomy in the sense described above. (Multiple realizability is also compatible with failure of autonomy if realization independence fails, as it can even when there is multiple realizability.) If, on the other hand, the psychological theory was empirically inadequate by itself and required extensive supplementation or correction by neurobio-

logical information in order to be adequate, the psychological theory would not be reducible to the neurobiological theory, but it wouldn't be autonomous either. What this shows is that the extent to which the relevant information in a lower-level theory is captured by the categories in an upper-level theory is very different from whatever is captured by Nagelian reduction.

29. This notion of computational opportunities arising from possibilities of avoiding modeling in detail various aspects of the systems with which we deal is developed in very rich detail in Wilson (2017).

30. See Wimsatt (1994), who attributes the phrase to Stuart Glennan.

31. As an illustration, consider the model of the causation of major depression in men developed in Kendler et al. (2006). This employs variables spanning many different levels: genetic risk, personality variables such as low self-esteem, conduct-related variables such as substance abuse, and social or environmental variables having to do with early parental loss, for example. None of these "screen off" the effects of the others on depression.

References

Bechtel, W., & Craver, C. (2007). Top-down causation without top-down causes. *Biology and Philosophy, 22*, 547–563.

Chalupka, K., Eberhardt, F., & Perona, P. (2017). Causal feature learning: An overview. *Behaviormetrika, 44*, 137–164.

Clark, S., & Lancaster, T. (2017). The use of downward causation in condensed matter physics. In M. Paleotti & F. Orilia (Eds.), *Philosophical and scientific perspectives on downward causation* (pp. 131–145). New York, NY: Routledge.

Ellis, G. (2016). *How can physics underlie the mind? Top-down causation in the human context.* Berlin, Germany: Springer.

Eronen, M. I. (2013). No levels, no problems: Downward causation in neuroscience. *Philosophy of Science, 80*(5), 1042–1052.

Fodor, J. (1974). Special sciences (Or: The disunity of science as a working hypothesis). *Synthese, 28*: 97–115.

Franklin-Hall, L. (2016). High level explanation and the interventionist's "variables problem." *British Journal for the Philosophy of Science, 67*, 553–577.

Green, S. (2018). Scale dependency and downward causation in biology. *Philosophy of Science, 85*(5), 998–1011.

Green, S., & Batterman, R. (2017). Biology meets physics: Reductionism and multi-scale modeling of morphogenesis. *Studies in History and Philosophy of the Biological and Biomedical Sciences, 61*, 20–34.

Heil, J. (2017). Downward causation. In M. Paoletti & F. Orilia (Eds.), *Philosophical and scientific perspectives on downward causation* (pp. 42–53). New York, NY: Routledge.

Herz, A., Gollisch, Y., Machens, C., & Jaeger, D. (2006). Modeling single-neuron dynamics and computations: A balance of detail and abstraction. *Science, 314*, 80–85.

Kendler, K., Gardner, C., & Prescott, C. (2006) Toward a comprehensive model for major depression in men? *American Journal of Psychiatry, 163*, 115–124.

Lewis, D. (2000). Causation as influence. In J. Collins, N. Hall, & L. Paul (Eds.), *Causation and counterfactuals* (pp. 75–106). Cambridge, MA: MIT Press.

Noble, D. (2006). *The music of life.* Oxford, UK: Oxford University Press.

Oppenheim, P., & Putnam, H. (1958). The unity of science as a working hypothesis. In H. Feigl, M. Scriven, & G. Maxwell (Eds.), *Concepts, theories, and the mind-body problem* (pp. 3–36). Minneapolis: University of Minnesota Press.

Sprites, P., & Scheines, R. 2004. Causal inference of ambiguous manipulations. *Philosophy of Science, 71*(5): 833–845.

Weslake, B. (2010). Explanatory depth. *Philosophy of Science, 77*, 273–294.

Wilson, M. (2017). *Physics avoidance: Essays in conceptual strategy.* Oxford, UK: Oxford University Press.

Wimsatt, W. (1994). The ontology of complex systems: Levels of organization, perspectives, and causal thickets. *Canadian Journal of Philosophy, 20*, 207–274.

Woodward, J. (2015). Interventionism and causal exclusion. *Philosophy and Phenomenological Research, 91*, 303–347.

Woodward, J. (2016). The problem of variable choice. *Synthese, 193*, 1047–1072.

Woodward, J. (in press). Explanatory autonomy: The role of proportionality, stability, and conditional irrelevance. *Synthese.*

10 Cancer beyond Genetics: On the Practical Implications of Downward Causation

Sara Green

Cancer is no more a disease of cells than a traffic jam is a disease of cars. A lifetime study of the internal combustion engine would not help anyone understand our traffic problems.
—Smithers (1962)

Overview

Discussions about reductionism and downward causation are often assumed to be primarily of interest to philosophers. Often, however, the question of whether multiscale systems can be understood "bottom-up" has important practical implications for scientific inquiry. Cancer research, I argue, is one such example. While the focus on genetic factors has intensified with recent investments in cancer genomics, the importance of biomechanical factors within the tumor microenvironment is increasingly acknowledged. I suggest that the role of solid-state tissue properties in tumor progression can be interpreted as a form of downward causation, understood as *constraining relations* between tissue-scale and microscale variables. Experimental demonstrations of these sorts of influences reveal limitations of reductionist accounts and expose the dangers of what Wimsatt calls *functional localization fallacies*. Such fallacies relate to the common bias of downgrading factors that—as a practical necessity—are left out of scientific analysis. Any heuristic, experimental or theoretical, involves foregrounding some aspects while ignoring others, and the complexity of cancer leaves room for the coexistence of many different partial perspectives. These perspectives are not reducible to one another, but neither do they in this case make up a neatly integrated "causal mosaic" of different influences. At present, the picture of cancer research looks more like a fragmented cubist painting in need of a more balanced attention to difference-making factors at higher levels or scales.

10.1   Introduction

Debates on the characteristics of cancer are as tangled today as they were in the beginning of the twentieth century (Baker, 2012; Plimmer, 1903). The question about the appropriate level of analysis is perhaps even more pressing in an era of large-scale investments in cancer genomics. Whereas some praise the potentials of genomics for understanding, stratifying, and treating different cancer types, others call for more attention to higher-level dynamics and organization of tissues. At stake are not only theoretical controversies but also issues with practical implications for experimental research and therapeutic interventions.

From a biological perspective, physics is often seen as a discipline aiming for reductionist (or fundamental) explanations. I challenge this view, and reductionism more generally, by focusing on the role of biomechanical features within the *tumor microenvironment* (TME) for cancer development (Laplane et al., 2018). I show how solid-state tissue properties in tumor progression support the importance of macroscale features of living systems. Moreover, I suggest that biomechanical constraints can shed light on the controversial notion of "downward causation" that is often used to distinguish between reductionist and anti-reductionist approaches in philosophy of biology (Campbell, 1974; Mossio et al., 2013; Wimsatt, 1994). An examination of downward causation in the context of cancer is of relevance also to broader questions about the explanatory scope of genetics.

Before we begin the analysis, it should be noted that "biological levels" are not unambiguously delineated (McGivern, 2008; Potochnik & McGill, 2012). Levels can be defined spatially, functionally, or temporally (DiFrisco, 2017), and the notion can be sensitive to circumstances of its use such as different disciplinary contexts (Brooks & Eronen, 2018; Wimsatt, 1972). In the following, I use the term "level" when referring explicitly to part–whole relations in organisms, such as genomes, cells, or tissues (see also Kaiser, 2015; Love, this volume; Wimsatt, 2007). However, the proposed view does not depend on any particular specification of levels. I prefer the term "scale" when more broadly discussing the influence of macroscale features, which also has implications for physics (Batterman, 2012).

I begin with some background for the discussion of whether a gene-centric view on cancer is adequate (sections 10.2 and 10.3). I then examine studies emphasizing the importance of macroscale biophysical features of the TME (section 10.4). I argue that influences of physical features such as matrix stiffness can be interpreted as a form of downward causation, understood as constraining relations that can be mathematically interpreted as boundary conditions (section 10.5). Sections 10.6 and 10.7 discuss implications for philosophical debates on reductionism and reflect on the possibility of a unified multiscale approach to cancer. Section 10.8 offers a summary and concluding remarks.

10.2   Background for the Debate

The dominant view of cancer presented in almost all cancer biology textbooks is that carcinogenesis is driven by genetic instability in terms of somatic mutations that alter cell signaling pathways. Historically, this view has been called the somatic mutation theory (Vaux, 2011) or the oncogene paradigm (Plutynski, 2018b). Cancer researchers today recognize that multiple factors influence the proliferation of cancer cells (Hanahan & Weinberg, 2011; Weinberg, 2007), but most research efforts still focus on genetic factors and molecular pathways as targets of intervention and explanation. Recently, investigation of genetic factors has intensified through large-scale projects such as The Cancer Genome Atlas (TCGA, 2005–2016), which utilized automated sequencing tools and algorithms to identify genetic variants associated with different tumor types. Identification of cancer biomarkers are expected not only to increase our understanding of the diversity of cancer types but also to allow for personalized therapies (Tomczak et al., 2015).

In contrast to the optimism concerning new potentials of cancer genomics, some see investments in large-scale genome projects as "one more misstep in the war on cancer" (Miklos, 2005). Despite decades of heavy investment in research on genetic markers and molecular pathways involved in cancer development, results in terms of improved understanding and clinical control have so far been disappointing (Lazebnik, 2002, 2010; Prasad et al., 2016). Some have questioned whether "cancer cells," understood as entities with distinct molecular properties, can be meaningfully delineated at all (Soto & Sonnenschein, 2011). The skepticism is motivated by a set of difficulties in distinguishing between mutational signatures in healthy and "cancerous" cells, as well as between malignant and benign tumors (Baker, 2013; Hanahan & Weinberg, 2011). Accordingly, some have argued that cancer is a disease that has to be characterized at higher scales or levels.

The importance of higher-level features for understanding cancer has been taken up by other philosophers, specifically in discussions of the tissue organization field theory (TOFT) (Sonnenschein & Soto, 1999; Soto & Sonnenschein, 2011). This approach characterizes cancer as a failure of tissue organization. Philosophical debates have centered on the extent to which the frameworks of SMT and TOFT can be combined (Bedessem & Ruphy, 2015, 2017) or whether these are theoretically incompatible due to conflicting ontological assumptions (Baker, 2013; Bizzarri & Cucina, 2016; Montévil & Pocheville, 2017). The debate provides intriguing insights into the implications of different ontological assumptions for our understanding of cancer. However, since the SMT and TOFT can be viewed as two extremes in a continuum of approaches to cancer, I shall focus more broadly on the role of the TME (see also Plutynski, 2018a).

The TME concept can encompass different experimental and theoretical approaches to cancer that go beyond TOFT (Laplane et al., 2018). The TME concept has a long and diverse history, as it has been highlighted by immunologists and researchers focusing on angiogenesis since the 1970s (e.g., Folkman, 1971; Witz, 2009).[1] What unites these is the view that the environment of the tumor cells influences or constrains the possibilities of cell growth and proliferation. Similar insights have been stressed by developmental biologists that have experimentally demonstrated the influence of physical properties of the extracellular matrix (ECM) on development of breast cancer (Lochter & Bissell, 1995). The framework of TME allows researchers to acknowledge top-down influences without denying the importance of genetic causation (Malaterre, 2011). Yet, as we shall see in the following, research on the role of the TME does challenge the explanatory scope of genetic approaches to cancer.

10.3 Postgenomic Puzzles and Recombination Experiments

The call for a higher-level perspective on cancer has a long history in biology but is currently also motivated by what Baker (2013) calls "postgenomic puzzles." The term refers to results of tumor sequencing that reveal unexpected complexity.

Efforts to understand cancer in terms of molecular mechanisms have been motivated by successful identification of genetic difference-makers from the 1970s and onward. Examples are the tumor suppressor gene P53 and BRCA markers used in risk analysis for breast cancer. Technologies such as genome-wide association studies have further resulted in the identification of a vast number of genetic factors and molecular pathways associated with cancer development. The genetic heterogeneity of tumors may even be so vast that each cancer is different. A widely cited quote from the U.S. National Institutes of Health with the launching of TCGA reads as follows:

Cancer is a disease of the genome and as more is learned about cancer tumors, the more we are finding that each tumor has its own set of genetic changes. Understanding the genetic changes that are in cancer cells is leading to more effective treatment strategies that are tailored to the genetic profile of each patient's cancer.[2]

While such quotes frame tumor heterogeneity as a potential for individualized treatments, others interpret such discoveries as a severe challenge to the idea of associating cancer types with stable genetic markers.

Many tumors have been found to have heterogeneous mutational patterns. A few mutations are highly frequent in many cancer types, but a long tail (graphically speaking) of diverse mutations has been discovered both *between* and *within* tumors. Mutational signatures have been found to vary when biopsies are taken at different spatial locations in the tumor or within the same location over time (Blanchard & Strand, 2017). Even more puzzling is that some tumors have normal mutation rates (Greenman et al., 2007; Qiu et al., 2008), and many cells with genetic lesions do not develop into tumors (Bissell & Hines, 2011). Such results have engendered skepticism about whether cancer development can be captured in simple mechanistic models at the scale of genetic or molecular interactions. Similar cancer phenotypes can be realized through multiple genetic and molecular pathways, suggesting that it may be useful to search for higher-scale influences that bring clarity to the diverse influences.

Stromal components, such as fibroblasts, immune cells, and the extracellular matrix, can make up a large percentage of a tumor (Plutynski, 2018a, p. 25). Whereas much genetic research has focused on molecular changes within epithelial cells and their resulting behavior, the following shows how recombination experiments shift the focus to the structural role of connective tissues. In *recombination experiments*, cells are transplanted to new host organisms (or new TMEs). Recombination experiments allow for observation of the effects of cell development in response to experimental interventions where epithelial cells or the surrounding tissue are exposed to a carcinogen. Such experiments draw attention to how philosophers should not only pay attention to how scientists *represent* levels but also how they *manipulate* or *intervene* on them (see also Love, this volume; Woodward, this volume).

Recombination experiments conducted in the 1950s suggested that carcinogen-induced "cancer cells" can be normalized if transplanted into normal tissue.[3] Similarly, neoplastic induction was observed when normal cells were inserted in carcinogenous tissue. From the 1980s, systematic experiments were conducted of how features at different scales influenced development and carcinogenesis. As a consequence, Bissell and colleagues challenged the view that intercellular structures merely constituted "passive conditions" or "housekeeping functions" for the maintenance of cells (Plutynski, 2018b).

Bissell's group showed that tissue-specific traits of relevance to the development of cancer were often lost in two-dimensional (2D) cultures (petri dishes with a single layer of cells). For instance, experiments demonstrated that the fate of cells with virus-activated oncogenes depended on whether the cells were cultured in a petri dish or left in a chick embryo (Dolberg & Bissell, 1984; Stoker et al., 1990). If the cells were removed from the embryo and cultured in a petri dish, they displayed the transformed phenotype within less than a day. However, if left in the embryo, the cells developed normally until much later. This suggests that the host environment plays an important role in buffering or dampening the effects of genetic lesions. More generally, the results suggest that the features of cancer cells are highly dependent on a context of reciprocal interactions with the microenvironment (see also Nelson & Bissell, 2005, 2006).

Other historically important experiments were conducted by Ana Soto, Carlos Sonnenschein, and colleagues. They transplanted tumor cells from a donor mouse to normal tissue in mice hosts of different ages and hormonal stages (Maffini et al., 2005). If genetic instructions determined cell growth, one would expect the hosts to develop the same rate of tumors. Instead, they observed that tumor incidence varied among the hosts and that in some cases, no tumors formed at all. Similarly, from a reductionist perspective, replacement of healthy fat pads with precultured tumor cells would be expected to result in tumor development. Yet, they observed normal development of tissue structures in all hosts. These experiments support the view that neoplastic cells can be normalized, depending on environmental conditions.

The cellular environment can also direct cell fates in the opposite direction: normal cells can undergo neoplastic induction if they are transplanted to tissue environments with carcinoma-exposed stroma. Maffini et al. (2004) investigated whether cancer originates in the stroma tissue or (bottom-up) with genetic mutations in epithelial cells. They surgically separated the stroma and epithelium such that each of these could be exposed to a chemical carcinogen. Interestingly, they observed that neoplastic transformation occurred only when the *stroma* had been exposed to the carcinogen. Similar results have been obtained also in experiments investigating tumor development around kidneys (Barclay et al., 2005). Additionally, studies within developmental biology show that the biomechanical context of the stroma can play essential roles in directing the development of the mammary gland and also influence the response to oncogenic mutations (Nelson & Bissell, 2006).

The ability of cancer cells to adapt or reprogram their expression profiles in response to their environment suggests that influences from the microenvironment in some cases can overrule genetic "instructions" and direct cell fate. The following sections examine how this has important practical as well as philosophical implications.

10.4   Moving Up Levels: The Tumor Microenvironment

In a widely cited *Nature* review, Bissell and Hines (2011) ask, "Why don't we get more cancer?" The question is spurred by the observation that there are far fewer incidents of cancer than we should expect from the number of harmful mutations identified in cancer genomics. This has important implications for understanding the problem of overdiagnosis in cancer, that is, for making sense of why some genetic or cellular abnormalities do not develop into symptomatically manifest disease (Green & Vogt, 2016). A perspective beyond genetic is therefore needed to better understand the context of cancer development, including how higher-scale physical cues influence hierarchical control of cell proliferation.

Novel technologies to experimentally measure and manipulate physical cues have increased the ability of researchers to identify the effects of tissue stiffness. For instance, the development of three-dimensional (3D) laminin-rich gels to simulate the TME has allowed for systematic studies of the plasticity of cell populations that enable tumor cell reversion, depending on microenvironmental signals and tissue architecture (Nelson & Bissell, 2005, 2006). A highly cited study by Paszek and colleagues (2005) illustrates this approach. The group used an electromechanical indentor to measure the relations between matrix stiffness and tumor development among colonies of mammary epithelial cells (MECs). The first step of the analysis compared normal breast tissues to malignant tissues in transgenic mice. The analysis showed that tumors and stroma attached to tumors are much stiffer on average (4,049±938 Pa) compared to a normal mammary gland (167±31 Pa).

To test whether stiff stroma was induced by genetic lesions or whether the influence could go also in the opposite direction, Paszek et al. (2005) cultured normal MECs on gels with varying rigidity. They found that increased matrix stiffness significantly perturbed the tissue architecture (figure 10.1).

The experiment by Paszek et al. (2005) suggested that increased matrix stiffness destabilizes adhesion of cell junctions and integrins. Integrins are transmembrane ECM receptors that function as force-sensitive mechano-transducers by triggering cellular pathways. Accordingly, the group hypothesized that integrin alteration could lead to altered activity of proteins involved in growth-related molecular pathways. The study confirmed that cells with the malignant phenotype have higher activities of the G-protein Rho and associated kinases (ROCK and MLC in figure 10.2).

Figure 10.2 highlights how integrin activation can trigger other molecular pathways stimulating cell proliferation, generating feedback that results in further stiffening of the matrix. The distinction between a "soluble" and "solid-state" signaling pathway emphasizes how studies of biochemical signaling pathways and of physical properties of the matrix, respectively, are considered equally important (Huang & Ingber, 2005).

The recombination experiments discussed in section 10.3 showed that transplantation of malignant cells to normal tissues can normalize cell growth and vice versa. Similarly, Paszek et al. (2005) demonstrated that the malignant phenotype can be strengthened or reverted, depending on the stiffness of the substrate. Approaches drawing the framework of condensed matter physics have similarly revealed that increased matrix stiffness correlates strongly with the development of breast cancer (Boyd et al., 2007) and that fluid and solid stresses influence the invasive potential of tumor cells (Rankin & Frankel, 2017; West et al., 2017) as well as treatment response (Stylianopoulus, 2017). Taken together, the results suggest an interpretation of the TME as a double-edged sword that can either protect from or promote cancer, depending its physical properties (Bissell & Hines, 2011).[4]

It has recently been suggested that the physical properties of the TME offer new potential targets for cancer treatments (Bissell & Hines, 2011; Hirata & Sahai, 2017; Rankin & Frankel, 2016; Stylianopoulus, 2017). Following an initiative by the U.S. National Cancer Institute in the late 1980s, antitumor drug screening is often carried out in 2D cultures. Yet, as highlighted in section 10.3, the fate of cancer cells in response to carcinogenic or anticarcinogenic influences may differ, depending on whether the cells are cultured in a petri dish or within the environmental context of the host organism. An important clinical implication of the experimental limitations of 2D cultures is that these cannot account for the development of drug resistance, which is very common in in vivo tumors. Limitations of 2D cultures have been suggested as part of the translational challenge that only 1–2 per 10,000 candidate drugs from antitumor drug screening currently make it further than phase III studies (Feng et al., 2013).

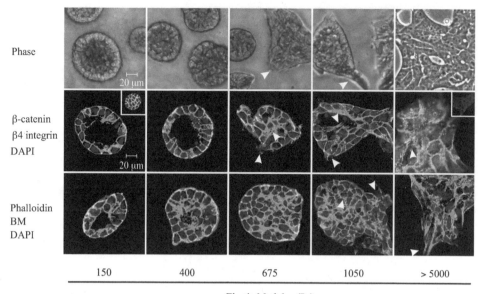

Elastic Modulus (Pa)

Figure 10.1
Phase contrast microscopy (top row) and confocal immunofluorescence (IF) images of mammary epithelium cell cultures on gels of varying stiffness, represented as increasing values for elastic modulus (a measure of how much a material will elastically deform when subjected to stress). The images show the colony morphology after twenty days. Nuclei were stained with DAPI (blue coloring). The two bottom rows show cultures grown on collagen gels and basement membrane (BM) gels, respectively, using different staining techniques. The arrows highlight how increasing matrix stiffness (left to right) leads to destabilization of adherence junctions (green staining for β-catenin) and adhesion of integrins (red staining). Figure reprinted from Paszek et al. (2005), with permission from Elsevier. A color version of the figure can be found in the online version of this chapter or in Paszek et al. (2005, figure 2C).

In contrast, 3D cultures can better account for the influence of biomechanical constraints on drug response (Nelson & Bissell, 2006; Soto et al., 2008). For instance, the response of breast cancer cell lines to two antitumor drugs, cisplatin and taxol, has been demonstrated to depend on substrate stiffness (Feng et al., 2013). Recently, 3D hyaluronan gels have been used to mimic the tumor ECM in cell invasion studies. This line of research is inspired by the discovery that the physical properties of stromal components of cancer-resistant naked mole rats differ compared to humans (reviewed in Rankin & Frankel, 2016). One of the important components of the microenvironment is hyaluronan, a glycosaminoglycan that can form gel-like structures influencing the density and fluid pressure of the ECM. Chemical modifications of hyaluronan could potentially allow for interventions on the physical properties of the ECM and subsequently affect cell proliferation and cell invasion. Elevated solid stress has also been shown to create a fluid flux from tumors to surrounding tissues, and stress-alleviating strategies have been proposed as a way to improve treatment response (Stylianopoulus, 2017).

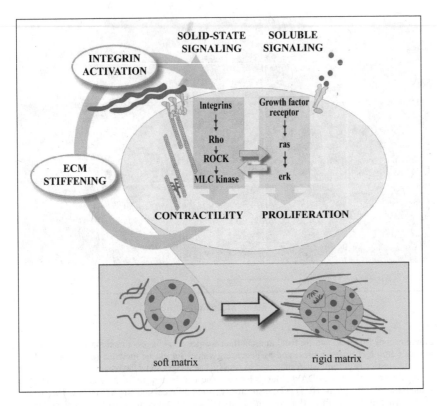

Figure 10.2
A mechanical autocrine loop that illustrates how ECM stiffening can lead to cancer development through a cascade of processes that further increases matrix rigidity. Reprinted from Huang and Ingber (2005) with permission from Elsevier.

Compared to 2D cultures, 3D cultures are more demanding in terms of experimental resources and have other limitations (see section 10.6). Yet, as argued in the following, they uncover aspects of cancer that may be interpreted as a form of downward causation.

10.5   Downward Causation

How can we make sense of the role of tissue-scale biomechanical factors in the context of cancer? I suggest that the role of tissue-scale biomechanics can be interpreted as an instance of downward causation. Importantly, however, the account defended is different from the account criticized by Kim (1998) and others, which entails strong ontological autonomy of higher-level wholes of lower-level parts. Instead, I consider downward causation as a *relation between system variables operating at different scales*. This suggestion is in line with Woodward's account (this volume) that highlights effects on lower-level variables through manipulation of high-level variables as the criteria for downward causation (see also Ellis, 2012). The experiments presented in sections 10.3 and 10.4 are examples of interventions on higher-scale factors (tissue stiffness) that influence the behavior of lower-scale variables (gene expression and cellular behaviors) and thus satisfy this criterion.

Top-down influences, in my view, are best understood as *constraining relations* that "select and delimit various types of the system's possible developments" (Emmeche et al., 2000, p. 25; see also Brooks & Eronen, 2018). This account is also inspired by systems biologist Denis Noble's suggestion that downward causation can be interpreted mathematically "as the influences of initial and boundary conditions on the solutions of the differential questions used to represent the lower level processes" (Noble, 2012, p. 55; see also Ellis, 2012). The account is developed in more detail in a separate paper on downward causation in the context of multiscale cardiac modeling (Green, 2018).[5] Here, I primarily wish to defend the explanatory importance of higher-level constraints.

By constraints, I understand features that delimit the degree of freedom of the dynamics within a given system (Hooker, 2013; Umerez & Mossio, 2013). Constraints not only limit possibilities through restraints on the possible system states but also enable certain states that would be impossible to reach for the unconstrained system. For instance, Noble highlights that the oscillating dynamics of the heart rhythm cannot be generated (or explained) without the boundary of cell structure, which creates a concentration gradient (cell voltage or cell potential) across the cell membrane. He argues that "without the downward causation from the cell potential, there is no rhythm" (Noble, 2012, p. 58).

Physical constraints are often mathematically represented as boundary conditions that impose limits on the domain of the model. These typically cannot be defined at the lowest explanatory scale. For instance, in Noble's example of the heart rhythm, the cell potential is a variable that cannot be understood or measured at the genetic level. It is a parameter that has to be measured at a higher scale (e.g., through microelectrode measurements of intact or coupled cells). Without imposing boundary conditions on the solutions of lower-scale models (in this context ordinary differential equations), the lower-scale models cannot be solved. Medium downward causation thus reveals important limitations to a reductionist perspective—by demonstrating the explanatory limitations of microscale details even for models at the lowest scale.

The requirement of higher-scale parameters is not unique to biology. Macroscale parameters are also indispensable when modeling physical behaviors, such as steady-state heat conduction in a rod (McGivern, 2008) or drop formation (Batterman, 2006; for other examples, see Batterman, 2012, 2018). In the context of multiscale cancer models, an important macroscale parameter is matrix stiffness, defined as Young's modulus (also called the elastic modulus), which outlines how much a material will deform in response to stress as a property evenly distributed over a given material's surface (Deisboeck & Stamatakos, 2011). The stiffness of a material, whether the material is steel or biological tissue, is a macroscale parameter that identifies elastic properties by treating structure as a larger continuum of matter (Green & Batterman, 2017). In our examples here, matrix stiffness is inherently a tissue-scale parameter because it depicts the physical forces acting on the *integrated effects* of cell populations that are constrained by certain geometrical structures (Davidson et al., 2009). Other examples of macroscale parameters used in continuum models of tumor dynamics are forces of cell-cell adhesion, migration velocity of cells, ECM fiber thickness, and length variation (Deisboeck & Stamatakos, 2011; West et al., 2017). Whenever the output of models or experiments targeting lower scales (e.g., gene regulatory networks) is influenced by changes in high-level variables such as tissue stiffness, I would argue that we have an instance of "downward" regulation of cell behavior.

The need for multiple approaches not only results from the complexity of biological systems but also from what, in the context of physics, has been called the "tyranny of scales" (Oden, 2006). The problem refers to the scale dependency of physical properties as well as the conceptual frameworks we understand these through. Batterman (2012) highlights that when modeling physical systems, such as a steel beam, it is not possible to describe all relevant factors in one overarching mathematical framework. Steel exhibits different physical properties at the atomic, intermediate, and macroscales, and capturing all aspects requires that different models are employed. For instance, continuum models describing elastic macroscale properties of steel treat the material as a continuum without discrete structures. But an examination of the structure of steel at a mesoscale reveals grain boundaries, cracks, and so on. Because physical behaviors and our concepts used to describe these are "multivalued" across scales (Wilson, 2012), researchers are inevitably forced to combine different mathematical frameworks to account for structures and processes across scales.

The situation is analogous in the context of cancer. Young's modulus is a continuum parameter that treats materials as if they were continuous rather than made up of discrete parts. Obviously, biological systems consist of very different parts with different properties, and such crude idealizations may seem misguided. But a striking insight is that coarse-grained models often work well, despite (and often *because of*) such simplifications (Batterman, 2018). Continuum models necessarily abstract from lower-scale details to identify parameters operating at higher scales. Importantly, and as further defended below, matrix stiffness cannot be understood simply as the aggregated effects of molecular actions (see also Batterman & Green, 2020). Relative autonomy of scales or levels is thus one of the key factors in support of downward causation. In the following, I consider and reply to potential objections to this point.

10.6   Potential Objections and Replies

A reductionist could object that since a tumor consists of nothing but molecular constituents, a more fundamental model targeting lower scales should *in principle* be sufficient. Such "in principle arguments" are, however, not particularly convincing or interesting if they are not accompanied by a clarification of how one would conduct such a reduction in practice (see also Batterman, 2018). Particularly, the reductionist has to clarify how upper-level variables are supposed to be reduced to lower-level variables.

One problem for the reduction of upper-level variables is *multiple realizability*, i.e., that a given upper-level variable has a number of different lower-level "realizers" (Woodward, this volume). An important consequence of multiple realizability is that much of the variation at a micro-level does not matter for the study of macroscale properties. Woodward refers to such cases as instances of *conditional independence*. In the context of cancer research, the dynamic state of a given cancer phenotype may be realized by multiple molecular states, and biochemical details do not always matter for prediction of the tissue-scale effects (West et al., 2017). The reductionist thus has to explain how it is possible to have this sort of explanatory autonomy (Batterman, 2018).[6]

Another important problem for the reductionist is that many higher-level variables are irreducible in the sense that they cannot be measured or conceptualized at lower scales

(Ellis, 2012). I began the chapter with a quote from Smithers (1962) arguing that cancer is no more a disease of cells than a traffic jam is a disease of cars. In both contexts, studying the constituents of the individual components, cars or cells, would not allow us to explain why the problem of cancer or traffic jams arises. Both are macroscale phenomena that we make sense of by looking at higher-level *relations* between the constituents, involving structural constraints such as roads and tissue boundaries. The examples are analogous to how the rules of a competitive sports game cannot be derived from detailed studies of the behavior of individuals (Sawyer, 2002) or how properties such as the temperature or pressure of a gas cannot be ascribed to individual gas molecules (Christiansen, 2000). The reductionist therefore has to clarify how we could possibly make sense of such macroscale phenomena without reference to higher-level concepts and parameters.

One strategy to respond to such examples in support of a reductionist view is to say that although constraints from higher-scale structures play some causally relevant role, it does not automatically grant top-down effects an *explanatory* role in science. A perspective giving explanatory priority to molecular mechanisms might, for instance, interpret the result of the experiments by Paszek et al. (2005) differently. It might be argued that the physical forces of tissue structures are just background conditions for the most important difference-makers, namely, the effects of increased Rho-activity on molecular pathways of cell cycle progression.[7]

Privileging genetic causation is motivated by an epistemic ideal in biology that emphasizes cell-autonomous or "instructive" molecular information. In cancer research, this ideal is reflected in the prioritization of "intrinsic" factors of carcinogenesis. Whereas this ideal has its merits for answering some explanatory questions, it may be counterproductive for addressing clinically relevant aspects of cancer. Methodological reduction (i.e., strategies to ignore or simplify some aspects in order to focus on others) is a necessary requirement for any scientific analysis of complex systems (Wimsatt, 2007). But if one is unaware of the existence of blind spots created through specific theoretical and experimental choices, there is a risk of committing what Wimsatt at the KLI meeting called *functional localization fallacies*. The term refers the common bias of downgrading factors that—as a practical necessity—are left out of scientific analysis. The rejection of the explanatory role of tissue biomechanics may partly result from a functional localization fallacy.

Biomechanical factors are often ignored or held fixed as a practical requirement for experimental intervention on genetic factors (Robert, 2004). Hence, the results of such experiments cannot be informative of the relative importance or (ir)relevance of biomechanical factors as these are excluded from the outset of the analysis. As mentioned, genetic difference-making is often studied through 2D cultures of cell populations with controlled and modified gene expression (e.g., via gene knockout experiments) (Vaux, 2011). Such studies, together with more recent tumor sequencing projects, have no doubt shed light on important genetic markers and pathways involved in carcinogenesis. But as it is now well established that epithelial cells behave differently if placed in a dish coated with basement membrane proteins or in cultures mimicking the TME rather than cultured on a plain plastic dish (Plutynski, 2018a, p. 41; 2018b), reference to the success of the gene-centric line of research cannot support the rejection of an explanatory role of physical features of the TME. Traditional gene-centric experimental designs are thus inherently limited when it comes to accounting for the role of macroscale physical factors in the natural environment of the cells (Soto & Sonnenschein, 2011; see Brigandt & Love, 2017, for a more general discussion of reductionism in biology).

As mentioned in section 10.4, alternative attempts involve building higher-dimensional models, such as a 3D model of the mammary gland and hyaluronan gels to mimic the TME (Nelson & Bissell, 2005; Rankin & Frankel, 2016). Moving to the tissue scale, however, typically requires that many molecular details are ignored, and the experiments are more laborious and difficult to automate. They would also become practically intractable for studying the effects, and combined effects, of hundreds or even thousands of genetic difference-makers. The key point is here that different experimental strategies foreground specific (and often complementary) aspects of a complex system, here ranging from the influence of biomechanical cues on cell differentiation to genetic difference making.

Cancer is characterized by what Wimsatt (1972) calls *interactional complexity*. Although different experimental and theoretical approaches are compatible, different ways of investigating the system often require distinct ways of demarcating and decomposing the phenomenon (Plutynski, 2018a). As a result, the different perspectives are not straightforwardly integrated. When Huang and Ingber (2005) distinguish between two forms of signaling (figure 10.2), it is because they wish to highlight that the physics perspective is not reducible to a biological perspective or vice versa. Whereas some would see the biomechanical constraints merely as background conditions, they emphasize that Paszek's interventions demonstrate the causal relevance of a "physical cue devoid of chemical specificity" (Huang & Ingber, 2005, p. 176).

The studies examined put forward the possibility that the architecture and mechanical properties of the tissues *can* be dominant over genotype, because malignant phenotypes can sometimes be reverted through interventions that alter microenvironmental factors (Bissell & Hines, 2011; Nelson & Bissel, 2006; Paszek et al., 2005). This suggests that physical factors of the TME cannot generally be considered merely as background conditions of limited explanatory and predictive value. A similar argument has been defended in the contexts of multiscale modeling in developmental biology (Green & Batterman, 2017) and systems biology (Green, 2018).

Multiscale models combine different models targeting different scales. These models are at the same time "explanatorily independent" and "epistemologically interdependent" (Potochnik, 2009). They are explanatorily independent in the sense that they legitimately ignore some aspects to address questions or properties at specific scales or levels. For instance, macroscale models are autonomous in the sense that they are often not improved by incorporation of all possible molecular details (Batterman, 2018). Similarly, microscale models typically black box many higher-scale features as these (autonomously) describe a process from a micro-level perspective.

At the same time, models targeting different scales in a multiscale modeling approach can be said to be "epistemologically interdependent" because they often rely on sources of information provided by other sources or models (Potochnik, 2009). A model at a characteristic scale often needs inputs or boundary conditions defined by models of a different spatial (or temporal) scale (Lesne, 2013). In cancer research, multiscale models must integrate continuum and discrete models to capture both large-scale volumetric tumor growth dynamics and behaviors of individual and heterogeneous cells (Deisboeck & Stamatakos, 2011). Because each framework presents a simplified picture of a system's dynamics, their applicability may depend upon factors that the models themselves do not represent or capture. Lower-scale models often require macroscale inputs and vice versa, and each perspective foregrounds some aspects while ignoring or idealizing others.

As a consequence of the tyranny of scales problem and interactional complexity, it seems futile to look for *the* origin of cancer or to debate what constitutes a "right level" of analysis. Different perspectives, addressing different scales using methods from different disciplines, can answer different types of explanatory questions. Each model addressing specific aspects at a given spatial (or temporal) scale will black box other questions or aspects at different scales. But if one is not aware of such "black boxing procedures," it may result in theoretical biases. It could be argued that the current privileging of genetic factors is an example of such a bias and that a more balanced view is needed.

10.7   Toward a Multiscale Perspective on Cancer?

Debates on cancer are often framed as a choice between alternatives (i.e., between views of cancer as a genetic or tissue-based disease). However, the concept of the TME can encompass several types of influences on cell differentiation, spanning from biochemical influences from the surrounding cells to mechanical and field forces (stiffness and surface tension) and topological geometry of the tissue (Bizzarri & Cucina, 2014; Laplane et al., 2018). Thus, an understanding of cancer across scales need not privilege any one level of causation (Bertolaso, 2011; Malaterre, 2011).

My aim in this chapter has not been to reject the merits of reductionist approaches. Not all explanatory questions require higher-level analysis, and gene-centric approaches have brought about many important insights to the development and treatment of cancer. Yet, the predictive capacities for clinical applications often depend on the willingness to embrace complexity. In other words, interactional complexity also has practical implications for how many theoretical perspectives are required to predict or explain the behavior of the system (Plutynski, 2018a; Wimsatt, 1972). Accepting a pluralistic stance should therefore not bar us from questioning the scope or sufficiency of the dominant gene-centric approach.

A consequence of specialization and complexity in science is often that narrow explanatory questions are pursued in different settings, with limited attempts to integrate the various research results. Thousands of papers are published each year on cancer, many with a focus on how specific genes or proteins contribute to a specific cancer-related pathway. But it has proven tremendously challenging to synthesize insights from these studies into clinically relevant applications (Lazebnik, 2002, 2010; Prasad et al., 2016). Accordingly, the debate on the relevant theoretical perspectives on cancer is not only philosophically interesting but also of practical importance to the design of experiments and prospects of funding strategies. As stated by one of key pioneers of the research program focused on the TME,

One of the first steps in our attempts to comprehend the big picture of tumor progression is to realize that single molecules or single signaling pathways are just solitary components of an immense network. This realization should lead to the abandonment of reductionism (which, I am afraid, is a difficult mission under the present culture of conducting science and its funding), and to the employment of approaches used in Systems Biology. (Witz, 2009, p. S13)

Similarly, Bizarri and Cucina (2014) highlight that the pharmaceutical industry, focused on the development of drugs targeting specific molecular components, is strengthening the bottom-up reductionist approach at the expense of other potentially useful approaches. Explanatory priorities regarding specific levels of analysis are therefore also related to the problem of opportunity cost (i.e., to the problem of how to prioritize investments in different solution strategies). An important job for philosophy of science could thus be to examine the merits of assumptions underlying such prioritizations of specific levels of analysis. Another could be to examine the potential or relevance of integrating different perspectives. Given the complexity of cancer, it cannot be assumed that the development of multiscale models will easily solve the translation problem.

Multiscale modeling is particularly challenging in biology, compared to modeling of many physical systems, because there are complex nonlinear feedback relations that cross-cut spatial and temporal scales. Moreover, biological systems have active boundaries that change over time as an organism responds to environmental stimuli, develops, and ages. This means that the causal effects of any type of difference-maker may be highly context sensitive. In the context of cancer, a given genetic instability, for instance, may have very different effects depending on the age stage of a person as aging involves changes in stiffness and elasticity of tissues (Bissell & Hines, 2011). Cancer is a "fast-moving target" that develops from a variety of causes, in a variety of ways, and at various levels (Bertolaso, 2011) and is hence especially challenging to model.

10.8   Summary and Concluding Remarks

Theoretical assumptions concerning the causal nature of cancer can influence choices of experimental designs to study cancer—and vice versa. Functional localization of specific features of complex systems necessarily foregrounds some aspects while backgrounding others. In cancer research, 2D cultures are efficient tools for studying effects of gene knockout and other molecular manipulations on cell proliferation. Yet, this approach is fundamentally limited in revealing environmental influences on tumor progression and drug response. Recombination experiments in vivo and studies of 3D cultures in vitro have led to important theoretical insights as well as to suggestions of new types of cancer treatment. Each framework is, however, only partial. Although they can be considered explanatorily independent (i.e., merited by particular questions of analysis), they may be considered epistemically interdependent in the sense that improving predictions of one model may rely on inputs from other models or other experimental sources.

This chapter has focused on the role of biomechanical features of the TME for cancer progression. I believe that the cases presented provide an interesting set of examples for philosophical discussions on the explanatory role and autonomy of macroscale features, as well as for discussions on the relative importance of biological and physical science approaches to cancer (see also Love et al., 2017). Contrary to the so-called layer-cake model of science, physical science approaches do not always play a theory-reducing role. On the contrary, in this context, physical science approaches highlight the importance of macroscale properties. I have argued that the influence of tissue-scale biomechanical constraints on gene expression and direction of cell fate can be interpreted as a form of downward causation. Specifically, I have defended an account of downward causation that can be mathematically interpreted as the influences of boundary conditions on the solutions of lower-scale models.

As a consequence of the interactional complexity of cancer, the results of the many efforts may not make up a neatly integrated *causal mosaic* of various factors (Love, 2017). The resulting picture may be more akin to a fragmented cubist painting where some aspects of the system are yet to become fully visible for scientific analysis.[8] While this may be seen as a disappointing conclusion, it does not mean that improvements cannot be reached. Multiscale approaches to cancer call for caution against the stubbornly persistent idea that genetic mutations constitute the most relevant or most reliable feature of cancer. A perspective from higher-scale tissue biomechanics allows for other visible patterns to materialize, which is not only theoretically intriguing but also of clinical relevance.

Acknowledgments

I thank Dan Brooks, James DiFrisco, and Bill Wimsatt for organizing the workshop on Hierarchy and Levels of Organization and the KLI for hosting a great event. A version of the chapter was presented at the Ghent-Brussels Seminars in Logic, History and Philosophy of Science, where I benefited particularly from comments from Leen de Vreese, Erik Weber, and Dingmar van Eck. Moreover, I received very helpful comments and suggestions from Bob Batterman, Anya Plutynski, Jim Woodward, Lauren Ross, Bill Bechtel, and the Philosophy of Science in Practice Group at UC San Diego. I also acknowledge The Carlsberg Foundation for research support (Semper Ardens grant CF17–0016).

Notes

1. Angiogenesis is a physiological process where new blood vessels are formed through branching from preexisting vessels. Blood supply is often a bottleneck in tumor development, and angiogenesis has recently become a target in cancer treatments.

2. The Cancer Genome Atlas Program: https://cancergenome.nih.gov/cancergenomics/impact (accessed July 30, 2018).

3. For a comprehensive review of important experiments, see Baker (2012) and Bizzarri and Cucina (2014).

4. For other studies appealing to factors beyond genetic instability, see Baker (2013), Bizzarri and Cucina (2014), Egeblad et al. (2010), Laplane et al. (2018), and Soto and Sonnenschein (2011).

5. As discussed by Mark Eronen at the meeting at the KLI, it can be debated whether one should insist on the term "downward causation" on this basis. I have no strong stance on this, but I hope to show that interpreting downward causation through constraining relations does not weaken the causal and explanatory power of top-down effects (Ellis, 2012).

6. As highlighted by Woodward, multiple realizability need not make higher-level explanations inferior, since explanatory relevance is dependent on epistemic aims that define the criteria for conditional relevance or irrelevance. Moreover, as illustrated by figure 10.2, one often needs to account for multiple causal factors operating in what Woodward calls causal cycles (i.e., where influences go in both directions). But it does present a challenge for the view that features at the lowest scale are sufficient.

7. Such an argument has been put forward in philosophical discussions about reductionism (e.g., by pointing out that macroscale features are merely background factors in molecular developmental biology) (Rosenberg, 1997).

8. I thank Leen de Vreese for suggesting the analogy of a cubist painting, inspired by the work of Caterina Marchionni, University of Helsinki.

References

Baker, S. G. (2012). Paradoxes in carcinogenesis should spur new avenues of research: An historical perspective. *Disruptive Science and Technology*, *1*(2), 100–107.

Baker, S. G. (2013). Paradox-driven cancer research. *Disruptive Science and Technology*, *1*(3), 143–148.

Barclay, W. W., Woodruff, R. D., Hall, M. C., & Cramer, S. D. (2005). A system for studying epithelial-stromal interactions reveals distinct inductive abilities of stromal cells from benign prostatic hyperplasia and prostate cancer. *Endocrinology*, *146*(1), 13–18.

Batterman, R. W. (2006). Hydrodynamics versus molecular dynamics: Intertheory relations in condensed matter physics. *Philosophy of Science*, *73*, 888–904.

Batterman, R. W. (2012). The tyranny of scales. In R. W. Batterman (Ed.), *Oxford handbook of philosophy of physics* (pp. 255–286). Oxford, UK: Oxford University Press.

Batterman, R. W. (2018). Autonomy of theories: An explanatory problem. *NOÛS*, *52*(4), 858–873.

Batterman, R. W., & Green, S. (2020). Steel and bone: Mesoscale modeling and middle-out strategies in physics and biology. *Synthese*.

Bedessem, B., & Ruphy, S. (2015). SMT or TOFT? How the two main theories of carcinogenesis are made (artificially) incompatible. *Acta Biotheoretica*, *63*(3), 257–267.

Bedessem, B., & Ruphy, S. (2017). SMT and TOFT integrable after all: A reply to Bizzarri and Cucina. *Acta Biotheoretica*, *65*(1), 81–85.

Bertolaso, M. (2011). Hierarchies and causal relationships in interpretative models of the neoplastic process. *History and Philosophy of the Life Sciences*, *33*, 515–536.

Bissell, M. J., & Hines, W. C. (2011). Why don't we get more cancer? A proposed role of the microenvironment in restraining cancer progression. *Nature Medicine*, *17*(3), 320–329.

Bizzarri, M., & Cucina, A. (2014). Tumor and the microenvironment: A chance to reframe the paradigm of carcinogenesis? *BioMed Research International*, *2014*, 934038.

Bizzarri, M., & Cucina, A. (2016). SMT and TOFT: Why and how they are opposite and incompatible paradigms. *Acta Biotheoretica*, *64*(3), 221–239.

Blanchard, A., & Strand, R. (Eds.). (2017). *Cancer biomarkers: Ethics, economics and society*. Kokstad, South Africa: Megaloceros.

Boyd, N. F., Guo, H., Martin, L. J., Sun, L., Stone, J., Fishell, E., Jog, R. A., Hislop, G., Chiarelli, A., Minkin, S., & Yaffe, M. J. (2007). Mammographic density and the risk and detection of breast cancer. *New England Journal of Medicine*, *356*(3), 227–236.

Brigandt, I., & Love, A. C. (2017). Reductionism in biology. In E. N. Zalta (Ed.), *Stanford encyclopedia of philosophy*. Retrieved from https://plato.stanford.edu/entries/reduction-biology/

Brooks, D., and M. I. Eronen. (2018). Levels of organization in biology. In E. N. Zalta (Ed.), *Stanford encyclopedia of philosophy*. Retrieved from https://plato.stanford.edu/entries/levels-org-biology/

Campbell, D. T. (1974). 'Downward causation' in hierarchically organised biological systems. In F. Ayala & T. Dobzhansky (Eds.), *Studies in the philosophy of biology* (pp. 179–186). Berkeley: University of California Press.

Christiansen, P. V. (2000). Macro and micro-levels in physics. In P. B. Andersen, C. Emmeche, N. O. Finnemann, & P. V. Christiansen (Eds.), *Downward causation: Minds, bodies and matter* (pp. 51–62). Aarhus, Denmark: Aarhus University Press.

Davidson, L., von Dassow, M., & Zhou, J. (2009). Multi-scale mechanics from molecules to morphogenesis. *International Journal of Biochemistry and Cell Biology*, *41*(11), 2147–2162.

Deisboeck, T., & Stamatakos, G. (Eds.). (2011). *Multiscale cancer modeling*. Boca Raton, FL: CRC Press.

DiFrisco, J. (2017). Time scales and levels of organization. *Erkenntnis*, *82*(4), 795–818.

Dolberg, D. S., & Bissell, M. J. (1984). Inability of Rous sarcoma virus to cause sarcomas in the avian embryo. *Nature*, *309*(5968), 552–556.

Egeblad, M., Nakasone, E. S., & Werb, Z. (2010). Tumors as organs: Complex tissues that interface with the entire organism. *Developmental Cell*, *18*(6), 884–901.

Ellis, G. F. R. (2012). Top-down causation and emergence: Some comments on mechanisms. *Interface Focus*, *2*(1), 126–140.

Emmeche, C., Køppe, S., & Stjernfelt, F. (2000). Levels, emergence, and three versions of downward causation. In P. B. Andersen, C. Emmeche, N. O. Finnemann, & P. V. Christiansen (Eds.), *Downward causation: Minds, bodies and matter* (pp. 13–34). Aarhus, Denmark: Aarhus University Press.

Feng, J., Tang, Y., Xu, Y., Sun, Q., Liao, F., & Han, D. (2013). Substrate stiffness influences the outcome of antitumor drug screening in vitro. *Clinical Hemorheology and Microcirculation, 55*, 121–131.

Folkman, J. (1971). Tumor angiogenesis: Therapeutic implications. *New England Journal of Medicine, 285*(21), 1182–1186.

Green, S. (2018). Scale-dependency and downward causation in biology. *Philosophy of Science, 85*(5), 998–1011.

Green, S., & Batterman, R. (2017). Biology meets physics: Reductionism and multi-scale modeling of morphogenesis. *Studies in History and Philosophy of the Biological and Biomedical Sciences, 61*, 20–34.

Green, S., & Vogt, H. (2016). Personalizing medicine: Disease prevention *in silico* and *in socio. Humana Mente Journal of Philosophical Studies, 9*(30), 105–145.

Greenman, C., Stephens, P., Smith, R., Dalgliesh, G. L., Hunter, C., Bignell, G., Davies, H., Teague, J. Butler, A., Stevens, C., Edkins, S. O'Meara, S., Vastrik, I., Schmidt, E. E., Avis, T., Barthorpe, S., Bhamra, G., Buck, G., Choudhury, B., Clements, J., Cole, J., Dicks, E., Forbes, S., Gray, K., Halliday, K., Harrison, R., Hills, K., Hinton, J., Jenkinson, A., Jones, D., Menzies, A., Mironenko, T., Perry, J., Raine, K., Richardson, D., Shepherd, R., Small, A., Tofts, C., Varian, J., Webb, T., West, S., Widaa, S., Yates, A., Cahill, D. P., Louis, D. N., Goldstraw, P., Nicholson, A. G., Brasseur, F., Looijenga, L., Weber, B. L., Chiew, Y.-E., deFazio, A., Greaves, M. F., Green, A. R., Campbell, P., Birney, E., Easton, D. F., Chenevix-Trench, G., Tan, M.H., Khoo, S. K., Teh, B. T., Yen, S. T., Leung, S. Y., Wooster, R., Futreal, P. A., &. Stratton, M. R. (2007). Patterns of somatic mutation in human cancer genomes. *Nature, 446*(7132), 153–158.

Hanahan, D., & Weinberg, R. A. (2011). Hallmarks of cancer: The next generation. *Cell, 144*(5), 646–674.

Hirata, E., & Sahai, E. (2017). Tumor microenvironment and differential responses to therapy. *Cold Spring Harbor Perspectives in Medicine, 7*(7), a026781.

Hooker, C. (2013). On the import of constraints in complex dynamical systems. *Foundations of Science, 18*(4), 757–780.

Huang, S., & Ingber, D. E. (2005). Cell tension, matrix mechanics, and cancer development. *Cancer Cell, 8*(3), 175–176.

Kaiser, M. (2015). *Reductive explanation in the biological sciences.* Cham, Switzerland: Springer.

Kim, J. (1998). *Mind in a physical world.* Cambridge, MA: MIT Press.

Laplane, L., Duluc, D., Larmonier, N., Pradeu, T., & Bikfalvi, A. (2018). The multiple layers of the tumor environment. *Trends in Cancer, 4*(1), 802–809.

Lazebnik, Y. (2002). Can a biologist fix a radio?—Or, what I learned while studying apoptosis. *Cancer Cell, 2*(3), 179–182.

Lazebnik, Y. (2010). What are the hallmarks of cancer? *Nature Comment, 10*, 232–233.

Lesne, A. (2013). Multiscale analysis of biological systems. *Acta Biotheoretica, 61*(1), 3–19.

Lochter, A., & Bissell, M. J. (1995). Involvement of extracellular matrix constituents in breast cancer. *Seminars in Cancer Biology, 3*(6), 165–173.

Love, A. C. (2012). Hierarchy, causation and explanation: Ubiquity, locality and pluralism. *Interface Focus, 2*(1), 115–125.

Love, A. C. (2017). Building integrated explanatory models of complex biological phenomena: From Mill's methods to a causal mosaic. *European Studies in Philosophy of Science, 5*, 221–232.

Love, A. C., Stewart, T. A., Wagner, G. P., & Newman, S. A. (2017). Perspectives on integrating genetic and physical explanations of evolution and development: An introduction to the symposium. *Integrative and Comparative Biology, 57*(6), 1258–1268.

Maffini, M. V., Calabro, J. M., Soto, A. M., & Sonnenschein, C. (2005). Stromal regulation of neoplastic development: Age-dependent normalization of neoplastic mammary cells by mammary stroma. *American Journal of Pathology, 167*(5), 1405–1410.

Maffini, M. V., Soto, A. M., Calabro, J. M., Ucci, A. A., & Sonnenschein, C. (2004). The stroma as a crucial target in rat mammary gland carcinogenesis. *Journal of Cell Science, 117*(8), 1495–1502.

Malaterre, C. (2011). Making sense of downward causation in manipulationism: Illustrations from cancer research. *History and Philosophy of the Life Sciences, 33*(4), 537–561.

McGivern, P. (2008). Levels of reality and scales of application. In B. Ellis, H. Sankey, & A. Bird (Eds.), *Properties, powers and structures: Issues in the metaphysics of realism* (pp. 45–60). New York, NY: Routledge.

Miklos, G. L. G. (2005). The human cancer genome project—one more misstep in the war on cancer. *Nature Biotechnology, 23*(5), 535–537.

Montévil, M., & Pocheville, A. (2017). The Hitchhiker's guide to the cancer galaxy: How two critics missed their destination. *Organisms: Journal of Biological Sciences, 1*(2), 37–46.

Mossio, M., Bich, L., & Moreno, A. (2013). Emergence, closure and inter-level causation in biological systems. *Erkenntnis, 78*(2), 153–178.

Nelson, C. M., & Bissell, M. J. (2005). Modeling dynamic reciprocity: Engineering three-dimensional culture models of breast architecture, function, and neoplastic transformation. *Seminars in Cancer Biology, 15*(5), 342–352.

Nelson, C. M., & Bissell, M. J. (2006). Of extracellular matrix, scaffolds, and signaling: Tissue architecture regulates development, homeostasis, and cancer. *Annual Review of Cell and Developmental Biology, 22*, 287–309.

Noble, D. (2012). A theory of biological relativity: No privileged level of causation. *Interface Focus, 2*(1), 55–64.

Oden, J. T. (Chair). (2006). Simulation based engineering science—An NSF Blue Ribbon Report. Retrieved from www.nsf.gov/pubs/reports/sbes_final_report.pdf

Paszek, M. J., Zahir, N., Johnson, K. R., Lakins, J. N., Rozenberg, G. I., Gefen, A., Reinhart-King, C. A., Margulies, S. S., Dembo, M., Boettiger, D., Hammer, D. A., & Weaver, V. M. (2005). Tensional homeostasis and the malignant phenotype. *Cancer Cell, 8*(3), 241–254.

Plimmer, H. G. (1903). The parasitic theory of cancer. *British Medical Journal, 2*(2241), 1511–1515.

Plutynski, A. (2018a). *Explaining cancer: Finding order in disorder*. New York, NY: Oxford University Press.

Plutynski, A. (2018b). The origins of 'dynamic reciprocity': Mina Bissell's expansive picture of cancer causation. In O. Harman & M. Dietrich (Eds.), *Dreamers, visionaries, and revolutionaries in the life sciences* (pp. 96–109). Chicago, IL: University of Chicago Press.

Potochnik, A. (2009). Explanatory independence and epistemic interdependence: A case study of the optimality approach. *British Society for the Philosophy of Science, 61*, 213–233.

Potochnik, A., & McGill, B. (2012). The limitations of hierarchical organization. *Philosophy of Science, 79*(1), 120–140.

Prasad, V., Fojo, T., & Brada, M. (2016). Precision oncology: Origins, optimism, and potential. *The Lancet Oncology, 17*(2), e81–e86.

Qiu, W., Hu, M., Sridhar, A., Opeskin, K., Fox, S., Shipitsin, M., Trivett, M., Thompson, E. R., Ramakrishna, M., Gorringe, K. L., Polyak, K., Haviv, I., & Campbell, I. G., (2008). No evidence of clonal somatic genetic alterations in cancer-associated fibroblasts from human breast and ovarian carcinomas. *Nature Genetics, 40*(5), 650.

Rankin, K. S., & Frankel, D. (2016). Hyaluronan in cancer—from the naked mole rat to nanoparticle therapy. *Soft Matter, 12*(17), 3841–3848.

Robert, J. S. (2004). *Embryology, epigenesis, and evolution: Taking development seriously*. New York, NY: Cambridge University Press.

Rosenberg, A. (1997). Reductionism redux: Computing the embryo. *Biology and Philosophy, 12*, 445–470.

Sawyer, K. R. (2002). Nonreductive individualism: Part I—Supervenience and wild disjunction. *Philosophy of the Social Sciences, 32*, 537–559.

Smithers, D. W. (1962). An attack on cytologism. *Lancet, 1*(7228), 493–499.

Sonnenschein, C., & Soto, A. (1999). *The society of cells: Cancer and control of cell proliferation*. New York, NY: Springer-Verlag.

Sonnenschein, C., Soto, A. M., Rangarajan, A., & Kulkarni, P. (2014). Competing views on cancer. *Journal of Bioscience, 39*(2), 281–302.

Soto, A., & Sonnenschein, C. (2011). The tissue organization field theory of cancer: A testable replacement for the somatic mutation theory. *BioEssays, 5*, 322–340.

Soto, A. M., Sonnenschein, C., & Miquel, P. A. (2008). On physicalism and downward causation in developmental and cancer biology. *Acta Biotheoretica, 56*(4), 257–274.

Stoker, A. W., Hatier, C., & Bissell, M. J. (1990). The embryonic environment strongly attenuates v-src oncogenesis in mesenchymal and epithelial tissues, but not in endothelia. *Journal of Cell Biology, 111*(1), 217–228.

Stylianopoulos, T. (2017). The solid mechanics of cancer and strategies for improved therapy. *Journal of Biomechanical Engineering, 139*(2), 021004.

Tomczak, K., Czerwińska, P., & Wiznerowicz, M. (2015). The Cancer Genome Atlas (TCGA), an immeasurable source of knowledge. *Contemporary Oncology, 19*(1A), A68.

Umerez, J., & Mossio, M. (2013). Constraint. In W. Dubitzky, O. Wolkenhauer, K.-H. Cho, & H. Yokota (Eds.), *Encyclopedia of systems biology* (pp. 490–493). New York, NY: Springer.

Vaux, D. L. (2011). In defense of the somatic mutation theory of cancer. *BioEssays, 33*(5), 341–343.

Weinberg, R. A. (2007). *The biology of cancer* (Vol. 1). New York, NY: Garland Science.

West, A. K. V., Wullkopf, L., Christensen, A., Leijnse, N., Tarp, J. M., Mathiesen, J., Erler, J. T., & Oddershede, L. B. (2017). Dynamics of cancerous tissue correlates with invasiveness. *Scientific Reports, 7*, 43800.

Wilson, M. (2012). What is classical mechanics anyway? In R. Batterman (Ed.), *Oxford handbook of philosophy of physics* (pp. 43–106). Oxford, UK: Oxford University Press.

Wimsatt, W. C. (1972). Complexity and organization. *PSA: Proceedings of the Biennial Meeting of the Philosophy of Science Association, 20*, 67–86.

Wimsatt, W. C. (1994). The ontology of complex systems: Levels of organization, perspectives, and causal thickets. *Canadian Journal of Philosophy, 20*, 207–274.

Wimsatt, W. C. (2007). *Re-engineering philosophy for limited beings: piecewise approximations to reality.* Cambridge, MA: Harvard University Press.

Witz, I. P. (2009). The tumor microenvironment: The making of a paradigm. *Cancer Microenvironment, 2*(1), 9–17.

11 Multiscale Modeling in Inactive and Active Materials

Robert W. Batterman

Overview

This chapter approaches the question of the nature of levels in biology from the point of view of materials science. I first examine multiscale investigations of materials such as steel beams (inactive materials). I compare these methods with recent work on the modeling of active materials—specifically, the behaviors and structures of the mitotic spindle. From this materials science perspective, scale separation allows for a conception of levels that does not necessarily track part–whole relations. Perhaps surprisingly, there are interesting and fruitful connections between multiscale modeling in physics and biology.

11.1 Introduction

Materials such as steel beams, fluids such as water, and schools of fish all exist at distinct "levels" or scales. By this I mean, at least in part, that they display distinct, dominant behaviors at different spatial and temporal scales. Water, for example, displays fluid properties at rather long (everyday) time scales. Thus, it is much easier to wade into a pool of water than it is to wade into a pool of hardened concrete. However, as anyone who has ever flubbed a dive into a swimming pool can attest, at short time scales, water can display solid-like behavior.

The fluid behavior of water is well described by the equations of hydrodynamics—the Navier–Stokes equations. These same equations describe the behavior of a host of other fluids that differ in their microscopic makeup. In other words, the equations describe upper-scale behavior that is largely autonomous from the theoretical characterization of the actual molecular and atomic constituents of the different fluids. This autonomy can be understood in terms of a claim of stability that can explain how the Navier–Stokes equations can be completely wrong about the actual molecular details of a fluid (these equations treat fluids as continua—as having no component structure whatsoever at any length scale), yet be extremely accurate in characterizing the behaviors at everyday length scales. The continuum behavior of a fluid is stable under perturbations of lower-scale molecular details.

One would like to understand the relationships between behaviors of a single system at different scales. It is the same system, but its distinct behaviors seem to require theories and models that are scale relative and applicable only at widely separated scales. For instance, the hydrodynamic equations are perfectly good for modeling the behaviors of fluid flow in a pipe, but they do not do as good a job in modeling the behavior of ink in an inkjet printer. For the latter, molecular dynamics is much more appropriate.[1]

The problem of modeling at different scales and the nature of the relationship between such models is actually at the heart of the age-old problem of the relationship between thermodynamics (a continuum theory) and statistical mechanics (a theory of collections of discrete particles). In this chapter, I will not be addressing this broad relationship between theories. Rather, I intend to discuss analogies and disanalogies that appear when attempting multiscale modeling of different kinds of systems.

In section 11.2, I consider how composite materials are modeled in materials science. Parameters that appear in continuum equations for the bending of metal bars differ in their values depending upon the type of metal—steel versus aluminum, for example. Those parameter values surely must reflect some lower-scale properties of the different materials. But in practice, the values are usually determined experimentally. This section considers the extent to which one can determine the values of those parameters theoretically. It provides a relatively detailed discussion of the modeling of materials at separate scales and the extent to which models at one scale can inform models at other scales. An important aspect to note here is that such modeling in materials science is fundamentally continuum modeling. Typically, details of the atomic or molecular level are not considered in any explicit fashion.

The next section addresses a similar problem in certain (primarily) biological systems that can be considered to be "active." Active systems differ from the materials discussed in the first section because they are inherently out of equilibrium and because the components of the systems independently burn their own energy.[2] Large-scale active systems include bird flocks and swarming bacteria. Inside cells, there are materials that are active in this sense as well. This section looks at the materials science of active matter and discusses differences and similarities with the materials science modeling of inactive materials. Can continuum models really be employed to understand the out-of-equilibrium, self-organizing structures that are characteristic of biological systems? One might be skeptical of such a possibility given the extreme complexity and nonequilibrium nature of such systems. It is, in fact, possible to study the material properties of such structures (the example discussed concerns the behaviors in the mitotic spindle) in a way quite analogous to the study of material properties in the cases of inactive matter like metal bars. I discuss briefly some of the commonalities between active and inactive material modeling and some of the differences.

The discussion suggests that a kind of "level" talk in the context of multiscale modeling yields a fruitful way of understanding the behaviors of various kinds of systems. Note that this conception of levels is *not* fundamentally based on a relation between parts and wholes. (Such a conception is common in various contexts, including biology.) Here the operative notion is that of scale separation. For the inactive materials studied in materials science, this way of thinking about levels has been very fruitful. Perhaps somewhat surprisingly, the same is true for active (typically biological) materials as well.

11.2 Inactive Materials

Steel is a composite material. At the level of atoms, it is primarily a lattice of carbon and iron atoms arranged in a certain order. At the level of railroad tracks, it appears as a homogeneous structure exhibiting no lower-scale structure at all. At this continuum scale, we can take the tracks to be roughly in (thermal) equilibrium yet subject to stresses when a train glides over them. When stressed, the track bends—it is perturbed from its equilibrium state—but (we hope) returns reasonably quickly to a quasi-equilibrium state after the train passes by. The way it returns to its equilibrium state depends on certain material parameters that characterize its bending and stretching behavior. From this continuum perspective, the equations that govern the behavior of the track under loading are called the Navier–Cauchy equations:

$$(\lambda + \mu)\nabla(\nabla \cdot \mathbf{u}) + \rho\nabla^2\mathbf{u} + \mathbf{f} = 0, \tag{1}$$

where $\mathbf{u}$ is the displacement vector, ρ is the material density, and $\mathbf{f}$ are the body forces (e.g., gravity) acting on the material. The material parameters, λ and μ, are the "Lamé" parameters. They codify the bending and stretching behaviors of the material. In order to model a system using these equations, values for such material parameters (along with a value for the material density ρ) need to be specified. Different values for these determine how steel differs in its loading behavior from, say, aluminum or wood. Typically, the values for the Lamé parameters are determined by *empirical experiments* on a workbench.

When the Navier–Cauchy equations were first written down, there was no consensus about whether materials like steel or iron had (atomic or molecular) structure at any scale below that of the continuum. In other words, the hypothesis that there are atoms and molecules was not remotely settled.[3] From our contemporary perspective, we know that materials do indeed have lower-scale structure. Steel, at atomic scales, is a lattice of iron and carbon atoms. And, at (meso) scales in between the atomic and the continuum, there are a host of other features. Surely, the material parameters, λ, μ, and ρ, must reflect some aspects of the lower-scale structures present in the material. In recent years, there has been considerable mathematical work on establishing appropriate connections between material structures at lower scales and *ranges* of allowed values for the material parameters. This work allows one, to a certain extent, to *upscale*—to determine values for the material parameters—in terms of structures at lower scales.

It turns out that the actual details at the level of the atomic lattice are largely irrelevant to the continuum scale behavior of the steel. This is important and reflects the *relative autonomy* of the continuum equations (the Navier–Cauchy equations and, for that matter, the equations of thermodynamics) from the detailed atomic and molecular configurations of the various systems. If we are interested in why the Navier–Cauchy equations work so well for designing safe bridges and buildings, we would need to demonstrate the stability of such structures under perturbation of lower-scale details. The functional form of the Navier–Cauchy equations gets things right at continuum scales. An important explanatory question is why? The answer, as noted, is in part that the important features in the modeling of material behavior are the structures that exist at mesoscales. We need to see how to talk about them.

11.2.1   Upscaling and RVEs

At everyday continuum scales, a steel bar looks reasonably homogeneous. If we look at it with our naked eyes or with a magnifying glass, we don't see much structure: it appears to be uniform. However if we zoom in, using high-powered microscopes or X-ray diffraction techniques, we will begin to see structures that are hidden at everyday lengths. In order to describe the main/dominant/important features of the steel at these shorter length scales, we employ the very important concept of a *representative volume element* (RVE). Consider the steel bar. A representative volume element is a material volume that is statistically representative of features of the steel at some particular spatial scale. In figure 11.1, the left side displays a material point, and it is surrounded by an infinitesimal material element. Structures in the material element—the voids, cracks, grain boundaries, and so on—are to be treated as the microstructure of that (macro) material element.

As is evident from the figure, the conception of the RVE involves the introduction of two length scales. There is the continuum or macroscale (D) by which the neighborhood of the material point is characterized, and there is a microscale (d) that represents the smallest microstructures whose properties (typically shapes are most important) are believed to directly influence the overall response to stresses and strains imposed upon the neighborhood surrounding the material point. These length scales must typically differ by orders of magnitude so that ($d/D \ll 1$). The requirement of this scale separation is independent of the nature of the distribution of the microstructures in the RVE. For example, they may be periodically or randomly distributed throughout the RVE. Of course, that difference can be important and may very well affect the properties displayed by the RVE under stresses and strains.

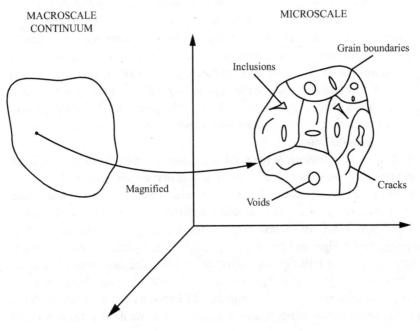

Figure 11.1
RVE.

It is important to note at least three features of RVEs:

First, the RVE concept is *scale relative*. The actual characteristic lengths of the structures in a RVE can vary considerably. As Nemat-Nasser and Hori (1999) note, the overall properties of a mass of compacted fine powder in powder-metallurgy can have grains of micron size, so that a neighborhood of 100 microns can very well serve as a RVE: "Whereas in characterizing an earth dam as a continuum, with aggregates [stones, sticks, clay etc.] of many centimeters in size, the absolute dimension of an RVE would be of the order of tens of meters" (p. 15).

Second, the RVE and the structures within it are all considered continua. That is to say, we are operating at scales much higher than that of the spacings in an atomic lattice but much lower than that of the scale where the material appears as a homogeneous continuum. So, there are at least *three* widely separated scales. Call the scale of the atomic lattice or the scale of intermolecular interactions between fluid molecules δ_a. Then, strictly speaking, we have the following relation:

$$\delta_a \ll d \ll D.$$

One typically eschews the very hard problem of upscaling from scales of order δ_a to D in favor of starting with those structures at the relevant RVE, namely, at scales of order d. This procedure, or this class of procedures, is known as "homogenization." In the next section, I provide a brief introduction to some of the mathematics.

Third, in employing RVEs to gain information about a material, one is implicitly ignoring boundaries or edges. In other words, one is essentially treating the material in bulk. We do not pay attention to what happens on the surface of the steel bar. This is, in effect, to treat the material as infinite in extent and is a natural and useful idealization.[4]

11.2.2 Upscaling

Our hypothesis is that the material parameters depend upon largely unknown aspects of the actual atomic or molecular makeup of the materials. Introducing material parameters such as the Láme parameters that code for lower-scale details actually introduces a level or scale intermediate between that of the continuum and the atomic.[5] The idea is that the nature of the important structures in the relevant RVE (which are at an intermediate scale d, $\delta_a \ll d \ll D$) largely determines the stable material parameters that characterize the material at the continuum scale. The appeal to an RVE, therefore, introduces a middle level or scale between the atomic and the continuum. It is difficult to determine the nature of the inclusions, cracks, grain boundaries, and so on that are present in an RVE. Recall figure 11.1. One can employ various imaging techniques to determine such structure. In addition, in analogy with work on so-called hydrodynamic descriptions of fluids (Kadanoff & Martin, 1963), one can attempt to employ correlation functions to try to map out structures. See also Torquato (2002) for a discussion of N-point correlation functions. More on this below in section 11.3.2.

We can see how this works by considering a composite material 50% of which is a good thermal conductor while the remaining 50% is an insulator. Suppose that the two distinct materials are well mixed. They might be periodic or randomly distributed, but suppose for illustrative purposes that for any reasonable RVE, the volume fractions of each are basically 1/2. If we tried to infer the conductivity of the material simply by averaging over

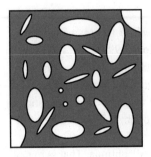

Figure 11.2
50–50 volume RVEs.

the volumes, we most likely will be grossly in error. If the dark material in figure 11.2 is the conductor and the configuration is as in the left RVE, then the material will be a very good conductor. But if the material is as in the right configuration, the material will be a terrible conductor.

A consequence of this simple example is that the sort of volume averaging typically assumed in statistical mechanics is not a good means for upscaling to determine continuum properties of systems. For instance, consider how, in typical statistical mechanical discussions, one might reasonably try to determine the density of an ideal gas. One would choose an RVE containing a large but finite number (N) of molecules and divide that number by the volume (V) of the RVE; then, taking limits ($N \rightarrow \infty$ and $V \rightarrow \infty$), one would arrive at a reasonable value for the density of the gas. Likewise, the identification of thermodynamic temperature with mean molecular kinetic energy also involves such simple averaging. But, these averages, as a rule, will only work for *homogeneous* systems as the example in figure 11.2 clearly shows. Therefore, making direct connections from the atomic/molecular level to the continuum scale is not as straightforward as many reductionist programs assume.

11.2.3   Homogenization

In fact, most upscaling starts at the mesoscale level and proceeds using sophisticated mathematics called "homogenization" to determine ranges of values or *bounds* for the various material parameters appearing in the continuum equations like equation (1). The simple example of figure 11.2 leads directly to the need for homogenization as opposed to averaging. Let's call the RVE in, say, the left figure Ω. Let's say light region is Ω_1, and the dark region is Ω_2. Let the thermal conductivity of the entire heterogeneous material in Ω be designated by γ. The conductivity of the material in Ω_1 is γ_1. Likewise, γ_2 is the conductivity for the material in Ω_2.[6] In other words,

$$\gamma(x) = \begin{cases} \gamma_1, & \text{if } x \in \Omega_1 \\ \gamma_2, & \text{if } x \in \Omega_2. \end{cases}$$

Denote the temperature at a point $x \in \Omega$, $u(x)$.[7] Since the conductivities differ in the two regions, we have

$$u(x) = \begin{cases} u_1(x), & \text{if } x \in \Omega_1 \\ u_2(x), & \text{if } x \in \Omega_2. \end{cases}$$

The flux of the temperature is given by

$$q = \begin{cases} q_1 = \gamma_1 \, \text{grad} \, u_1 & \text{in } \Omega_1 \\ q_2 = \gamma_2 \, \text{grad} \, u_2 & \text{in } \Omega_2. \end{cases}$$

We assume the continuity of the temperature and of the flux at the interfaces between the two materials:

$$\begin{cases} u_1 = u_2, & \text{on } \partial\Omega_1 \cap \partial\Omega_2 \\ q_1 \cdot n_1 = q_2 \cdot n_2, & \text{on } \partial\Omega_1 \cap \partial\Omega_2 \end{cases} \tag{2}$$

where the n_i are outward normal unit vectors on the boundaries $\partial\Omega_i$, ($i = 1, 2$) and $n_1 = -n_2$ on the boundary $\partial\Omega_1 \cap \partial\Omega_2$.

If we take $f(x)$ to be a heat source, then the heat conduction problem (to find the temperature $u(x)$ given the heat source and the above boundary conditions) can be written as follows:

$$\begin{cases} -\text{div}(\gamma(x)\text{grad}\, u(x)) = f(x) & \text{in } \Omega_1 \cup \Omega_2 \\ u = 0, & \text{on } \partial\Omega \\ u_1 = u_2, & \text{on } \partial\Omega_1 \cap \partial\Omega_2 \\ q_1 \cdot n_1 = q_2 \cdot n_2, & \text{on } \partial\Omega_1 \cap \partial\Omega_2. \end{cases} \tag{3}$$

Had the region Ω contained an isotropic single material with constant conductivity, this problem is an easily solvable elliptical partial differential equation. However, because we are considering a composite that satisfies the conditions (2), this means both that the gradient of the temperature u is *discontinuous* and that the flux q is *not differentiable*. Thus, to solve the problem (3) for the composite presents considerable difficulties. In fact, the problem requires the mathematics of weak solutions, with a weak notion of derivative that involves a formulation on a particular of space of functions known as a Sobolev space, H. Without going into detail, the heat conduction problem expressed in (3) is given a *variational* formulation:

Find $u \in H$ such that[8]

$$\sum_{i=1}^{N} \int_\Omega \gamma(x) \frac{\partial u}{\partial x_i} \frac{\partial v}{\partial x_i} \, dx = \int_\Omega f \, v \, dx, \quad \forall v \in H. \tag{4}$$

In many cases of physical interest, it is reasonable to assume (contrary to our 50–50 example) that one of the materials has a much smaller volume than the other. So let us suppose that Ω_1 is much smaller in volume than Ω_2, and then we can proceed as follows. Further simplifying (again as is often reasonable), suppose that the heterogeneities of material in Ω_1 with conductivity γ_1 are periodically distributed throughout the RVE. See figure 11.3 for an idea of what this means.

The periodicity is represented by the small parameter ε. Given this, the conductivity γ for the composite as a whole becomes a function of ε, and problem (4) now has the following formulation:

Find $u^\varepsilon \in H$ such that

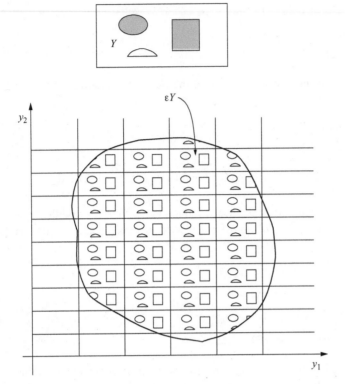

Figure 11.3
Periodic heterogeneities with period ε (Cioranescu & Donato, 1999, p. 5).

$$\sum_{i=1}^{N} \int_{\Omega} \gamma^{\varepsilon}(x) \frac{\partial u^{\varepsilon}}{\partial x_i} \frac{\partial v}{\partial x_i} \, dx = \int_{\Omega} f \, v \, dx, \quad \forall v \in H. \tag{5}$$

To represent the periodicity explicitly is reasonable to define γ^{ε} as follows:

$$\gamma^{\varepsilon} = \gamma \left(\frac{x}{\varepsilon} \right) \text{ almost everywhere on } \mathbb{R}^{N}. \tag{6}$$

We let γ be a periodic function with period Y; then by (6), the heterogeneities in Ω having size of order ε are periodic with period equal to εY. Now our problem (5) can be rewritten as follows:

Find $u^{\varepsilon} \varepsilon H$ such that

$$\sum_{i=1}^{N} \int_{\Omega} \gamma \left(\frac{x}{\varepsilon} \right) \frac{\partial u^{\varepsilon}}{\partial x_i} \frac{\partial v}{\partial x_i} \, dx = \int_{\Omega} f \, v \, dx, \quad \forall v \in H. \tag{7}$$

Two scales now appear in the problem. There is a macroscopic scale x and a microscopic scale $\frac{x}{\varepsilon}$. Now, speaking metaphorically (but in a way that can be made mathematically precise), we note that as $\varepsilon \to 0$, $\frac{x}{\varepsilon}$ will get larger (hence we are upscaling). Furthermore, the heterogeneities will get ever smaller, and we can hope to get a smoother, more *homo-*

geneous mixture in that limit.[9] This is a very different way to upscale than is the typical volume averaging employed in statistical mechanics. It allows one to find effective values for continuum parameters that will more accurately reflect the structures present at meso-scales in various materials.

11.2.4 Bounds

A desideratum for upscaling would be to demonstrate that there are determinate ranges of values for the continuum scale Lamé parameters or for the thermal conductivity, γ, given some minimal facts about the nature of composite materials. If all we know is the volume fraction of the various components (of, e.g., the materials with conductivities γ_1 and γ_2) but we lack information about the geometry or distribution of the materials in the RVE, can we still delimit bounds within which the values of the effective parameters will be guaranteed to reside?

In fact, some rigorous mathematical results allow us to do exactly that. These results are particularly useful for determining the range of responses to composite materials under stresses.[10]

11.2.5 Summary Remarks

The main upshot of this discussion is that for heterogeneous materials and for nonhomo-geneous (nonideal) gases, the volume averaging techniques typically assumed in philosophical assertions of a direct/reductionist connection between statistical mechanics and thermodynamics will almost surely fail. Intermediate, mesoscale structures characterized by appropriate RVEs are crucial to upscaling. In fact, such mesoscale structures are typically the best places to start to bridge between models of materials at different scales.

Why was this section titled "Inactive Materials?" The answer is because by and large, the composite materials considered here are assumed to be (essentially) in thermal equilibrium. Among other things, this assumption allows one to focus on bulk continuum behaviors of the systems, effectively ignoring any boundaries that might exist. But not all materials are like this. In fact, living systems are definitely not like this. There is a relatively new field at the intersection of biology and physics that has come to be known as the study of "active matter." These systems include swarms of bacteria, schools of fish, flocking birds, and processes in living cells. Understanding the behaviors of such systems requires some interesting twists to multiscale modeling. In the next section, I consider the nature of upscaling for active systems.

11.3 Active Materials

A crucial assumption of the discussion of upscaling in the last section (upscaling for what I call "inactive" materials) is that the system is a composite that is essentially in a state of equilibrium or quasi-equilibrium. However, the world is not static, and sometimes one needs to consider the upper-scale behavior of systems that are not in equilibrium. Relatively standard physics can be brought to bear on studying out-of-equilibrium behavior in *steady states*. A paradigm example is Rayleigh–Bénard convection. In this phenomenon, a fluid is contained between two plates with the distance between the plates considerably less than that of the extent of the plates. Initially, the temperatures of the two plates are the same, and the

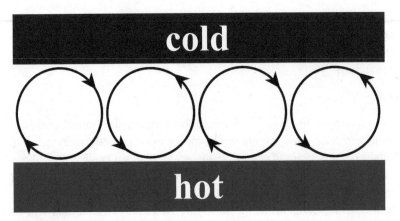

Figure 11.4
Cartoon of Rayleigh–Bénard convection (Research Group Felix Otto, n.d.).

fluid is at thermal equilibrium with the plates. Upon increasing, slightly, the temperature of the lower plate, the system will no longer be in thermal equilibrium and a temperature gradient between the plates will be established. As a result of this gradient, there will appear alternating clockwise/counterclockwise motions, as displayed in figure 11.4.

This behavior is stable under reasonably small perturbations and will continue as long as the temperature differential between the top and bottom plates is maintained. Note, though, that this out-of-equilibrium steady-state flow *requires* that thermal energy be introduced to the fluid-plate system exogenously. This is very interesting behavior, but it does not present any real modeling challenges.

On the other hand, the swarms, flocks, schools, and so on mentioned above are behaviors that do present new challenges. The difference is that the energy involved is not exogenous. Each component of the swarm, flock, school—each bacterium, bird, or fish—has its own energy source. Considered as microscopic degrees of freedom, they burn their own internal energy. This can lead to (emergent/collective) behaviors quite distinct from collections of passive or inactive components. Needleman and Dogic (2017, p. 136) say that active particles can "self-organize," whereas passive particles "self-assemble."[11] The collective behavior of the fluid molecules in the Rayleigh–Bénard cells is an example, in their terminology, of self-assembly.

As a rough guide, we can say that active matter has the following unifying features:

They are composed of self-driven units … each capable of converting stored or ambient free energy into systematic movement. The interaction of the active particles with each other, and with the medium they live in, give rise to highly correlated collective motion and mechanical stress. Active particles are generally elongated and their direction of self-propulsion is set by their own anisotropy, rather than fixed by an external field. (Marchetti et al., 2013, p. 1144)

11.3.1   Materials Science of the Mitotic Spindle

One area of rather intense research on the (bio)physics of active particles concerns the behavior of the mitotic spindle within the cytoskeleton of eukaryotic cells. The spindle (figures 11.5 and 11.7) is a mechanism that segregates chromosomes in cell division and is an ensemble of filament structures called microtubules, molecular motors, and other proteins.

Recent research approaches the behavior of the spindle from the point of view of materials science.[12] The molecular motors "walk" along the filament structures that are often polarized—oriented with plus and minus ends. In fact, these molecular motors often crosslink the microtubules and in the process cause the filaments to slide past one another. See figure 11.6. The motors are fueled by ATP (Adenosine Triphosphate) and so are out-of-equilibrium components. Their motions and their effects on the microtubules allow the spindle to segregate chromosomes into daughter cells during mitosis. See figure 11.6. These sliding motions allow for the mesoscale formation of different structures within the cytoskeleton.[13] In fact, the nonequilibrium nature of the spindle allows for it to do some really quite remarkable things.

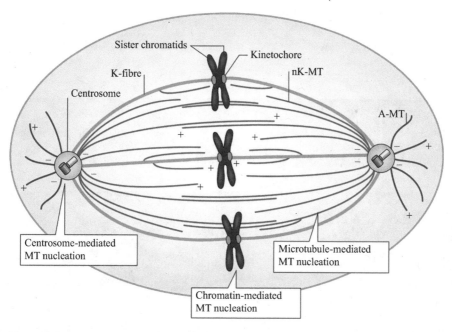

Figure 11.5
Mitotic spindle (Prosser & Pelletier, 2017, p. 188).

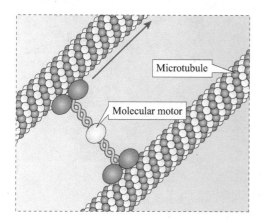

Figure 11.6
Microtubules and molecular motors (Needleman & Dogic, 2017).

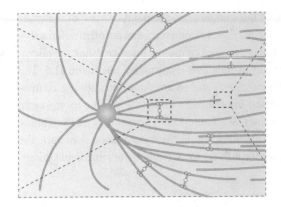

Figure 11.7
Spindle (Needleman & Dogic, 2017).

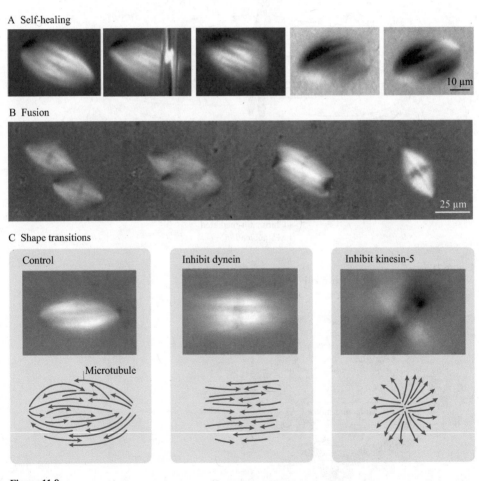

Figure 11.8
Mesoscale properties of spindles (Needleman & Dogic, 2017).

In vitro studies provide evidence for this. In figure 11.8, a spindle is cut and heals itself. In other instances, two spindles can fuse into one. Furthermore, by inhibiting certain molecular motors, one can change the shape of the spindle structure. All of this provides evidence of how larger-scale material structures can result from self-organizing active matter.

11.3.2 Mesoscale Descriptions of Active Materials

In section 11.2, we saw that material parameters reflect crucial features of composite (or noncomposite) inactive materials at scales of the relevant RVE. As noted, they introduce a third level or scale between the atomic and the macroscopic. Similarly, in condensed matter physics, another type of parameter, an *order* parameter, plays a crucial role in describing the continuum behaviors of fluids, magnets, and superconductors.[14]

In the mitotic spindle, as noted, some of the filaments are polar. One can study these structures at a mesoscale noting that different structures can be classified by different symmetries. If there is no relative orientation of the tubules with respect to one another, one will have isotropic structure at large length scales. If the tubules are polar yet oriented along an axis but pointing in different directions, one has essentially a nematic liquid crystal. If they are polar, axially oriented, and pointing in the same direction, one has a polar liquid crystal. See figure 11.9. These structures are well understood in (soft) equilibrium condensed matter physics. Their responses to external pressures and stresses are also well understood.

In active materials, things become more complicated but nevertheless allow for the mesoscopic study of stresses and strains in the relevant RVEs. In fact, adding active stresses resulting from the actions of molecular motors genuinely leads to new behaviors and different material properties. Dipolar active stresses preserve volume and therefore can initiate

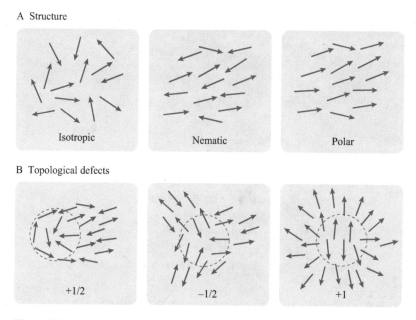

Figure 11.9
Symmetries of polar objects (Needleman & Dogic, 2017).

flowing behavior in different directions, thereby helping to explain some of the larger-scale shape shifting one sees during mitosis. Figure 11.10 provides an idea of how isotropic and dipolar stresses can effect shape changes and induce motion or flow in the relevant cellular RVEs.[15] Isotropic stresses (such as uniform pressure) can change the volume (expanding or contracting). Dipolar stresses can induce motion.

So far, we are describing bulk behaviors as we did in the cases of inactive matter. That is, we have tacitly assumed that the active materials studied are not spatially confined or walled in. But, an important feature of some active materials (including what goes on inside the cell) is the spatial confinement of the processes. This confinement or spatially bounded behavior can have dramatic effects on behaviors and classifications of the active materials we have been considering. Needleman and Dogic (2017) say that

chaotic dynamics is a generic feature of intrinsically unstable active nematics. By contrast, although the mitotic spindle is well described as an active nematic system, it retains a uniform alignment throughout its entire structure because of its finite size. This observation clearly demonstrates that boundary effects also strongly influence the behaviours of active nematics. (p. 9)

In "Physical Basis of Spindle Self-Organization," Brugués and Needleman (2014a) develop a (minimal) liquid crystal model for understanding the self-organizing processes inside the spindle. The model assumes that the orientation of the microtubules and their density are cross-correlated or coupled. As such, the model allows for a higher-scale description of the dynamics inside the spindle. They employ sophisticated experimental techniques to determine values for *spatiotemporal correlation functions* of microtubule density, of microtubule orientation, and of stresses. Their model reproduces the experimentally determined values of these correlation functions. It is a continuum model described by nematic

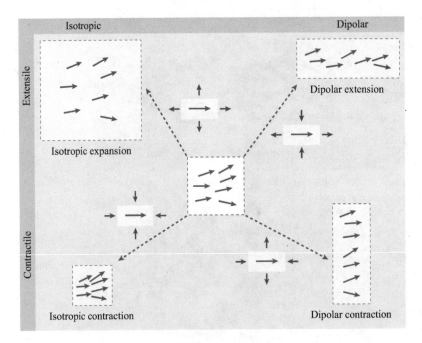

Figure 11.10
Stresses in active materials (Needleman & Dogic, 2017).

and polar fields, which result from the collective effects of local microtubule interactions that are determined by lower-scale effects of motor proteins and cross-linkers. Thus, they construct a mesoscale model that allows for qualitative and quantitative predictions of spindle behaviors, some of which are pictured in figures 11.8a, 11.8b, and 11.8c. Specifically, their measurements of the various correlation functions and of the generation and propagation of stress in the spindle are "consistent with spindle self-organization arising from the local interactions of microtubles, mediated by crosslinkers and motors, and microtubule polymerization dynamics" (Brugués & Needleman, 2014a, p. 18499).

In this model, the orientation of the microtubules is determined by nematic interactions, and the polar field convects the microtubule concentration and polarity magnitude. The magnitude of the nematic field is taken to be constant throughout the spindle, while the magnitude of the polarity field depends on motor activity and self-advection. (Brugués & Needleman, 2014b, p. 1)

The takeaway here is that continuum models of active materials, just as with inactive materials, can be very fruitful. Mesoscale parameters and fields can code for complex molecular scale behaviors, and the continuum minimal models can be shown to be quantitatively and qualitatively accurate. It remains to be seen, I believe, whether justifications of such models of the kind provided by homogenization theory (and the renormalization group) will be possible in the active matter case. Further questions include describing the role of boundaries and confinement in active matter systems. In the case of the spindle discussed here, the ordered structures seem to depend upon the constraints that exist within the cytoskeleton. However, flocks of birds and schools of fish appear to generate their own boundaries (amorphous as they may be). Are there ways of explaining the emergence of such features? Does it make sense to say that for some active matter systems, boundaries are emergent?

I do think answers to these questions are forthcoming and depend upon a phenomenon quite common in condensed matter physics—the important fact that long-range correlations can form even among systems whose components interact only locally. Early flocking models (e.g., Vicsek et al., 1995) modeled individuals interacting in response to their immediate neighbors. Such "agent-based" models (akin to Ising models) were able to simulate quite interesting behaviors, including phase transitions from disordered to ordered states. More recently, in the vein of Brugués and Needleman's approach, continuum hydrodynamic descriptions of flocks have been presented with an important focus on correlations (see Toner et al., 2005). *The key to successful mesoscopic continuum modeling of both active and inactive systems is to have an understanding of the nature of the relevant RVE.* Spatial and temporal correlation functions allow for just such an understanding.

11.4 Conclusion

The conception of biological levels of organization is problematic, as witnessed by (at least some of) the chapters in this volume. But a conception of properties or dominant behaviors at various spatial (and temporal) scales appears to be widespread across various sciences.[16] An important aspect of such a conception is the fact that when behaviors appear at widely separated scales, it is possible to model them largely independently of features that exist at different scales. Multiscale modelers take advantage of this relative autonomy of behaviors at separated scales to gain explanatory insight into the nature of different systems.

Much work has been done to understand how such scale-relative behaviors can relate to one another in single systems. In section 11.2, I have characterized some of this work by focusing on the extremely important concept of a representative volume element (RVE). This is a mesoscale concept that can be exploited to try to determine the continuum scale behaviors of materials whether they be samurai swords or earthen dams. For the latter, the makeup of the RVE can easily be determined by digging with a shovel. For the former, various forms of microscopy will likely be necessary. In the case of fluids (as studied in statistical mechanics and thermodynamics), information about the RVEs can be theoretically probed using various spatial and temporal correlation functions whose values can be subjected to empirical tests.[17] It turns out that some of the most important information about RVEs for upscaling is geometrical and topological, which is exactly the kind of information encoded in the correlation functions.

It is extremely interesting that one can actually, in certain contexts, treat biological structures and processes using some of the same techniques. The field of active matter is brand new and brings with it a new set of modeling challenges. In the last section, I provided an all-too-brief discussion of the multiscale modeling of active materials. The specific focus was on understanding the behaviors of the mitotic spindle using a continuum approach that is quite analogous to that discussed in the section on inactive materials. The analog of atoms or molecules is microtubules and molecular motor proteins. The analog of RVEs with their encoding of geometric and topological features of inactive materials is the nematic and polar fields within a spindle. Active materials can generate their own stresses and strains (unlike inactive materials). As a result, they are typically more complex than inactive materials. Yet, because of scale separation, the resulting upper-scale behaviors can be nicely modeled without detailed knowledge of the lower-scale behaviors of microtubules and molecular motors. The multiscale methods of materials science generalize nicely to the cases of active materials despite the radically different behaviors active materials can exhibit.

Acknowledgments

I thank Sara Green for helpful comments and discussions. I also thank Bill Wimsatt for many stimulating conversations about levels and emergence over the years. My thoughts about scales and levels have been deeply influenced by his work.

Notes

1. The continuum hydrodynamic equations are partial differential equations. The equations of molecular dynamics are ordinary differential equations. These are of completely different mathematical type. See Batterman (2009) for more details.

2. Of course, this energy is typically provided by the environment.

3. See Batterman (2013, section 3, pp. 269–273) for a discussion of this situation and the controversy surrounding the proper derivation of the Navier–Cauchy equations.

4. It is also an idealization that sometimes is not warranted.

5. For a discussion concerning the similarities between this feature of material parameters and so-called order parameters that are introduced to account for symmetry behaviors of condensed matter systems, see Batterman (2019).

6. χ, γ_i are *material parameters*. The following discussion shows how one can connect between such continuum parameters and mesoscale properties of RVEs.

7. The ensuing discussion follows Cioranescu and Donato (1999).

8. $v \in H$ are knowns as "test functions."

9. A lot of difficult mathematics is involved in demonstrating that this limit exists, that the values for the temperatures u^ϵ converge to an effective temperature, and that there exists an effective limit conductivity γ_{eff} to which the γ^ϵ converges. This mathematics is called "homogenization theory."

10. It is beyond the scope of the present chapter to address these interesting investigations in any further detail. See Milton (2002, chap. 21) for a brief overview.

11. Clearly, these are loaded terms. We need to see whether the phenomena described truly warrant such descriptors.

12. The discussion here largely follows Needleman and Dogic (2017).

13. There are three types of cytoskeletal motors—myosin, kinesin, and dynein. They direct different types of motion (yielding different kinds of shapes), as can be seen in figure 11.8.

14. For a ferromagnet, the order parameter is the net magnetization of the material. As the ferromagnet is heated, the net magnetization (reflecting a preferred spatial direction) weakens as the spins begin to point in all directions. Above a critical temperature, the system is paramagnetic with full rotational symmetry.

15. Needleman and Dogic (2017, p. 6) note that this figure is schematic in that the individual filaments appear to be the units experiencing the active stresses. In cells, the units might be pairs of filaments, asters, or bundles of different kinds.

16. Suggestions along these lines can be found in the groundbreaking work of Bill Wimsatt. See Wimsatt (this volume) for a summary and discussion of many issues relating to levels and scales.

17. Such experimental tests involve X-ray and neutron scattering, among other techniques.

References

Batterman, R. W. (2009). Idealization and modeling. *Synthese, 169*, 427–446.

Batterman, R. W. (2013). The tyranny of scales. In R. W. Batternman (Ed.), *The Oxford handbook of philosophy of physics* (pp. 255–286). Oxford, UK: Oxford University Press.

Batterman, R. W. (forthcoming). *A middle way: A non-fundamental approach to many-body physics.* Oxford, UK: Oxford University Press.

Brugués, J., & Needleman, D. (2014a). Physical basis of spindle self-organization. *Proceedings of the National Academy of Sciences, 111*(52), 18496–18500.

Brugués, J., & Needleman, D. (2014b). *Proceedings of the National Academy of Sciences,* Supporting Information.

Cioranescu, D., & Donato, P. (1999). *An introduction to homogenization.* Oxford, UK: Oxford University Press.

Kadanoff, L. P., & Martin, P. C. (1963). Hydrodynamic equations and correlation functions. *Annals of Physics, 24*, 419–469.

Marchetti, M. C., Joanny, J. F., Ramaswamy, S., Liverpool, T. B., Prost, J., Rao, M., & Aditi Simha, R. (2013). Hydrodynamics of soft active matter. *Reviews of Modern Physics, 85*(3), 1143–1189.

Milton, G. (2002). *The theory of composites.* Cambridge, UK: Cambridge University Press.

Needleman, D., & Dogic, Z. (2017). Active matter at the interface between materials science and cell biology. *Nature Reviews Materials, 2*(17048), 1–14.

Nemat-Nasser, S., & Hori, M. (1999). *Micromechanics: Overall properties of heterogeneous materials* (2nd ed.). Amsterdam, Netherlands: Elsevier.

Research Group Felix Otto. (n.d.). Rayleigh–Bénard convection. Retrieved from https://www.mis.mpg.de/applan/research/rayleigh.html

Prosser, S. L., & Pelletier, L. (2017). Mitotic spindle assembly in animal cells: A fine balancing act. *Nature Reviews: Molecular Cell Biology, 18*, 187–201.

Toner, J., Tu, Y., & Ramaswamy, S. (2005). Hydrodynamics and phases of flocks. *Annals of Physics, 318*, 170–244.

Torquato, S. (2002). *Random heterogeneous materials: Microstructure and macroscopic properties.* New York, NY: Springer.

Vicsek, T., Czirók, A., Ben-Jacob, E., Cohen, I., & Shochet, O. (1995). Novel type of phase transition in a system of self-driven particles. *Physical Review Letters, 75*(6), 1226–1229.

12 Using Compositional Explanations to Understand Compositional Levels: An Integrative Account

Carl Gillett

Overview

Scientists talk of "levels" where they offer compositional models and explanations. Once we appreciate both the plural, and integrated, character of such compositional models/ explanations, I show we can finally provide a more adequate account of compositional levels that accommodates scientific practice. I outline how scientists use "level" to refer to the often large ontological commitments of their integrated compositional models in what I term the "Integrative Account" of levels. And I outline how, contrary to the claims of philosophical critics, such "levels" are precise and rigorous in character and perform useful roles for scientists that cannot be replicated using the notion of "scale."

The *integration* of function across many *levels of organization* is a special focus of physiology. (To *integrate* means to bring varied elements together to create a unified whole.)
—Silverthorn (2013, p. 2)

The concept of integrative levels of organization is a general description of matter... new levels of complexity are superimposed on the individual units by the organization and integration of these units into a single system. What were wholes on one level become parts on a higher one.... Each level of organization possesses unique properties of structure and behavior.
—Novikoff (1945 p. 204)

Compositional levels of organization... are constituted of families of entities usually of comparable size and dynamical properties, which characteristically interact primarily with one another, and which, taken together, give an apparent rough closure over a range of phenomena.
—Wimsatt (1994, p. 225)

In areas of science where researchers talk about "levels," these researchers also commonly give what I term "compositional" models and explanations where we explain by taking entities to compose each other. For instance, figure 12.1 frames the basis of the famous "sliding filaments" account, and associated explanations/models, of skeletal muscle contraction using compositional relations to activities of the parts of the muscle, shown in the figure, at what are termed by scientists various "levels."

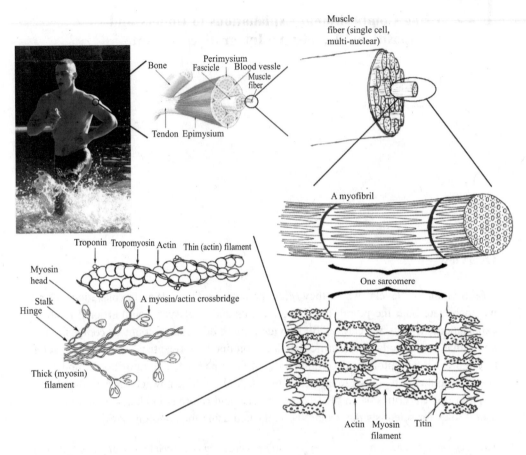

Figure 12.1
The basis of the famous sliding filaments model of skeletal muscle contraction. (Wikimedia commons image created by Raul654 distributed under CC-BY 3.0 license: https://en.wikipedia.org/wiki/Muscle_contraction#/media/File:Skeletal_muscle.jpg)

My goal here is to sketch a positive account of such levels building from a better appreciation of various species of compositional models/explanations and their integration with each other. Once we appreciate the ontological commitments of plural, integrated compositional models/explanations, I show we can finally provide a more adequate account of compositional levels that accommodates scientific practice while also overcoming the objections of critics.

I begin the chapter, in section 12.1, by briefly clarifying my project, assumptions, and methodology. Then, in section 12.2, I use examples from molecular biology, cell biology, and physiology to survey the various species of compositional model to show they extend beyond the so-called constitutive mechanistic explanations highlighted by the New Mechanists that explain an *activity* of a whole using compositional relations to *activities* of parts. In addition, I highlight the species of compositional explanation that explains a *property* of a whole, in what I call "Standing" models/explanations, and also a species of model that posits a compositional relation between a *whole* and its *parts*, in what I term an "Analytic" compositional model. Just as importantly, I highlight how the species of compositional model are usually *highly integrated* in their ontological posits in an interesting example of what Sandra Mitchell (2002, 2003) has termed "integrative pluralism." Where their plural models are ontologically integrated, we can expect that scientists may often consequently have larger ontological commitments, going beyond those of any one model, and I show that this is true in ways that ultimately illuminate what scientists refer to as "levels."

To explore these larger commitments of integrated models, in section 12.3, I show that compositional models/explanations are committed to what I term "activity closure." Thus, roughly put, in the sliding filaments model, framed in figure 12.1, cells act on other cells, organelles act on other organelles, and macromolecules act on macromolecules, but cells do not act on organelles or macromolecules or the muscle itself, or vice versa. I provide reasons to conclude we have such activity closure of the parts and wholes posited across integrated species of compositional models/explanations—that is, these parts and wholes cannot engage in activities with effects upon each other.

I then outline, in section 12.4, the still larger ontological commitments of plural, integrated compositional models incorporating such activity closure *and* the compositional relations posited in specific models. Consider, for example, the commitments of the plural models integrated with the sliding filaments model. Here we have groups of individuals that act only on other members of their groups (i.e., cells acting on cells, organelles acting on organelles, and macromolecules acting on macromolecules). But, in addition, the individuals in each group are related as parts and wholes (macromolecules composing organelles that compose cells that compose muscles), their activities are composed by the activities of individuals in other groups (activities of myosin and actin composing the contraction of myofibrils where these activities compose the cells' activities of contracting, which in turn compose the muscle's contracting), and the properties of wholes are also composed by properties of their parts.

The plural, integrated models of scientists are thus committed to layers of compositionally related entities that only act upon each other, where I term each of these layers a "compositional array." That is, roughly put, a group of individuals, and their properties and activities, that are (1) individuals that are all parts of the relevant terminal whole, (2) individuals that productively act upon other individuals in the group, but (3) individuals such that they do not act upon, and are not acted upon by, the terminal whole or individuals in other such groups. Once we appreciate such ontological commitments, then we see that scientists plausibly use "level" to refer to such compositional arrays, and I offer what I term the "Integrative Account" of levels to capture this insight. Most important, I show the Integrative Account captures key features of levels and scientific practices of ascribing such levels, such as the "cellular," "organelle," and "molecular" levels ascribed in the sliding filaments model.

One longstanding objection of philosophers of science, such as Philip Kitcher and Kenneth Schaffner, is that "levels" are inherently vague and imprecise.[1] However, in section 12.5, I show the Integrative Account allows us to precisely frame when entities are in a level, in the same level, and in a higher or lower level relative to each other, thus rebutting Kitcher and Schaffner's objection.

Using the Integrative Account, in section 12.6, I sketch some obvious practical reasons why scientists find it useful to talk of "levels" in the face of various forms of complexity. And I use this work, in section 12.7, to assess a new wave of philosophical skepticism about "levels," pressed by Marcus Eronen, Angela Potochnik, and others, who have argued that "scale" should replace "level" in actual practice.[2] However, I illuminate why "scale" cannot successfully perform the work of, or hence replace, "level" in actual scientific practice once we finally appreciate the character of this work.

12.1   Clarifying the Project and Some Assumptions

It is important to clarify what I am, and am not, seeking to do here, since my focus is superficially similar to some very different projects. To this end, it is most useful to start by distinguishing what I term "internal" and "ultimate" ontology.[3] I take certain models and explanations, that is, certain representations, to be intended to represent various ontological categories that I term their "ontological posits." My descriptive work here is focused on what we may term "internal ontology"—that is, the study of such ontological posits of successful scientific products in certain models and explanations. This contrasts with what we may term "ultimate ontology," which is focused on the nature of reality itself. Elsewhere I have discussed at length the ongoing scientific debates over what conclusions in ultimate ontology we may draw, if any, from our compositional models/explanations given their internal ontology, but here I focus solely on the descriptive project of internal ontology.[4]

I am an explicit pluralist both about scientific models, explanations and practices, and also about the concepts, including ontological notions, these various models, explanations, and practices deploy. Here I take myself to be describing only the ontological notions specific to certain compositional models and explanations where the same words are often used to express different ontological concepts in different models, explanations, and practices, so let me clarify my stance on some key issues.

First, for example, it is important to note that the entities posited in compositional models are *ahistorically individuated*—that is, these individuals, properties, and activities are individuated by what they do now.[5] The notions of "individual," "property," and "activity" in compositional models are therefore different from, for instance, those often used in evolutionary biology in models, explanations, and practices focused on *historically individuated* entities.

Second, when I talk of "part" and "whole," I mean the notions posited in compositional explanations whose nature I briefly sketch below.[6] Other notions of part are used in other kinds of product in other areas of the sciences (Healey, 2013). It is also unsafe to blithely assume that notions of "part" used in analytic metaphysics, whether from standard mereology or otherwise, track any of these scientific notions.

Third, I am a pluralist about the notions of "level" deployed in the sciences that come in explanatory, methodological, and ontological varieties, among others.[7] Once more, here I am only focused on the ontological notion of a "level" associated with compositional models/explanations. In ending the chapter, I briefly return to how some of the various notions of "level" might be related.

Fourth, and most important for my discussion, I need to carefully highlight my assumption that at least two different families of "causal" concepts are used in different models, explanations, and practices. As we shall see, central to compositional explanations is the thick causal concept of an "activity," or what I also term a "productive" relation, which is a behavior or doing of an individual resulting in effects upon another individual (or itself). In contrast, other models posit thin causal notions such as that of "manipulability" between individuals. I assume thick and thin notions of causation are distinct. Manipulability is often posited between two individuals *without* any such activity between these individuals. And activity plausibly requires more (in doings resulting in effects) than manipulability.[8] I illustrate these differences between activity and manipulability at various points in my discussion and also the importance of separating the thick and thin concepts.

12.2   Compositional Models/Explanations, Their Species, and Integration

More descriptive and theoretical work is needed for compositional models and explanations than I can devote to them here, since my focus is on the nature of "levels."[9] However, to illuminate the levels posited by scientists, I survey cases of compositional explanation from human physiology, and related areas of cellular and molecular biology, focusing primarily on our opening example. In 12.2.1, I look at models/explanations of an activity of a whole using a compositional relation to activities of parts. Then, in 12.2.2, I more briefly look at models/explanations of a property of a whole using a compositional relation to properties of parts. Most briefly, in 12.2.3, I note models of a whole itself using a compositional relation to other individuals that are its parts. I highlight the selective nature of such models, in 12.2.4, but more important, their integration with each other. In 12.2.5, I draw the findings of my brief survey together and highlight how my work offers potential insight about levels.

12.2.1   Explaining an Activity of a Whole: Dynamic Compositional Models/Explanations

In response to the question "Why did the muscle contract?" one good answer, in certain contexts, is, "The myosin crawled along the actin," where this is based around the sliding filament model framed in a textbook example in figure 12.2. Other good answers focused on what scientists explicitly term other "levels" in this model are that "the cell fibers contracted," at the cellular level, or, at the organelle level, that "the myofibrils contracted." These are all examples of the species of compositional explanation widely acknowledged by philosophers of science in what the New Mechanists term "constitutive mechanistic explanations" and which I term "Dynamic" compositional explanations and models.

Let us look more carefully at the type of model we find in figure 12.2 that underlies such an explanation. We explain the muscle's contraction at some time using a compositional relation, at that time, to the behavior of various individuals at a number of what are termed by researchers "levels," where these activities are taken to be compositionally related to each other and, transitively, to the muscle's contracting.

Working from the bottom of the diagram, the muscle's contracting is taken to be composed by many contracting cells (or "cell fibers") that are interconnected, or "organized," so that as each cell contracts, it pulls on the cells to which it is connected and which are also contracting. Hence, the contracting cells compose (or what I term "implement") and explain the muscle's contracting. Turning to the next pair of "levels," the cells fibers contracting are taken to be composed by their many contracting myofibrils, which are so

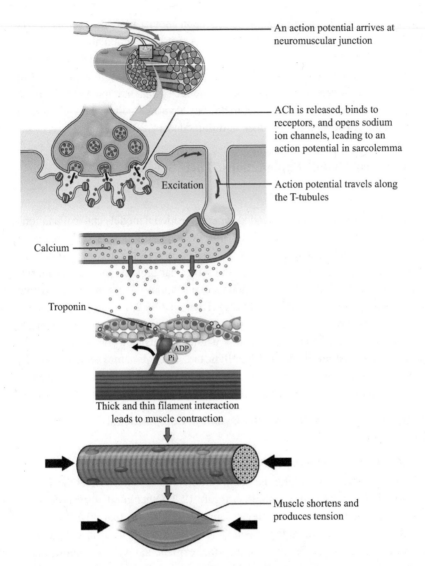

An action potential arrives at
neuromuscular junction

ACh is released, binds to
receptors, and opens sodium
ion channels, leading to an
action potential in sarcolemma

Action potential travels along
the T-tubules

Excitation

Calcium

Troponin

ADP
Pi

Thick and thin filament interaction
leads to muscle contraction

Muscle shortens and
produces tension

Figure 12.2
A textbook diagram of the sliding filament model of muscle contraction and a Dynamic compositional model.
(From Betts, 2013, chap. 10, section 10.3, figure 1)

organized, through their interconnection and alignment, that their contracting implements
and explains the cell fiber's contracting. Lastly, many myosin proteins crawling along actin
filaments implement and explain the myofibrils contracting. (This last pair of "levels" is
represented in one layer in the diagram since a myofibril is represented alongside its
component myosin and actin.)

   As has been widely noted by New Mechanist philosophers of science, these are explanations of an activity of an individual, in a skeletal muscle contracting, using a compositional
relation to activities of other individuals, in either cells, organelles like myofibrils, or
proteins of actin and myosin.[10] The latter is a descriptive claim in internal ontology about
the posits of these explanations/models—namely, that they posit activities of individuals

that are doings or behaviors that result in effects on themselves or other individuals. These effects involve changes in the motions of masses, and hence, unsurprisingly, these activities involve transfers of energy in one direction or the other.

Notice that in this latter observation about the posited activities involving transfer of energy, I am not making a claim about how to analyze the scientific concepts of particular activities or of the general notions of either "activity" or "causation." Rather, I am making an observation about our empirical findings for such activities—they always involve transfer of energy. I rely on such empirical findings below.

In addition, as is also widely accepted, the individual whose activity is explained and the individuals whose activities are used to explain it are taken by the researchers offering these explanations to be related as what they explicitly term *parts* and *whole*.[11] All of these parts and wholes are also parts of what I term a "terminal whole" in a whole not taken to be a part of further whole—in this case, the human body, as the model in figure 12.1 makes clear.

As textbooks and specific academic papers also routinely lay out, scientists take the proteins and their activities to be among the parts and activities of the skeletal muscle at what is termed the "molecular level," the myofibrils and their activities are among its parts and activities at the "organelle level," the cells and their activities are among its parts and activities at the "cellular level" and within the body itself, and the skeletal muscle is taken to be at the "organ" level.[12] So scientists routinely refer to "levels" in their Dynamic compositional models and explanations.

Lastly, it is worth noting the names of the latter levels only reflect the main known parts at each level—thus scientists take the molecular level in their models to include other entities than proteins, and the cellular level includes other individuals than cells. For brevity, scientists have named each level after the main kind of individual found at that level, whilst knowing there are other kinds of individuals at that level. In section 12.4.2 below, I outline the reasons why this situation arises of often having a mix of kinds, and even scales, of individual at a certain level.

12.2.2   Explaining a Property of a Whole: Standing Compositional Models/Explanations

There are plausibly other species of compositional explanation that even the New Mechanists have overlooked. For example, there are what I have elsewhere termed "Standing" compositional explanations and models that explain a property of a whole using a compositional relation, what is termed "realization," to properties of parts.[13]

Prominent among such Standing explanations in physiology, cell biology, and molecular biology are those where we explain the energy of a whole (i.e., one of its properties) using a compositional relation to the energy of its parts at a certain level (i.e., to properties of its parts). For example, we explain the energy of a skeletal muscle at a time using a compositional relation to the combined energies of all of its parts at the cellular level, or to all of its parts at the organelle level, at that time. In my coming discussion, I again rely on these well-confirmed compositional explanations of the energy of a whole.

Let me briefly point to another example focused on our primary case. In response to the question "Why is the muscle strong?" good answers, in the relevant contexts, include "Because the cells each have a certain strength" or "Because myosin has the property of exerting a certain force as it crawls along an actin filament" or an answer focused on properties of myofibrils. Here we explain the strength of the muscle, a property of a whole, using a compositional relation of realization to properties of its parts at what scientists again term the cellular, organelle, and molecular "levels."

12.2.3   Part–Whole Relations and Another Species of Compositional Model

Lastly, we should note that when asked "What is a skeletal muscle?" a number of good answers, in the relevant contexts, can be built from the model framed in our starting example in figure 12.1, including "bundled muscle fibers/cells," "organized macromolecules," or an answer built around organelles. It is also worth noting that this larger model combines smaller models of this kind, including the models in figure 12.3 framing parts of muscles at the tissue and cellular level, figure 12.4 framing parts of the cell at the organelle level, and figure 12.5 framing parts of a region of a myofibril at the molecular level.

Here we again plausibly have a type of model. But in these cases, there are models of a certain whole, and they posit a compositional relation to individuals that together compose (or as I shall say "constitute") this whole. Once more, working scientists explicitly refer to tissue, cellular, organelle, and molecular "levels" in this context with reference to such groups of constituent individuals. I term these "Analytic" compositional models.

Elsewhere, and more contentiously, I have suggested that this type of model and the answers noted are examples of another species of compositional explanation where the explanandum is a certain whole (i.e., an individual), the explanans is some group of individuals that are

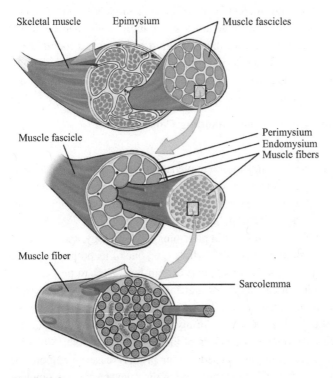

Figure 12.3
A textbook diagram of the composition of a skeletal muscle at tissue and cellular levels in muscles cells/fibers and hence an Analytic model of it. (From Betts et al., 2013, chap. 10, section 10.2, figure 1)

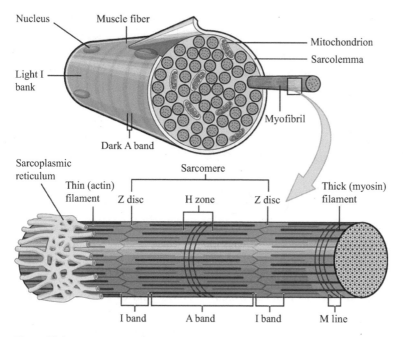

Figure 12.4
A textbook diagram of the composition of a muscle cell/fiber at the organelle level and thus an Analytic model of it. (From Betts et al., 2013, chap. 10, section 10.2, figure 2)

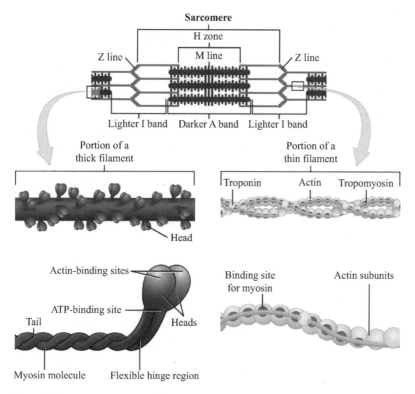

Figure 12.5
Textbook diagram of the composition of a myofibril at the molecular level and an Analytic model of it. (From Betts et al., 2013, chap. 10, section 10.2, figure 3)

its parts (at a certain "level"), and the backing relation is the constitution relation between these individuals.[14] Here I simply assume we have Analytic models and leave to one side the more contentious issue of whether they are also ever explanations.

It is worth noting that the individuals taken to be parts in such Analytic models are just those individuals whose activities compose activities of the whole or whose properties realize properties of the whole. The individuals highlighted by Dynamic and Standing models/explanations are thus those that are taken to be parts in Analytic compositional models/explanations. And, in the reverse direction, the individuals taken to be parts of the whole in Dynamic and Standing models/explanations are typically the individuals posted in successful Analytic models.

12.2.4   Selective Ontological Representations and Integrative Pluralism

The literature on models is large and familiar to philosophers of science, so let me briefly note a common feature that these compositional models share with other models.[15] The subject matter of physiology, cell biology, and molecular biology is highly complex, in many ways. To make such complex phenomena cognitively tractable, scientists follow the strategy made familiar in work on models, which is "Divide and Cognize," rather than "Divide and Conquer."

To see this, consider our compositional models focused on the skeletal muscle, its activity of contracting, and related properties, that we have just surveyed. In each of the species of compositional model, just *one* category of entity is taken as explanadum and just *one* instance of that category. For example, in our Dynamic model, just *one* activity is the explanadum; in our Standing models, just *one* property is the explanadum; and in our Analytic model, *just the individual itself* is the focus, rather than any of its activities or properties. Similar points hold about the selectivity of the explanans represented in each model: Dynamic models just use activities of parts, Standing models just use properties of parts, and Analytic models just use parts. In fact, the categories explicitly represented in these models are usually limited in these ways; thus, Dynamic models do not represent properties, Standing models do not represent activities, and Analytic models usually represent neither properties nor activities.[16]

What compositional relation is represented, and is the backing relation of the model/explanation, is also selective. Each of the species of compositional model we have looked at has just *one* kind of compositional relation explicitly represented in the relevant model. Thus, a Dynamic model explicitly represents just an implementation relation between activities of parts and whole, a Standing model just explicitly represents a realization relation to properties of parts and whole, and an Analytic model explicitly represents just a constitution relation to individuals.

Scientists are well aware, for example, that there are many other compositional relations or that the phenomenon is an individual with many activities and also many properties. But compositional models are like so many models in being *selective* representations in these various ways. Such models are presumably also often idealizations, and/or abstractions, and/or have other features, but their nature as selective representations suffices for my purposes here. More important for my purposes is to highlight how, alongside the manner in which scientists "Divide and Cognize," we also find significant ways that the resulting models are also integrated.

First, and most bluntly, we find that larger compositional models, like that in figure 12.1, incorporate less encompassing models like those in figures 12.3, 12.4, and 12.5. Here, Analytic models of individuals and their parts at lower levels are combined into an Analytic model encompassing all of these individuals and their relations. Similar examples are found with Dynamic and Standing models and hence with the activities and properties of parts and wholes.

Second, if we just focus on the relations across the species of compositional model associated with skeletal muscles, their contraction and the relevant properties, we see other forms of integration. For notice the systematic, and consistent, integration among the individuals posited in these models. Focusing on the organ, organelle, and molecular levels, our Dynamic, Standing, and Analytic models, for instance, all posit a skeletal muscle, myofibrils, and actin and myosin proteins, among other individuals. So a crude point is that the internal ontology across the models is integrated by positing overlapping kinds of individuals at various levels.

Third, we find such systematic integration of the internal ontology of models even with posited entities that do not overlap across these models. For example, although Dynamic models posit activities and no properties, and Standing models posit properties and no activities, the internal ontology of these models is still systematically integrated in another way. The properties posited in parts and wholes, at various levels, in Standing models are those that manifest themselves in the activities posited, at various levels, in the relevant Dynamic models. For instance, in a skeletal muscle energy and strength are the properties that result in the activity of contracting in the muscle when manifested, while having a certain energy and exerting a certain force in myosin proteins are the properties that result in the myosin's activity of moving down a chain of actin when manifested. Similar points hold about the properties and activities of myofibrils at the organelle level. The properties posited in Standing models, and the activities posited in Dynamic models, are thus systematically integrated. Consequently, the internal ontology of Standing and Dynamic models is integrated not just through individuals that each posits, but even with those entities, like properties and activities, that these models do not overlap in positing.

Fourth, as I noted earlier in relation to Analytic compositional models, the individuals counted as parts of a certain whole are plausibly those individuals whose properties and/ or activities are successfully posited in Standing and Dynamic models as composing properties and activities of the relevant whole. And Standing and Dynamic models count as parts of those individuals figuring in Analytic models. So the three species of compositional models exhibit another kind of ontological integration between their posited backing relations.

There are still further forms of such ontological integration (i.e., integration in the models' internal ontology), but just these examples illustrate how the species of compositional model/explanation about some phenomenon are often highly integrated, and so too are specific examples of each species of model.

We should briefly note how this latter kind ontological integration is not limited to compositional models but also *extends to certain types of causal models* such as the "etiological mechanistic explanations" that the New Mechanists have documented as also positing activities of an individual to explain certain effects.[17] Thus, for example, we explain a certain bone moving by positing an activity of contracting in a muscle. Or we explain a change in the length of a myofibril by positing an activity in it of contracting. Notice that such causal models posit activities and individuals overlapping those in our Dynamic compositional models. And such etiological models are also bluntly incorporated into Dynamic compositional models. Given the ontological integration of Dynamic and other compositional models, we can thus see that causal models are also plausibly systematically ontologically integrated with these compositional models.

Overall, our survey thus documents how compositional models and explanations, and related causal models and explanations, are plausibly another example of what Sandra Mitchell (2002, 2003) has termed "integrative pluralism" in cases where a variety of different kinds of models and explanations are integrated. The kind of integration here is philosophically novel because it involves the systematic integration of the internal ontology of a variety of distinct species of model. So, in the following sections, I explore the potential implications of such ontological integration for understanding what scientists mean by "level" when talking about their integrated compositional models.

12.2.5   A Neglected Family of Models/Explanations and a Novel Form of Integrative Pluralism

Let me briefly summarize our findings. All of the compositional explanations we have surveyed are what I shall term "ontic" explanations that work by representing an ontological relation between entities in the world, the explicitly represented backing relation of the explanation, whose ontological nature drives these explanations. Furthermore, these explanations are all backed by compositional, rather than causal, relations, since their backing relations all share common ontological features lacking in causal relations. For example, among other singular features, their backing relations are all synchronous relations, between entities that are in some sense the same (although not identical), and which involve synchronous changes in their relata.[18] So we can thus see that these are plausibly not causal explanations.[19]

We thus have another, philosophically neglected, family of ontic explanations of singular facts or events in the sciences in addition to causal explanations—hence supporting a pluralism about ontic explanation.[20] I term these "compositional" models or explanations, rather than "mechanistic" models or explanations, since Standing and Analytic explanations are not mechanistic because they do not have activities as their explanans or explanandum.[21] And my brief survey highlights how pluralism is also true about this family of models/explanations itself, since we have found at least three species that differ, in the ways framed in table 12.1, in the categories of entity they represent, their explanans, explanandum, and their backing relation.

As we have just seen, species of compositional models/explanations are systematically integrated in their internal ontology in a way that extends to related causal models, too. This is interesting, since we have seen that working scientists also routinely use the term "level" to refer to all of the integrated species of compositional model/explanation—and in a systematic and consistent manner across such integrated models. Does what scientists mean by "level" relate in some way to the ontological integration of compositional and causal models? In coming sections, I explore this idea in stages by looking at whether such integrated models consequently have larger ontological commitments that have previously been overlooked by philosophers.

Table 12.1
Species of compositional model/explanation and their differing characteristics

	Categories represented	Explanandum	Explanans	Backing compositional relation
Analytic compositional model (and/or explanation)	Individuals	An individual whole	Individual parts	Part–whole relations (i.e., *constitution*) between individuals
Standing compositional model and explanation	Properties and individuals	A property of a whole	Properties of parts	*Realization* of property of whole by properties of parts
Dynamic compositional model and explanation (a.k.a. constitutive mechanistic explanation)	Activities and individuals	An activity of a whole	Activities of parts	*Implementation* of activity of whole by activities of parts

12.3   The Activity Closure of Parts and Wholes in Compositional Models/Explanations

In this section, I look at whether our integrated compositional models have either of two forms of what I term, in 12.3.1, "activity closure" of parts and wholes—that is, parts and wholes that do not act upon each other. In 12.3.2, I outline descriptive evidence from our examples that scientists *do* respect activity closure in their compositional models/explanations. Next, in 12.3.3, I present arguments, using features of our integrated models, and related empirical findings, to that show scientists *ought* to accept activity closure. I conclude, in 12.3.4, that scientists do have larger ontological commitments from their groups of integrated compositional models. For example, scientists consequently do, and ought to, take compositional explanations to embody activity closure of thick causal relations of parts and wholes across such distinct but integrated compositional models. However, I also carefully mark that scientists have other models illuminating thin causal relations that apparently can hold over time between such parts and wholes. Scientists thus have a complex set of commitments to track.

12.3.1   Two Forms of Activity Closure

It is important to frame, and distinguish, two closure claims about the activities of parts and wholes in integrated compositional models. First, there is the claim that individuals directly related to each other as part and whole cannot act upon each other. This is what I term "Narrow Activity Closure":

> (Narrow Activity Closure) An individual that is a scientific part of some other individual cannot engage in activities that have effects upon this whole or vice versa.

This thesis is a "closure" claim about activities, since it claims that individuals directly related as scientific parts and wholes are closed with regard to the activities of each other. Thus, a cell that is part of a skeletal muscle cannot engage in an activity with an effect on this muscle or vice versa.

Second, we have a far wider form of activity closure—namely, that parts and wholes in the same terminal whole cannot act upon each other regardless of whether these individuals are themselves related to each other as part and whole. Call this wider thesis "Broad Activity Closure":

> (Broad Activity Closure) Two individuals, s_a and s_b, that are each parts of some individual s^*, but where s_a and s_b are not related as part and whole, and where s_a and s_b are at different levels in s^*, cannot engage in activities that have effects upon each other.

Here we have the claim that any individuals at what scientists term different "levels" within the compositional hierarchy of the same terminal whole, and even though not directly related as part and whole, cannot act upon each other.[22]

12.3.2   Justifying Activity Closure (I): Descriptive Evidence

I have sketched these theses about activity closure since it appears integrated compositional models are systematically committed to activity closure. But what is the descriptive evidence that scientists respect either form of activity closure? To appreciate this evidence, let us look at the Dynamic compositional explanation of muscle contraction from figure 12.1 or 12.2.

For instance, the explanation posits cells that act on other cells, but not cellular activities with effects on the muscle, organelles like myofibrils, or macromolecules, nor any activity of the muscle, organelles, or macromolecules with effects on the cells. We thus find that this Dynamic model respects both Narrow and Broad Closure, where this is a descriptive claim about the posits of the explanation and its underlying model in figure 12.1 or 12.2. And similar points hold for other Dynamic models and explanations.

In addition, consider the internal ontology of our examples of Standing compositional models and explanations where we find properties of cells that manifest in activities on other cells but not in activities with effects on the muscle, organelles like myofibrils, or macromolecules. And in such Standing models, we do not find properties posited in the muscle, organelles, or macromolecules that manifest in activities with effects on the cells. So we see that these Standing compositional models and explanations also plausibly respect both Narrow and Broad Closure, and again, we plausibly find this feature in other Standing models/explanations.

Lastly, it is worth noting that given the systematic integration of the internal ontology of Dynamic and Standing compositional models, as well as to Analytic and causal models, we thus plausibly find consistent commitments to activity closure of the same entities across distinct, but integrated, models.

12.3.3   Justifying Activity Closure (II): Energetic Arguments

We have found plausible descriptive evidence that in practice, working scientists *do* respect both Narrow and Broad Activity Closure in their integrated compositional models/explanations. But we can still ask whether they *ought* to accept either Closure claim. To answer this question, I want to explore the joint commitments of our integrated Dynamic and Standing compositional explanations, along with other empirical evidence, in a kind of reductio ad absurdum argument, what I dub an "Energetic Argument," focused on the energy of parts and wholes.[23]

Let us start with a reductio argument for Narrow Closure. So consider the muscle cell, s_{b1}, and a molecule of myosin that is one of its parts, s_{a1}, and for the sake of reductio, assume the myosin productively acts on the cell between t1 and t2 to change the cell.

Given our empirical evidence about activities, I assume that the latter activity and change involves a transfer of energy. So the myosin s_{a1} transfers energy Y to the cell s_{b1} by t2. But, at t1, we know from our Standing explanations that the energy of the cell equals the combined energy of all the proteins that are its parts, in $s_{a1}-s_{an}$, including our myosin molecule, and let this equal N. So at t1, the energy of the cell is N. But, given the transfer of energy by t2 from s_{a1}, we can conclude the energy of the cell at t2 is (N+Y). In similar fashion, given this transfer, we can also conclude the energy of the molecular parts $s_{a1}-s_{an}$ at t2 is (N−Y). But, at t2, the cell's energy is the combined energy of the parts. So we can also conclude that at t2, the energy of the cell is (N−Y). Therefore, the energy of the cell at t2 is, and is not, (N+Y)—a contradiction.

This Energetic Argument takes premised form thus:

(1) All activity involves transfer of energy.

(2) The energy of the cell s_{b1} at time t equals the combined energies of the proteins, $s_{a1}-s_{an}$, that are its parts at the molecular level at t.

(3) At t1, the energy of the proteins $s_{a1}-s_{an}$ equals N.

(4) Between t1 and t2, the myosin protein s_{a1} acts upon the cell s_{b1}.
      From (2) and (3):

(5) At t1, the energy of the cell s_{b1} equals N.
      From (1) and (4):

(6) By t2, the protein s_{a1} transfers Y energy to the cell s_{b1}.
      From (5) and (6):

(7) At t2, the energy of the cell s_{b1} is N+Y.
      From (3) and (6):

(8) At t2, the energy of the proteins $s_{a1}-s_{an}$ is N−Y.
      From (2) and (8):

(9) At t2, the energy of the cell s_{b1} is N−Y.
      From (7) and (9):

(10) At t2, the energy of the cell s_{b1} is, and is not, N+Y.

Here some premise must be false. Premise (1) is the empirical claim that all activities of individuals, of the relevant kind, involve transfer of energy. In premise (2), we have the empirical finding, from Standing compositional explanations, that the energy of a whole cell is equal to the combined energies of its parts at the molecular level. Premise (3) is a simple assumption, again empirically supported, about the starting energy of the proteins. So the only plausible candidate to be false is premise (4) and the claim that a part acts upon its own whole.

The very same kind of argument can be run against the claim that a whole can act upon its own part. Thus, Energetic Arguments support Narrow Activity Closure—that is, that in compositional models/explanations, a part cannot act upon its own whole or vice versa. However, we can press such Energetic Arguments further to support Broad Activity Closure and the claim that parts and wholes, at different levels in the same terminal whole, cannot act upon each other despite not themselves related as part and whole.

To see this, consider our skeletal muscle, s^*, which has as parts at the cellular level certain cells, s_{b1}–s_{bm}, and has as parts at the molecular level certain proteins, s_{a1}–s_{ax}, where the proteins are also parts of the respective cells. Assume that a myosin protein, s_{a1}, is a part of the cell s_{b1} but is not a part of the cell s_{b2}. In this situation, assume for reductio's sake that between t1 and t2, our myosin protein, s_{a1}, acts upon the cell s_{b2} of which it is not a part. So, from t1 to t2, s_{a1} through this activity transfers energy Z to s_{b2}. At t1, the combined energy of all the cells s_{b1}–s_{bm} that are parts of s* was M. Thus, we can conclude that at t2, given the transfer from s_{a1} to s_{b2}, the energy of the cells s_{b1}–s_{bm} is (M+Z). We may assume that the combined energy of the proteins that are parts of the muscle s* at t1 was also M. So we can also conclude, given the transfer from s_{a1} to s_{b2}, that at t2 the combined energy of the proteins s_{a1}–s_{ax} is (M–Z). However, the energy of the muscle just is the combined energies of its parts at a level. So, we can conclude that at t2, the muscle s* does, and does not, have an energy of M+Z.

This Energetic Argument can be framed as follows:

(11) All activity involves transfer of energy.

(12) The energy of a whole at a time equals the combined energies of its parts at a certain level at that time.

(13) At t1, the energy of all the proteins s_{a1}–s_{ax}, which are parts of the muscle s* at the molecular level, equals M.

(14) At t1, the energy of all the cells s_{b1}–s_{bm}, which are parts of the muscle s* at the cellular level, equals M.

(15) Between t1 and t2, the myosin protein s_{a1} acts upon the cell s_{b2}.
     From (11) and (15):

(16) By t2, the protein s_{a1} transfers Z energy to the cell $s_{b2.}$
     From (14) and (16):

(17) At t2, the energy of all the cells s_{b1}–s_{bm}, which are parts of the muscle s*, is M+Z.
     From (13) and (16):

(18) At t2, the energy of all the proteins s_{a1}–s_{ax}, which are parts of the muscle s*, is M–Z.
     From (12) and (17):

(19) At t2, the energy of the muscle, s*, is M+Z.
     From (12) and (18):

(20) At t2, the energy of the muscle, s*, is M–Z.
     From (19) and (20):

(21) At t2, the energy of the muscle, s*, is, and is not, M+Z.

Once more, one of the premises leading to this contradiction must be false. Here, premises (11) and (12) are supported by the same empirical findings supporting the premises of our simple argument, while (13) and (14) also simply frame well-supported empirical claims. Thus, premise (15) is plausibly the false premise. Similar arguments again apply to activities of a whole on a lower-level part. So a part cannot act upon a whole, or vice versa, even when they are only in the same compositional hierarchy at different levels and not themselves directly related as part and whole. We thus have an independent argument, built upon empirical findings, supporting Broad Activity Closure.

12.3.4   Part–Whole Causation in Compositional Models/Explanations?
Thick Relations of Activity, No; Thin Relations, Yes

My work in this section supports both Narrow and Broad Activity Closure about the internal ontology of compositional models/explanations, but it is worth exploring a potential objection since it brings out more clearly the scope of my conclusions and some implications for wider questions.

Someone might object that my arguments face a crippling problem because they prove too much by showing there is no causation between parts and wholes. But, the objection continues, working scientists routinely, and apparently highly successfully, illuminate causation between parts and wholes, and their activities and properties, using interlevel experiments. So, concludes the objection, something has gone badly wrong with my arguments about activity closure.

However, my arguments in this section *only apply to activities*, and hence *thick* causal relations, but not to manipulability, difference-making, and *thin* causal relations. My arguments show that compositional models/explanations do not, and should not, posit activities—and thick causal relations—between parts and wholes and/or their activities and properties. But my arguments do not apply to, and do not show anything about, whether there are thin causal relations of manipulability or difference-making over time between parts and wholes and/or their activities and properties. Plausibly, interlevel experiments establish that such thin causal relations often hold, over time, between the parts and wholes in a certain terminal whole.

My arguments therefore do not show too much, nor do their conclusions conflict with results from interlevel experiments.[24] What the objection does usefully highlight is that scientists inherit a complex set of commitments from their various models. From their successful compositional models, scientists are committed to parts that productively interact with other parts and wholes that act upon other wholes, but where there is no activity between them. From their plethora of models derived from interlevel experiments, scientists are committed to various thin causal relations over time, whether manipulability or difference-making relations, between these same parts and wholes, and their activities or properties.

Scientists thus have a complex set of ontological commitments, framed in figure 12.6, as a result of their various models. Pretty obviously, it will be practically challenging for researchers to track such commitments to both thick and thin causal relations. In subsequent sections, I explore how use of the term "level" might aid scientists in navigating this complexity.

12.4   The Integrative Account of Compositional Levels

We have found that we have a variety of species of compositional models that are systematically integrated in their internal ontologies and where such models posit compositional relations, but also embody activity closure. Drawing these findings together, in 12.4.1, I outline how our integrated compositional models are consequently committed to what I term "compositional arrays" of individuals, properties, and activities that are compositionally

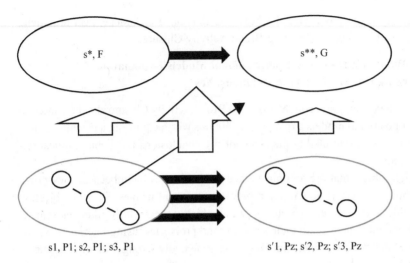

s1, P1; s2, P1; s3, P1   s'1, Pz; s'2, Pz; s'3, Pz

Figure 12.6
Diagram of the combined ontological commitments of scientists when incorporating their compositional models
and models positing thin causal relations between parts and wholes. Circles are individuals, thick horizontal lines
are activities, thin diagonal lines are thin causal relations, and vertical hollow arrows are compositional relations.

related and embody activity closure. I sketch what I term the "Integrative Account" of levels
that takes scientists to use the term "level" to refer to these compositional arrays. I then
show the Integrative Account captures scientific practices of level ascription and does better
in this respect than extant accounts such as Craver (2007). Furthermore, in 12.4.2, I show
how the Integrative Account illuminates various features of levels, including the ways in
which compositional levels are local and their nuanced relationship to scales.

12.4.1   Integrated Ontological Commitments, Compositional Arrays, and Integrative "Levels"

Our primary example, in the integrated compositional models concerning skeletal muscles,
illustrates the commitment of these models to both activity closure and compositional relations. The compositional models about skeletal muscles we have looked at, roughly put, take
proteins to act on other proteins and macromolecules, organelles that act on other organelles,
and cells that act on other cells, but none of the individuals in these groups act upon individuals in other groups or the muscle itself—thus, the models embody activity closure.

In addition, these integrated compositional models take the individuals in these groups,
and their activities and properties, to be compositionally related. Proteins together compose
organelles that together compose cells that together compose the muscle. And the models
take activities of myosin and actin to compose the contraction of myofibrils where these
activities compose the cells' activities of contracting, which in turn compose the muscle's
contracting. Similar points hold about the properties of wholes that we have seen are also
composed by properties of their parts.

Scientists thus have complex, and highly ordered ontological commitments from their
plural, integrated compositional models in which, across these groups of systematically
integrated models, we find closed layers of individual components that only act upon each
but do not act on individuals in other closed layers. I coin the neutral term "compositional

array" to refer to these individual layers.[25] A compositional array is a group of individuals (and their activities and properties) such that (1) individuals in the group are all parts of the relevant terminal whole, (2) individuals in the group productively act upon other individuals in their group, but (3) individuals in the group do not act upon, and are not acted on by, the terminal whole or individuals in other such groups.

Appreciating that plural, integrated compositional models are committed to compositional arrays is important in itself. But what should also be striking in our examples of such compositional models is that scientists apparently use the term "level" to refer to just such compositional arrays. What I call the "Integrative Account" of levels frames this idea that "level" is used by scientists in these areas to refer to the compositional arrays to which their integrated compositional models are committed. Later I provide a precise framework for level ascriptions under the Integrative Account. But let us now see how the Account does in accommodating actual ascriptions of "levels" and their features.

We can quickly confirm that the Integrative Account matches actual scientific practice. To see this, consider what the Account says about the levels of various entities in the sliding filament model and associated models. First, the Integrative Account places cells at the same level, since cells are components and productively interact with each other. And the Account places the properties and activities of cells at this level, too. Similarly, the Integrative Account places interacting component proteins and macromolecules, as well as their activities and properties, at the same level. And it places interacting component organelles, as well as their activities and properties, at the same level. And the Integrative Account places cells, or organelles, or proteins, or their properties and activities, at different levels. All of these ascriptions match the actual practices we have seen in our examples.

Let us briefly look at what the account says about cases where scientists *withhold* level ascriptions. Consider a virus newly entered into a cell. Here the Integrative Account says that the virus is at no level, since our models take the activities and properties of the virus to compose no activities or properties of the cell or the body—thus the virus is in no compositional array and hence no level. Again, this matches actual practice, since researchers do not take the virus to be a component or at a level.

The Integrative Account of levels thus appears to fit well with actual practices of ascribing/withholding "levels" in compositional models and explanations, and associated accounts. So let us change gears to critically compare the Integrative Account to the best extant treatment of compositional levels in Carl Craver's (2007) "Levels of Mechanisms Theory." Craver's framework is *solely* based on Dynamic compositional explanations, what he terms constitutive mechanistic explanations, since this is the only species of compositional model/explanation so far recognized by the New Mechanists—hence explaining why Craver's account is based around compositional relations of just the activities and individuals we find in this one species of model. Famously, Craver's Levels of Mechanisms Theory has two internal problems (Eronen, 2015): first, Craver's view cannot say when entities are at the same level, and second, Craver's account cannot provide level ascriptions across the entities posited in distinct models.

The Integrative Account obviously contrasts in its basis from Craver's treatment, and we can now see that this allows the Integrative Account to overcome the difficulties of the Levels of Mechanisms Theory. Crucially, the Integrative Account acknowledges Standing and Analytic compositional models, as well as Dynamic models, and takes them all to be integrated about certain phenomena. So the Integrative Account does not take "level" to refer to the internal ontology of a single compositional model or even a single species. Instead, the Integrative Account takes "level" to refer to the ontological commitments of plural, integrated compositional models and often causal models, too.

Given these differences, first, the Integrative Account does underwrite ascriptions of when entities are at the same level, since this is driven by entities being in the same compositional array due to their productive and compositional relations. In section 12.5, I provide a precise framework for when entities are at the same level under the Integrative Account that confirms this point.

Second, it should also be obvious that the Integrative Account makes level ascriptions across entities posited in different models and hence also across different categories of entity, including properties as well as activities and individuals. Thus, for example, the Integrative Account places an activity of myosin, posited in our Dynamic compositional explanation, as at the same level as the property of the myosin of exerting a certain force, posited in our Standing compositional model, since both entities are in the same compositional array. The framework offered in section 12.5 again confirms these points. The Integrative Account of compositional levels thus avoids the main problems of Craver's Theory.

It is also worth emphasizing at this point that, for simplicity of exposition, I have focused throughout the chapter on just a handful of integrated compositional models focused on one group of phenomena in those associated with skeletal muscles and their contraction. But the muscle has many other activities than contracting and many other properties than strength, and the compositional models for these activities and properties are also ontologically integrated with the few models I have looked at here.

Furthermore, the muscle and its various activities and properties themselves figure as components in many other compositional explanations of the body, and its activities and properties, where these models involve various other organs and their activities and properties, as well as their components, their activities, and their properties at lower levels. Once more, these compositional models and explanations of the body are plausibly also ontologically integrated with each other and with compositional models of components like the skeletal muscle, cells, organelles, and so on, including those I have looked at.

Given the latter points, it is clear that the extent of our integrated compositional models driving level ascriptions under the Integrative Account are often far greater than the handful of compositional models I explicitly looked at in the case in section 12.2. The precise extent of such ontological integration across models, and hence the extent of various "levels," is an important avenue for future research, so I wanted to mark it for the reader.

12.4.2   Their Local Nature, Nuanced Relations to Scale, and Other Features of Levels

The Integrative Account of levels can also be shown to accommodate, and illuminate, many of the features of compositional levels highlighted in the descriptive work of writers like Silverthorn, Novikoff, or Wimsatt quoted in the starting passages of this chapter. For example, reflecting on features of compositional relations themselves highlights why compositional levels have the feature Novikoff notes of involving qualitatively distinct individuals, properties, and activities (Gillett, 2016a, chap. 2). Earlier sections on activity closure have also already highlighted why the Integrative Account is built around levels that are what Wimsatt (1994) terms "families of entities ... which characteristically interact primarily with one another, and which, taken together, give an apparent rough closure over a range of phenomena" (p. 225). And we have now seen that the Account obviously incorporates the centrality of integration that Silverthorn emphasizes.

Below I also detail how the Integrative Account illuminates why Wimsatt is also right that levels concern "families of entities usually of comparable size and dynamical properties." But let us build up to that point by first highlighting how the Integrative Account supports the claim that compositional levels are local as writers like Craver (2007) and others have emphasized. For when a part acts beyond the boundary of its terminal whole, then a part is acting on a nonpart and we do not have activity closure, a compositional array, or a compositional level. So we can see that under the Integrative Account, we do not have global levels, but only levels that are local to the boundaries, both external and internal, of the relevant terminal whole.

The levels supported by compositional explanations are also local in another sense. If different kinds of individual organize, and productively interact, in diverse ways in distinct kinds of individual, then we have different types and/or numbers of level across different kinds of terminal whole. And this appears to be what we find. For example, there is no cellular, tissue, or organ level in rocks or glaciers. So the extent of a level is local to its terminal whole, but so too are the kind, and hence numbers, of level.

Critics of "levels," like Eronen and others, have pointed to putative problem cases where we sometimes have entities of the same kind, and hence the same scale, at different levels. But the Integrative Account illuminates why these are not problematic cases and also illuminates the relation of levels and scales.

If one individual of a kind productively interacts with the individuals in one compositional array, while another individual of the same kind interacts with individuals in another compositional array, then the individuals are in distinct compositional arrays and the Integrative Account illuminates why scientists hence place the different individuals of the same kind, and hence the same scale, at distinct levels.

The Integrative Account consequently allows us to better understand the relation of scale and compositional level. The key point to appreciate is that specific composed and component entities, i.e. entities related compositionally to each other, are always of different *relative* scales, since the compositional relations posited in compositional models/explanations are many–one, with many components and one composed entity (Gillett, 2016a, chap. 2).[26] Thus, many activities of proteins implement the muscle's contracting, many properties of proteins realize the muscle's strength, and many proteins constitute a muscle. Given the many–one character of scientific composition, the composed (whether individual, property, or activity) will be many times its components in size, mass, energy, duration, and so on— hence leading to the *relative* difference in scale of composed and components entities at different levels.

If we combine the latter point with the reasons why we can end up with entities of different scales in the same level, due to their particular productive interactions, then the Integrative Account illuminates why Wimsatt is correct in carefully emphasizing that entities at the same level are only *usually* of the same scale and that entities at higher and lower levels are only *usually* of a different scale, since neither is *absolutely* the case. Although each entity at a level is of a different relative scale than its own components, there is no guarantee that this entity is of the same scale as other entities at the relevant level, since entities of different scales may end up in the same compositional array due to their productive relations with each other. In fact, what we sometimes end up with is a motley of individuals at a level that all productively interact though being of different kinds and even different scales. As I noted in 12.2.1 above, for simplicity researchers often, for example, refer to the "molecular level" but they are well aware that there are many other kinds of individuals than proteins that are parts at this level (i.e., ions, lipids, sugars, nucleic acids, and many other nonproteins).[27] Similar points apply to other levels.

12.5   Same, Higher, and Lower Level under the Integrative Account: Rebutting the Kitcher–Schaffner Objection

The wider scope of "compositional level" under the Integrative Account that spans many categories and models, particularly in contrast to Craver's framework, might raise concerns that we have a vague notion of "level." As I noted, earlier critics, like Kitcher and Schaffner, dismissed the notion of a "compositional level" because they claimed that no precise answers about levels can be given.

However, the Integrative Account is not intrinsically vague in its ascriptions and shows that the Kitcher–Schaffner objection is simply mistaken. Under the Integrative Account, using the ideas of earlier sections, we can now precisely frame what it is to be in a level, and at the same level, as follows:

(Level) Aside from the terminal whole w, which is at its own level, individuals s_{b1}–s_{bn}, and their properties and activities, are in the same level, and hence in a level, within the whole w, under conditions \$ at time t, *if and only if*, at t under \$, (i) individuals s_{b1}–s_{bn}, and their properties and activities, bear compositional relations to w and its properties and activities; (ii) the individuals s_{b1}–s_{bn} can or do productively interact with each other; and (iii) if individuals s_{b1}–s_{bn}, and their properties and activities, bear compositional relations to individuals s_{a1}–s_{am}, and their properties and activities, then individuals s_{b1}–s_{bn}, and their properties and activities, do not productively interact with individuals s_{a1}–s_{am}, and their properties and activities, or vice versa.

Here we have the main features of compositional arrays, in (i) to (iii), that we have already sketched. The resulting account of a level successfully provides precise, same-level ascriptions cross-categorially for individuals, activities, and properties across distinct, but integrated, models. This confirms that the Integrative Account, unlike Craver's Levels of Mechanisms framework, makes ascriptions about when entities are at the same level. (As an aside, note that the terminal whole is often taken by scientists to be at a level, although not in a group of interacting components, so I have treated it as special case, but condition (iii) still entails we have activity closure with the terminal whole and its parts.)

Furthermore, building on the latter, we can precisely articulate under the Integrative Account when for any two individuals, or their properties and activities, they are at higher or lower levels relative to each other as follows:

(Higher/Lower Level) An individual s, or its property or activity, is in a lower [higher] level relative to an individual s_x, or its property or activity, in whole w, under \$ at time t, *if and only if*, under \$ at time t, (i) s is in a level in w, (ii) s_x is in a level in w, and (iii) s or one of its activities or properties composes [is composed by] an entity, whether an individual, property, or activity, in the same level as s_x.

Again, this definition works cross-categorially for individuals, or their activities and properties, and it also applies regardless of whether two entities bear compositional relations to each other. So we get precise cross-categorial ascriptions of being higher or lower level that once more apply not just to one model but across various integrated models of however great an extent.

We can therefore see that Kitcher and Schaffner's claims about the vagueness and imprecision of levels is mistaken. Rather than levels being in some way problematic, it appears that the neglect of compositional models/explanations precluded philosophers from understanding compositional levels in all their complexity and sophistication.

12.6   Some Reasons Why Scientists Talk about "Levels"

Can we now offer any ideas about why working scientists might find it useful to talk of levels? The most obvious utility of using the term "level" derives from the complexity of compositional arrays. Scientists *could* talk about groups of productively interacting individuals, and their activities and properties, that compose (and are composed by) individuals, and their activities and properties, in other such groups, where these groups of individuals, and their activities and properties, do not act upon each other. But that is a very large mouthful! Instead, researchers have coined a term, in "level," that allows them to concisely talk about these complex commitments.

On top of this simple reason, our earlier observations also highlight why the term "level" is practically useful in other ways. Compositional explanations and models posit parts and wholes that are closed to each other's thick causal relations of activity. However, interlevel experiments generate other models positing thin causal relations between the same entities over time, whether manipulability or difference-making relations. In this situation, active researchers face numerous practical challenges in tracking "causal" relations. Using the term "level," in the manner framed by the Integrative Account, aids researchers with these difficulties by allowing them to more swiftly track, and/or communicate, which entities are in which arrays. This is important as researchers sift through the significance of their various models. For instance, talking of "levels" allows scientists to track which thin causal relations revealed by experiments pick-out activities consistent with the commitments of their compositional models/explanations.

There are plausibly other reasons why it is useful to talk in this way of "levels."[28] But just the two kinds of reason I have outlined suffice to highlight the utility of talking about "levels" in areas of science where plural, integrated compositional models/explanations are common.

12.7   The New Philosophical Skepticism about Levels: Why Scale Fails as a Replacement for Level in Scientific Practice

Although we now have a strong case that the concept of a "compositional level" plays a useful role in actual practice, some philosophers of sciences have argued that this practice should be revised. A number of writers have recently advocated replacing the concept of "level" with the notion of "scale," including Potochnik (2010, 2017; Potochnik & McGill, 2012) and Eronen (2013, 2015), among others. So we need to briefly assess this proposal.

When we seek a replacement for some successful scientific notion X, then the main work one needs to do is to show how the broached replacement serves the purpose of notion X just as well or better. One can therefore feel some sympathy for the recent proponents of scale as a replacement for level because we have not had a clear idea of either the nature of a level or the purpose levels serve for working scientists. And so it has not been easy to make the required case that scale serves the purpose of level just as well or better.[29]

Fortunately, our earlier work about the nature and purpose of levels means we can more carefully assess whether scale can indeed replace level. We have now seen that "level" plausibly refers to the compositional arrays to which integrated compositional explanations/models are committed. And we saw that talking of "levels" allows scientists to efficiently communicate about arrays and to track what entities are in arrays, among other useful roles. Focusing on these purposes that the notion of "level" serves, we can thus better assess how well "scale" does as a replacement.

Unfortunately, our work highlights a couple of foundational reasons why "scale" fails as a successful replacement for "level" given the purposes of scientists. First, in 12.4.2, we saw that proponents of the scale idea, like Eronen, highlight cases where we find individuals of the same kind, and hence same scale, in *different* compositional arrays of the same integrated compositional models/explanations. Consequently, the scale of some entity does not always track which compositional array that entity is in, and we outlined the deeper reasons for this mismatch above in 12.4.2. So "scale" cannot successfully replace the notion of "level" for scientists in tracking the commitments of their plural, integrated compositional models/explanations.

Second, and more significantly, an entity being at some scale does not even suffice for this entity to be a component of some terminal whole or to be in any compositional array at all. As we have seen, a virus can be inside a cell and be of the same scale as its subcellular parts, but the virus is not a part of the cell at all or hence in any compositional array. Why is that so? Plausibly because an individual is a part of another, in the sense used in compositional models/explanations, when the individual's properties or activities realize or implement properties or activities of the relevant whole.[30] Simply being of a certain scale thus does not leave an individual in *any* compositional array. The notion of "scale" is therefore too blunt a tool for scientists to track whether an entity is even in a compositional array at all, let alone which of them it is in.

Overall, we thus find that the concept of "scale" fails for scientists in key roles that the concept of "level" plays for them. And similar points can be used to show that "scale" also fails to allow scientists to communicate about compositional arrays as well, again unlike "level." But "scale" can only replace "level" in actual practice if "scale" serves the purposes of "level" as well or better—so "scale" plausibly cannot replace "level" in actual practice. This does not mean it is never useful to use the concept of "scale" as well as, or even instead of, the notion of "level," since the purposes of researchers are many. But it scuppers the claim that we can replace level with scale in scientific practice, hence undercutting the Eronen–Potochnik proposal.

12.8 Conclusion

Compositional or organizational levels in the sciences can only be properly appreciated once we finally begin to understand compositional models and explanations, since the

Integrative Account highlights how scientists use the term "level" to talk about the complex ontological commitments of their plural, integrated compositional models/explanations. I have sketched why scientists use the term "level" in this way for a variety of practical reasons that cannot be duplicated by the concept of "scale"—hence rebutting the recent Eronen–Potochnik proposal. Furthermore, the Integrative Account frames precise ascriptions of same, higher, and lower level in ways that capture actual scientific practice, thus also rebutting the Kitcher–Schaffner objection to levels.

As we grapple with the broader scientific practices associated with integration between different areas of science, and their products, we now need to explore how the Integrative notion of level, used with compositional models, is related to other notions of "level" used with different products, including those used in these other areas of the sciences.[31] For instance, evolutionary biology uses explanations, practices, and associated notions like "levels of selection" that are often focused on *historically individuated* entities—and hence not the *ahistorically individuated* entities of compositional models/explanations. However, an interesting option to explore is whether the Integrative notion of a compositional level in some way informs the notion(s) of a "level of selection" at play in evolutionary biology.[32] One can only be excited at such new research questions that are revealed once we finally take compositional models/explanations seriously.

Acknowledgments

I am grateful to the audience at the KLI workshop on levels where I presented a draft of this chapter. I am also grateful to Ken Aizawa, James DiFrisco, Toby Friend, and Varun Ravikumar for comments on earlier drafts.

Notes

1. See Kitcher (1984, p. 25, note 3) and Schaffner (1993, p. 287).

2. Eronen (2013, 2015), Potochnik (2010, 2017), and Potochnik and McGill (2012).

3. Cf. the distinction drawn in Godfrey-Smith (2010).

4. See Gillett (2016a) for detailed discussion of these scientific debates over reduction and emergence.

5. Amundsen and Lauder (1994).

6. We now have promising accounts of such scientific part–whole relations; see, for example, Gillett (2007, 2013, 2016a) and Kaiser (2018).

7. See the useful survey of different notions of "level" in Craver (2007, chap. 5).

8. Cf. Machamer (2004).

9. Elsewhere I have offered more detailed treatments of such explanations and the compositional relations they posit (Gillett, 2016a, chap. 2). Barberis (2017) persuasively argues the latter accounts do not fit with extant philosophical treatments of "levels." However, the present chapter outlines a notion of "levels" that is used in actual scientific practice and that is compatible with my accounts.

10. For example, Machamer et al. (2000), Craver (2007), and Glennan (2017).

11. See the extended treatment of this point in Craver (2007, chap. 4).

12. See, for example, Betts et al. (2013) or Silverthorn (2013). The latter list of levels is not taken to be exhaustive, since there is a tissue and chemical level, but these are the levels cited in our exemplar models.

13. For a discussion, and an account, of such realization relations see, for example, Gillett (2002) and (2003).

14. Aizawa and Gillett (2019).

15. See, for example, Giere (2006).

16. These features are usually, but not always, the case.

17. See, for instance, Machamer et al. (2000), Craver (2007), and Glennan (2017).

18. Elsewhere I have highlighted still further differences between the features of such compositional relations and causal relations. See Gillett (2016a, chap. 2). For a survey of wider philosophical discussions about, and the variety of accounts of, vertical relations, and which of these accounts are appropriate to scientific cases, see Aizawa and Gillett (2016a) and (2016b).

19. More detailed arguments can be given that the backing relations of these explanations cannot be either ontologically "thin" causal relations, like manipulability (Baumgartner & Gebharter, 2016; Gillett, 2020; Romero, 2015), or ontologically "thick" causal relations, such as activities (Gillett, 2020).

20. See Gillett (2020) for a more detailed defense of this pluralist conclusion about ontic explanation.

21. I also avoid the term "part–whole" explanation (Love & Huttemann, 2011), since part–whole relations are not the backing relations of Standing or Dynamic explanations.

22. The wider thesis is stated in terms of "levels," but since we have seen this is common in scientific practice, it is hard to avoid and does not pose a problem for my later arguments.

23. To simplify presentation of the Energetic Arguments I have made a number of assumptions that the reader should bear in mind. This kind of argument can be given in a more complex form without such assumptions, but for ease of exposition I look at the simpler version here. The reader should thus note, for example, that I have assumed, first, that the parts of the cell remain constant across the relevant time from t1–t2 and, second, that the whole and the relevant parts are not constantly exchanging energy with the surroundings, or other parts/wholes, from t1–t2. Obviously, both simplifying assumptions are incorrect, since we know both that individuals, like cells, constantly change parts and also that individuals, like cells and proteins, are constantly exchanging energy with other parts and the surroundings. However, such details can be added to the argument at the cost of complexity while leaving the conclusion untouched, so I leave them aside having noted them. Thanks to James DeFrisco and Toby Friend for pressing me to make some of the assumptions explicit.

24. My findings therefore potentially support the important earlier conclusion of Craver and Bechtel (2008) that there is "no inter-level causation" in Dynamic compositional explanations if this is intended to solely refer to productive relations, rather than also precluding manipulability relations over time, between such parts and wholes.

25. Why do we need another term? Why not just call these "levels"? It is a descriptive finding in itself that integrated compositional models have such commitments and all accounts of "levels" need to accommodate this evidence. It begs the question against other theories of "level" to simply assume that compositional arrays are what scientists refer to by "level." I provide arguments for that further claim below.

26. Cf. Craver (2007, chap. 5).

27. Thanks to James DiFrisco for pressing me to make this point clear.

28. For example, Brooks and Eronen (2018) provide a plausible account of how a broader notion of "level" frames problems and aids problem solving. This purpose is also plausibly served by the narrower "levels" described by the Integrative Account.

29. This may explain why critics often focus on *philosophical* concepts of level (Brooks, 2017), rather than the notions of level found in scientific practice.

30. I noted the circumstantial descriptive evidence supporting this point in section 12.2 above, but for a more detailed defense, see Gillett (2013; 2016, chap. 2).

31. Helpful starting points are Brooks and Eronen (2018) and Brooks (this volume).

32. For discussion of closely related questions, see DiFrisco (this volume).

References

Aizawa, K., & Gillett, C. (Eds.). (2016a). *Scientific composition and metaphysical grounding.* New York, NY: Palgrave MacMillan.

Aizawa, K., & Gillett, C. (2016b). Vertical relations in science, philosophy and the world. In K. Aizawa & C. Gillett (Eds.), *Scientific composition and metaphysical grounding* (pp.1–38). New York, NY: Palgrave MacMillan.

Aizawa, K., & Gillett, C. (2019). Defending pluralism about compositional explanations. *Studies in the History and Philosophy of Science, Part C, 78,* 101202.

Amundsen, R., & Lauder, R. (1994). Function without purpose. *Biology and Philosophy, 9,* 443–69.

Barberis, S. (2017). Multiple realization, levels, and mechanisms. *Teorema, 36,* 53–68.

Baumgartner, M., & Gebharter, A. (2016). Constitutive relevance, mutual manipulability and fat handedness. *British Journal for Philosophy of Science, 67,* 731–756.

Betts, J. G., Young, K., Wise, J., Johnson, E., Poe, B., Kruse, D., Oksana, K.,Johnson, J., Womble, M., & Desaix, P. (Eds.). (2013). *Anatomy and physiology.* Retrieved from https://openstax.org/books/anatomy-and- physiology/pages/1-introduction

Brooks, D. (2017). In defense of levels: Layer cakes and guilt by association. *Biological Theory, 12*, 142–156.

Brooks, D. S., & Eronen, M. I. (2018). The significance of levels of organization for scientific research: A heuristic approach. *Studies in History and Philosophy of Biological and Biomedical Sciences, 68–69*, 34–41.

Craver, C. (2007). *Explaining the brain*. Oxford, UK: Oxford University Press.

Craver, C. F., & Bechtel, W. (2007). Top-down causation without Top-down Causes. *Biology and Philosophy, 20*, 715–734.

Eronen, M. I. (2013). No levels, no problems: Downward causation in neuroscience. *Philosophy of Science, 80*(5), 1042–1052.

Eronen, M. I. (2015). Levels of organization: A deflationary account. *Biology and Philosophy, 30*(1), 39–58.

Giere, R. (2006). *Scientific perspectivism*. Chicago, IL: University of Chicago Press.

Gillett, C. (2002). The dimensions of realization: A critique of the standard view. *Analysis, 62*, 316–323.

Gillett, C. (2003). The metaphysics of realization, multiple realizability and the special sciences. *Journal of Philosophy, 100*(11), 591–603.

Gillett, C. (2007). Hyper-extending the mind? Setting boundaries in the special sciences. *Philosophical Topics, 351*, 161–188.

Gillett, C. (2010). Moving beyond the subset model of realization: The problem of qualitative distinctness in the metaphysics of science. *Synthese, 177*, 165–192.

Gillett, C. (2013). Constitution, and multiple constitution, in the sciences: Using the neuron to construct a starting framework. *Minds and Machines, 23*, 309–337.

Gillett, C. (2016a). *Reduction and emergence in science and philosophy*. New York, NY: Cambridge University Press.

Gillett, C. (2016b). The metaphysics of nature, science, and the rules of engagement. In K. Aizawa & C. Gillett (Eds.), *Scientific composition and metaphysical grounding* (pp. 205–248). New York, NY: Palgrave MacMillan.

Gillett, C. (2020). Why constitutive mechanistic explanation cannot be causal. *American Philosophical Quarterly, 57*, 31–50.

Glennan, S. (2017). *The new mechanical philosophy*. New York, NY: Oxford University Press.

Godfrey-Smith, P. (2010). Causal pluralism. In H. Beebee, C. Hitchcock, & P. Menzies (Eds.), *Oxford handbook of causation* (pp. 326–337). New York, NY: Oxford University Press.

Healey, R. (2013). Physical composition. *Studies in History and Philosophy of Science Part B: Studies in History and Philosophy of Modern Physics, 44*, 48–62.

Kaiser, M. (2018). Individuating part-whole relations in the biological world. In O. Bueno, E. Chen, & M. Fagan (Eds.), *Individuation across experimental and theoretical sciences*. New York, NY: Oxford University Press.

Kitcher, P. (1984). 1953 and all that: A tale of two sciences. *Philosophical Review, 93*, 335–373. Reprinted in his *In Mendel's Mirror*. New York, NY: Oxford University Press. All references are to the reprint.

Love, A. C., & Huttemann, A. (2011). Comparing part-whole reductive explanations in biology and physics. In D. Dieks, W. Gonzalez, S. Hartmann, T. Uebel and M. Weber (Eds.), *Explanation, prediction, and confirmation* (pp. 183–202). New York, NY: Springer.

Machamer, P. (2004). Causation and activities. *International Studies in the Philosophy of Science, 18*, 27–39.

Machamer, P., Darden, L., & Craver, C. (2000). Thinking about mechanisms. *Philosophy of Science, 67*, 1–25.

Mitchell, S. (2002). Integrative pluralism. *Biology and Philosophy, 17*, 55–70.

Mitchell, S. (2003). *Biological complexity and integrative pluralism*. New York, NY: Cambridge University Press.

Novikoff, A. (1945). The concept of integrative levels and biology. *Science, 101*, 209–215.

Potochnik, A. (2010). Levels of explanation reconceived. *Philosophy of Science, 77*(1), 59–72.

Potochnik, A. (2017). *Idealization and the aims of science*. Chicago, IL: University of Chicago Press.

Potochnik, A., & McGill, B. (2012). The limitations of hierarchical organization. *Philosophy of Science, 79*(1), 120–140.

Romero, F. (2015). Why there is not inter-level causation in mechanisms. *Synthese, 192*, 3731–3755.

Schaffner, K. (1993). *Discovery and explanation in biology and medicine*. Chicago, IL: University of Chicago Press.

Silverthorn, D. (2013). *Human physiology: An integrated approach*. San Francisco, CA: Pearson.

Wimsatt, W. (1994). The ontology of complex systems: Levels of organization, perspectives and causal thickets. *Canadian Journal of Philosophy, 20*, 207–274.

13   Functional Kinds and the Metaphysics of Functional Levels: In What Sense Are Functionally Defined Kinds and Levels Nonarbitrary?

Thomas A. C. Reydon

Overview

This chapter explores the metaphysical connections between kinds and levels of organization. By applying a newly developed account of natural kinds, the Grounded Functionality Account, to the case of kinds of genes located at specific levels of organization, it is shown how the nonarbitrariness of kinds and levels can be understood in terms of metaphysical grounding—how kinds at different levels of organization can be metaphysically connected and how such a connection between kinds can result in derivative metaphysical connections between levels. The case of gene kinds shows that such grounding does not necessarily consist of higher levels depending on lower levels but can also manifest itself the other way around: genes are metaphysically dependent on genomes. Using the metaphysical notion of grounding provides a picture in which kinds at one level of organization are grounded in kinds at another level of organization and in which levels of organization are grounded in other levels in a derivative way (dependent upon the kinds at one level being grounded in the kinds of another level).

13.1   Introduction

This chapter explores the metaphysical connections between practices of grouping entities into kinds and locating those kinds at particular levels of organization. Both of these practices constitute central elements of investigation throughout the sciences. It is safe to say that every area of science involves, if not an exhaustive classification of all entities in the domain under study, at least some set of basic assumptions regarding what kinds of entities should be distinguished and on what basis entities should be allocated to kinds.[1] Such classificatory assumptions are not limited to views on how to group entities into the kinds that make up the "furniture" of the domain under study. Typically, they also include conjectures regarding how these kinds and entities relate to each other hierarchically—that is, how entities at one level are composed of entities at lower levels (Eronen, 2015, pp. 39–40; Wimsatt, 1994, p. 222) and how higher-level kinds contain lower-level kinds.

The central notions in these practices and in the assumptions underwriting them, "kind" and "level of organization," stand at the focus of strikingly similar philosophical debates.[2]

Regarding kinds and in particular the narrower notion of "natural kinds," there is persistent disagreement on a number of central questions. What makes an entity a member of its kind? Are there fundamental kinds that determine what entities essentially are? Do kinds have an independent existence as parts of the world's structure, or are they fundamentally dependent on human classificatory and investigative interests? If they depend to some extent on human interests, in what sense can kinds be said to be nonarbitrary groupings of things? And so on. The disagreement on these matters runs so deep that several authors have suggested that the quest for answers is hopeless, and we should abandon the notion of "natural kind" as well as the project of developing an account of the epistemology and metaphysics of natural kinds (e.g., Hacking, 2007; Ludwig, 2018). With respect to levels of organization, a similar debate exists. There is a persistent lack of agreement on such issues as whether levels of organization are real (i.e., whether they exist independently as parts of the world's structure or are fundamentally dependent on how we analyze systems under study), how levels of organization are to be characterized, in what sense (if at all) levels can be said to be nonarbitrary, and so on. In a similar way as in the case of kinds, skepticism about the notion of "levels of organization" has emerged, involving calls to abandon the notion altogether.[3]

One aim of the present chapter is to counterweigh skepticism about "levels thinking" by exploring how a concise concept of levels of organization could be formulated in terms of the metaphysical dependence between entities and kinds at different levels, using a newly developed account of kinds. Both "kinds thinking" and "levels thinking" are important aspects of scientific practice, I hold, and achieving clarity about their epistemological and metaphysical commitments is important for our understanding of how science works. Elsewhere, I have argued that skepticism about "kinds thinking" is unwarranted and that a useful notion of "natural kinds" is achievable (Ereshefsky & Reydon, 2015, in press; MacLeod & Reydon, 2013; Reydon, 2010, 2016). The present chapter builds on this to develop a useful notion of "levels."

13.2   Kinds and Levels as Intertwined Problems

The debates on levels and kinds have so far been largely disconnected, but connections can and should be made. Views about what kinds of entities should be recognized in a particular domain of investigation constrain what levels of organization are acceptable. A level of organization is at least in part defined by the entities (material objects, processes, functional components, etc.) that exist there, and it is unclear what to make of levels at which no kinds of things would be located.[4] Conversely, views about which levels should be recognized in a domain constrain which kinds are acceptable, as "level-less" kinds are difficult to conceive of—when a kind is recognized that cannot be allocated to any of the recognized levels of organization in the domain under consideration, this seems a good reason to add a level to the hierarchy. Therefore, any philosophical project examining the kinds structure of a particular domain of investigation must answer the question how the various kinds in that domain relate to each other—whether there are higher-level kinds that contain lower-level ones, for instance. Conversely, any inquiry into the levels of organization in a particular domain is faced with the question which kinds of entities are characteristic for the various levels in that domain. Clearly, the discussions on kinds and levels are mutually relevant, and new developments in one discussion will affect the other. I will try to forge connections between the two discussions by bringing a new account of kinds to bear on the question of levels of organization.

The connections I want to forge are metaphysical, because both "kinds thinking" and "levels thinking" have metaphysical import. The sets of principles and the orderings that follow from them are commonly seen as telling us something about what there is in the world—what kinds of things exist and how entities make up other entities. As such, the kinds and levels that are recognized in a domain of study cannot be completely arbitrary orderings. For, if the grouping of entities into kinds and the allocation of kinds to levels in a particular domain were completely arbitrary, how could those kinds and levels successfully feature in scientific investigations, in explanations of phenomena, in knowledge claims, and so on? Nonarbitrariness (in the sense of not being *entirely* arbitrary) can thus be thought of as a minimal criterion that kinds and levels should meet in order to be useful for scientific purposes (Ereshefsky & Reydon, 2015; Ludwig, 2018).[5]

I will explore the nonarbitrariness of kinds and levels by focusing on functional kinds and functionally defined levels in the life sciences. In what sense can we say that functional kinds and functionally defined levels of organization are nonarbitrary? While this question arises for all kinds of kinds and levels, functional kinds and functionally defined levels are highly suited to explore this question. When it comes to the question of nonarbitrariness, they constitute a more difficult topic than traditional natural kinds and associated levels, such that exploring the nonarbitrariness of functional kinds and levels can teach us much about the nonarbitrariness of kinds and levels generally. In contrast to natural kinds, functional kinds are often thought to depend strongly on human interests, such that it is difficult to see them as nonarbitrary groupings. However, natural kinds, too, are thoroughly dependent on our practices and interests, such that there is no fundamental difference between natural and functional kinds in this respect (Ereshefsky & Reydon, 2015, in press; Reydon, 2009a, 2016). But in order to be able to perform epistemic roles in the sciences, kinds must also have *some* foundation in the world as it exists independently of us. For both natural and functional kinds and their associated levels, this foundation is what makes them natural and nonarbitrary, I will argue.[6] I will explore how this grounding of kinds and levels in nature is realized and present a view of levels as grounded on kinds, which in turn are grounded on connections between epistemic and investigative interests, on the one hand, and relevant features of the world, on the other hand.

I will proceed as follows. First, I will briefly discuss the question of the (non)arbitrariness of functional kinds and levels by considering the example of gene kinds. The next section then presents a novel account of natural kinds that is sufficiently naturalistic to cover all kinds that feature in the various areas of science (including the social sciences and technological disciplines) while still having enough normative force to provide criteria to distinguish between natural kinds and other groupings. I then use this account to clarify in what sense we can say that functionally defined levels of organization in biology are nonarbitrary, looking at kinds of genes, kinds of genomes, and the levels of organization at which these are located.

13.3   Functional Kinds and Functional Levels in Biology

Practices of grouping entities into kinds and locating these at particular levels of organization often rest on at least two straightforward, but problematic, assumptions about the world. The first is that entities—conceived broadly as material objects, substances, natural processes, events, and so on—come in kinds. The assumption is that every entity is an entity of some kind or other, belongs to only one fundamental kind (membership of which tells us something about what it is and what its characteristic properties are), and may belong to multiple higher-order kinds as long as lower-level kinds are neatly nested into higher-level kinds. Perhaps the most prominent example in the life sciences is the Linnaean system, which is a hierarchically structured classificatory system with a fundamental level of kinds (the species level) and higher-level kinds containing lower-level kinds. On the traditional interpretation, in the Linnaean system, the allocation of an organism to a species tells us what the organism *is*, with lower-level kinds (subspecies, varieties) and higher-level kinds (genera, families, orders, etc.) providing additional information about the traits that the organism will probably exhibit.[7] The second assumption is that the entities in a particular domain are related by way of part–whole relations (Wimsatt, 1994, p. 222). For example, genes are parts of genomes, genomes are parts of cells, cells are parts of organisms, organisms are parts of populations, and so on. Accordingly, kinds of cells are located at a higher level of organization than kinds of genes *because* genes are parts of genomes, which in turn are parts of cells, and so on.

If these two assumptions would hold, they would support a straightforward view of the world as having a clear-cut structure of kinds located at various levels and in which every entity has its proper place as a member of a particular kind located at a particular level of organization. They would also support an account on which the levels featuring in one hierarchy are connected by two kinds of relations—part–whole relations that hold between the entities at neighboring levels and relations of containing that hold between the kinds located at these levels. And as both kinds of relations are transitive, such relations would be what keeps the structure of levels together.

Both assumptions are highly problematic, however. They involve a realist stance toward kinds and levels: if kinds and levels do not correspond to aspects of the structure of the world as it is independently of us, these assumptions seem difficult to uphold. But realism about kinds is a problem in itself, and even when one assumes realism about natural kinds, functional kinds are still problematic. Realism about kinds is the view that natural kinds are aspects of the natural order of things, as it exists independently of us.[8] This motif is found throughout the history of the discussion on the topic and is often expressed by referring to the Platonic slogan that natural kinds "carve nature at her joints" or by talking of "real kinds." Bird and Tobin's (2018) review article expresses it as follows:

Scientific disciplines frequently divide the particulars they study into *kinds* and theorize about those kinds. To say that a kind is *natural* is to say that it corresponds to a grouping that reflects the structure of the natural world rather than the interests and actions of human beings. We tend to assume that science is often successful in revealing these kinds; it is a corollary of scientific realism that when all goes well the classifications and taxonomies employed by science correspond to the real kinds in nature. The existence of these real and independent kinds of things is held to justify our scientific inferences and practices.

The idea is that natural kinds either have an independent existence *as kinds* in reality (which is an implausibly strong claim) or that natural kinds *represent* the fundamental structure of the world. Both versions of this idea are manifest in the traditional assumption that natural kinds (or the aspects of the world's structure they represent) are mind independent.[9] On such a view, it is not difficult to see what the nonarbitrariness of natural kinds would consist of. Also, as kinds are located at various levels of organization with entities of a particular kind at one level being composed of entities of various lower-level kinds, an argument for the nonarbitrariness or robustness of such levels (Wimsatt, 1994, p. 225; this volume) seems possible on the basis of the alleged nonarbitrariness of natural kinds. However, the question remains how the mind-independent existence of natural kinds should be conceived of. The various accounts of natural kinds available in the philosophical literature encompass attempts to answer this question, so far, however, without having resulted in a final answer or even a widely accepted majority view.

For the natural kinds realist, functional kinds are problematic. The problems are well known. Functions are multiply realizable in the sense that often the same function can be realized by materially and structurally very different entities, and entities often are multifunctional in the sense that one entity can realize very different functions in different contexts. Moreover, the individuation of functions depends crucially on the way in which we analyze the workings of a system. Functions aren't simply found in the world—they depend on how investigators analyze and decompose a system or process into functional units. As functional decompositions can be done in multiple ways, depending on the investigators' interests, functional kinds seem mind dependent in ways that natural kinds are not. Natural kinds and associated levels of organization thus seem nonarbitrary in a sense in which functional kinds and functionally defined levels are not. So, an argument for nonarbitrariness of functionally defined levels of organization on the basis of the nonarbitrariness of functional kinds seems more difficult to make than for natural kinds and levels.

Consider kinds of genes.[10] Genes come in kinds: *Drosophila melanogaster* fruit fly genomes include, among others, *dachshund*, *diaphanous*, *wingless*, *windbeutel*, and *ken and barbie* genes, and humans have *BRCA1*, *DIAPH1*, and *PAX6* genes. But genes are not simply found in nature as discrete material entities waiting to be classified. Individual genes, such as a particular *PAX6* gene in a particular cell of my body, are both individuated *as genes* (i.e., as members of the category of genes) and as members of particular kinds of genes on the basis of their functions. Entirely nonfunctional stretches of a genome are not individuated as genes. Genes are, as Griffiths and Stotz (2006, p. 500; 2013, p. 75) aptly put it, "things you can do with your genome" or "ways in which cells utilize available template resources to create the biomolecules that are needed in a specific place at a specific time." But not all functions that can be realized by parts of an organism's genome are considered to individuate genes. The functions of coding for a particular amino acid such as tryptophan or lysine, of being a start or stop signal, of functioning as a promoter, and so on, are not taken as individuating genes. Codons are functional parts of genes and appear as functional elements in genetics and genomics, but codons themselves aren't genes. Similarly, the developmental "function" of producing an entire organism is realized at the molecular level by a whole genome, and whole genomes are not themselves members of the gene category. The function that is used to individuate genes is that of being used as a template for the production of macromolecular products of particular kinds (functional polypeptides or RNA molecules) (Griffiths & Stotz,

2013, chap. 4; Portin, 1993; Waters, 1994). Genes are individuated by reasoning backward from functional macromolecules, metabolic functions, or traits, to what may be thought of as the counterparts on the genome of these macromolecules, functions, or traits, for the production of which they function as templates (Gerstein et al., 2007; Griffiths & Stotz, 2007, p. 100; Griffiths & Stotz, 2013, p. 75).

Individuating genes by reasoning backward isn't a straightforward matter, though, as there is no simple one-on-one mapping of a species' proteome—let alone of higher-level functions or organismal traits—onto its genome (Harrison et al., 2002). A number of processes that can occur in DNA transcription and translation, such as alternative splicing, frameshifting, or RNA editing, usually make a unique mapping of macromolecular products (and downstream effects) onto particular stretches of DNA impossible. Due to such processes, a particular segment of an organism's genome may be involved in the production of multiple expression products, and conversely, an expression product may involve several separate stretches of DNA (Gerstein et al., 2007; Griffiths & Stotz, 2006, 2007, 2013, pp. 71–75, 84–97; Reydon, 2009b, 2016; Waters, 1994; Wieben, 2003). In addition, the same stretch of DNA can count as a gene in some contexts and not count as a gene in other contexts. For example, a nonfunctional homolog in the genome of a species B of a functional sequence in the genome of species A does not count as a gene for species B because of its nonfunctionality, notwithstanding the large degree of structural similarity between the homologous sequences. Such nonfunctional homologs—pseudogenes—are "defunct genomic loci with sequence similarity to functional genes but lacking coding potential" (Pei et al., 2012, p. 1). In the tunicate species *Molgula occulta*, for example, several functional genes in the Tyrosinase family that are found in other species of the genus *Molgula*, as well as in other tunicate genera, are in the process of losing functionality and becoming pseudogenes (Racioppi et al., 2017).

In the face of the preceding discussion, how can we say that genes and gene kinds are real? The example suggests that functional kinds are fundamentally dependent on what investigators count as relevant functions and how they individuate functions and their realizers—and thus cannot be seen as representing aspects of the structure of the world in the way natural kinds are usually thought to do.

Gene kinds thus seem to better fit a conventionalist or antirealist view of kinds (Bird & Tobin, 2018, section 1.1.2). On such a view, the kinds that feature in the sciences are useful instruments for various purposes, but we cannot think of them as representing the natural order of things. According to the conventionalist, all classifications are fundamentally dependent on the perspective of those who did the classifying, and no classification is privileged. As Bird and Tobin (2018) put it, "The classifications of botanists do not carve nature at its joints any more than the classifications of cooks." While this way of thinking about kinds may avoid the difficulties confronted by natural kind realism, an important problem is the question *why* certain ways of grouping entities into kinds are useful for various purposes. Whenever we find that a particular set of kinds is successfully used for, say, supporting inferences or explanations, the question arises what makes these kinds suitable for these particular purposes rather than a completely different way of grouping the same entities. Conventionalist views of kinds fail to answer this question. They do not go beyond the observation that what works, works—a stance that is both epistemologically and metaphysically unsatisfactory.

Functional kinds thus remain problematic. For realists about kinds, functional kinds are not acceptable and should be thought of as conventional groupings that do not represent the world's structure. For antirealists or conventionalists about kinds, functional kinds are as acceptable as natural kinds, but in both cases, this comes at the price of a lack of explanation of epistemic success. Either way, functional kinds and the associated levels of organization seem to have an ineliminable aspect of arbitrariness that seems to make them ill-suited for scientific purposes.

This view of functional kinds and levels, however, does not sit well with scientific practice. Practices of grouping genes into kinds are not wholly arbitrary, as gene kinds must match functions that are actually realized in cells. Moreover, gene kinds as well as functional kinds in general do play important epistemic roles in the sciences, and their success in playing these roles is in need of explanation (Ereshefsky & Reydon, 2015). Such an explanation should involve an account of the nonarbitrariness of functional kinds. In the following section, I will present an account of kinds—the *Grounded Functionality Account* (GFA)—that is able to cover functional kinds. I will use this account to link functional kinds to functionally defined levels of organization and clarify in what sense we can say that functionally defined levels of organization in biology are nonarbitrary.

13.4   The Grounded Functionality Account of Natural Kinds

The GFA was developed in response to the failure of available accounts of natural kinds to cover the diversity of kinds that play important roles in the sciences.[11] The motivating criticism was that available accounts impose overly strict criteria for attributing natural kind status to groupings of entities and, in doing so, exclude various kinds of kinds from being covered by the account on a priori grounds. Available accounts tend to single out one factor as fundamental when it comes to supporting natural kinds (e.g., microcompositional essences, homeostatic mechanisms, causal networks), assuming that this is *the* aspect of the world's fundamental structure that underwrites all kinds. But ultimately we do not know anything about the world's fundamental structure, such that the best position is to remain agnostic: the world might be well ordered or it may be an amorphous mess, but either way, we do not have good reasons to single out one particular factor as the fundamental, distinguishing characteristic of natural kinds. The criticism of available accounts thus is that despite aspirations to be naturalistic accounts of what natural kinds are, available accounts are not naturalistic enough (Ereshesfky & Reydon, 2015, in press).

A starting point, then, is agnosticism about the world's fundamental structure, that is, the view that we cannot assume that the world has a particular fundamental structure and, *a fortiori*, that we cannot use claims about the world's structure as a basis for philosophical accounts.[12] For the metaphysics of science, this means that we should resist trying to formulate overall metaphysical frameworks and instead focus on the *local* metaphysics of the various areas of scientific practice instead. Which aspects of the world are in focus in a particular area of work to some extent depends on the epistemic aims that scientists in that area pursue, and so do the metaphysical commitments of the natural kinds they use (i.e., the aspects of the world that the kinds in that area are thought to latch onto). Therefore, we should not approach the metaphysics of classification by using one single condition for attributing natural kind status, as available accounts do, but instead assume a thoroughly pluralist stance that recognizes that different areas of scientific investigation latch onto different aspects of the world. The metaphysical project of the GFA thus is very different from traditional work on natural kinds.

Consider how Hacking phrased the metaphysical question in the debate on natural kinds: "are there natural kinds—real or true kinds found in or made by nature?" (Hacking, 1990, p. 135). From the perspective of the GFA, asking whether there *are* kinds in nature, kinds *made by* nature, or kinds *found in* nature does not make much sense. We do not simply find kinds of things in the world, nor does nature make kinds. Rather, we group things into kinds because we believe such groupings represent important and interesting features of the world. Thus, the question should be in what ways the kinds that feature in the various areas of scientific investigation are supported by aspects of the world that render them suitable for the purposes for which they are being used. Looking at the matter in this way allows us to bring functionally defined kinds into the domain of natural kinds, as these successfully feature in the sciences too.

The GFA addresses the above question using the notion of "classificatory programs" that was introduced by Ereshefsky (2001; see also Ereshefsky & Reydon, 2015, in press). Classificatory programs are the parts of research programs that are aimed at constructing classifications of the entities under study and consist of three elements: sorting principles, motivating principles, and classifications. Sorting principles sort entities into kinds. Motivating principles are the epistemic aims of a classificatory program and motivate why that program should sort entities a particular way. The classifications produced by a classificatory program identify putative natural kinds—the assumption is that if a classificatory program is successful in producing a classification that is useful as a basis for research activities, its success is explained as successfully having identified groupings that represent relevant aspects of the world. According to the GFA, a natural kind classification should be judged by how well it satisfies the aims of the specific classificatory program that produced it.

Note that the GFA does not presuppose any particular epistemic aims as characteristic for classificatory programs that identify natural kinds. Science involves a variety of aims, including explanation, understanding, prediction, control, inference, description, and so on. The aims of classificatory programs are found in the programs themselves and can vary between programs. Whether a program can be said to have produced natural kinds depends on how well its classification achieves the program's particular aims. Natural kinds thus are *functional* in the sense of being useful for achieving certain aims.

In addition to this functionality aspect, the GFA imposes a "*grounding* condition" for natural kinds. This condition can be formulated by slightly modifying Bird and Tobin's (2018; quoted above) condition on what it is to be a natural kind. According to the grounding condition, to say a kind is *natural* is to say that it corresponds to a grouping that reflects an aspect of the world rather than *merely* the interests and actions of human beings (Ereshefsky & Reydon, in press). This condition demands that for a classification to involve natural kinds, the kinds should in part depend on the world and not exclusively on our conceptions of what the world might be like. The grounding condition clarifies how natural kinds can be tools for obtaining knowledge about the world: while they are made by us in the context of classificatory programs that have specific aims, to be used to achieve those aims, natural kinds and the classifications in which they feature should also depend on those aspects of the world that are under investigation and not merely on our conceptions of it.

The GFA uses the term "grounding" to indicate the straightforward point that for a kind to be useful, it must in *some* way be anchored to aspects of the world, be supported by such aspects, or latch onto such aspects, or something else along those lines. The GFA does not presuppose any particular way in which kinds are grounded in the world but is neutral on this point because it is intended as a thoroughly naturalistic account of kinds that allows for different ways in which kinds can be supported by the world. The sciences investigate different domains of reality, and there is no particular reason to assume that throughout all these domains, kinds of entities should be grounded in the same way. Saying that a kind is grounded in aspects of the world thus amounts to saying that it is connected to aspects of the world that explain the successful use of the kind for particular purposes in the context of a particular research program. Often this will involve explaining the similarities that obtain between the members of the kind, but this is not necessarily always the case.

As used in the GFA, "grounding" thus refers to a relation of metaphysical dependence of a kind upon aspects of the world, where this relation can be manifested very differently in different cases. In contemporary analytic metaphysics, the term "grounding" is used to distinguish between causal relations and relations of metaphysical dependence (e.g., Bliss & Trogdon, 2016; Correia & Schnieder, 2012; Trogdon, 2013). Here, "causal relations" should be read in a strict sense as referring to cases in which an entity at one level of organization causes the presence of an entity or occurrence of a phenomenon at another level—think, for instance, of how the presence of a particular allele in an organism's genome causes the presence of a trait. In cases of metaphysical dependence, in contrast, the presence of an entity or occurrence of a phenomenon is explained noncausally by specifying ways in which the entity or event depends on (but is not caused by) aspects of the world. Such explanations thus are seen as constituting a different mode of explanation from causal explanations (Bernstein, 2016; Bliss & Trogdon, 2016; Fine, 2012, p. 37; Trogdon, 2013, pp. 97–98).

Note that the notion of "grounding" in the GFA should not be aligned with any strong view of what grounding consists of. The notion of "grounding" is highly contested, and very different interpretations exist (Correia & Schnieder, 2012; Bliss & Trogdon, 2016), but the GFA does not encompass a specific metaphysical view of how grounding should be understood. The notion in the context of the GFA can be clarified by contrasting it with Hacking's (1990, p. 135) metaphysical question about natural kinds, quoted above. Hacking asks whether there are kinds the existence of which is caused by factors in nature—that is, Hacking asks about kinds as aspects of the world. The GFA, in contrast, asks about kinds that are rooted in *both* our practices and the world, that is, about how groupings made by us for certain purposes depend on the world. The GFA's general thesis is that a kind's successful use in a particular context of investigation is explained by its depending in *some* way on the world, where this dependence is not causal (the way the world is does not cause a kind's existence, nor does it cause a kind's successful use in scientific investigation). The successful use of a kind in a particular context of investigation is explained as being *in virtue of* a particular aspect of the way the world is (Trogdon, 2013, p. 97).[13] The GFA thus constitutes a functional approach to kinds while at the same time requiring that kinds are grounded in the world.

13.5   The GFA in Practice: Genes and Genomes in Different Research Contexts

Let me now turn to the application of the GFA to a concrete case of "kind thinking" and "levels thinking." As the grounding of kinds is the core of the GFA, I will examine how grounding occurs here. Consider again the example discussed above, kinds of genes and kinds of genomes. As the GFA conceives of kinds as explicitly context dependent, the context of investigation (or, classificatory program) under consideration must be specified. I will first look at the research context of developmental biology.

Developmental biology is a context of investigation in which gene kinds are firmly embedded in a structure of entities located at different levels of organization. In descriptions of organismal development, at minimum, the gene level, the genome level, the level of molecular expression products (functional RNA molecules, proteins), the cell level, and the organism level must be distinguished. The levels structure is not linear, though: while genes are parts of genomes, which are parts of cells, which are parts of organisms, expression products are parts of cells but do not stand in part–whole relations with genes and genomes. Also, it is easy to recognize additional levels, such as the tissue level, the organ level, or the organelle level. In any case, part of the explanation of organismal development is an account of the activation and suppression of gene expression in the various cells in the organism at different stages of development. The identification of functional parts of the genomes of various species of organisms, as well as the study of how these functional parts control the formation of traits at the organism level, play central roles. An important part of clarifying how genes control development is to perform comparative studies by examining how differences in functional parts of genomes are related to differences in traits, either between organisms of the same species or between organisms of different species. In gene knockout experiments such as those famously carried out in mice (Hall et al., 2009), for example, comparisons are made between organisms with the "normal" functional allele of a particular gene and organisms in which the gene is silenced.

In this sense, developmental biology clearly exemplifies how biologists reason backward from genomes to genes. This reverse reasoning is not merely an epistemological aspect of research but reflects the metaphysical relation between genes and genomes, between gene kinds and genome kinds, and between the gene level and the genome level. Genes in the research context of developmental biology are metaphysically dependent on genomes, as they are individuated as kinds of functional parts of genomes connected to the control of developmental processes. The same holds for gene kinds and kinds of genomes: the various kinds of genes found in the human genome—*BRCA1*, *DIAPH1*, *PAX6*, and so on—are metaphysically dependent on the relevant genome kind—the *Homo sapiens* genome—as they are individuated only as functional kinds in explicit relation to the *H. sapiens* genome. The same holds for the various kinds of genes in other species, such as *Drosophila melanogaster*. Kinds of *Drosophila* genes, such as *dachshund*, *diaphanous*, *wingless*, *windbeutel*, and *ken and barbie*, exist only as parts of the *D. melanogaster* genome. This is of course not to say that the discovery of gene kinds must temporally follow the discovery of genome kinds. Clearly, this is not the case, as genetics is a much older field than genome sequencing. Large amounts of information about *D. melanogaster* genes were available long before the *D. melanogaster* genome was sequenced in 2000 (Adams et al., 2000), and much was known about human genes long before the sequencing of the human genome a year later (International Human Genome Sequencing Consortium, 2001; Venter et al., 2001). Much can be discovered about parts of an organism's genome before the genome as a whole is known. Still, ontologically, these parts depend on the whole, because they only exist as parts of the whole: genes exist as loci on an organism's genome.

Strictly speaking, then, there are no *dachshund* genes *simpliciter*, but only *D. melanogaster dachshund* genes. To be sure, biologists often find that organisms of different, sometimes distantly related, species have the same genes. For example, a recent press statement on the EurekAlert! website, the news service of the American Association for the Advancement of Science, was titled, "Genes that aid spinal cord healing in lamprey also present in humans, MBL team discovers."[14] But such talk of the same genes occurring in different species is for reasons of simplicity—what is meant is that organisms of two different species have homologous genes that are structurally similar (but not identical) and may or may not perform similar functions. The actual usage of gene names often only implicitly distinguishes between homologous gene kinds.[15] In principle, a particular name is used to uniquely denote a locus in the genome of one particular species. Often, though, the same or a very similar name is used to denote homologous loci on the genomes of distinct species, and biologists speak loosely about "the same" gene occurring in different species. In such cases, the gene name is explicitly or implicitly linked to the species under consideration. For example, the human gene *DIAPH1* (also: *DFNA1*), a gene that in humans apparently is involved in nonsyndromic hearing loss, is an ortholog of the *Drosophila melanogaster* gene *diaphanous* (Lynch et al., 1997), and sometimes the gene is simply referred to as "the human *diaphanous* gene." Although "*diaphanous*" thus is used as the informal name for a number of distinct orthologous gene kinds found on the genomes of different species, biologists use gene names in such a way as to make clear in relation to which species a statement involving "*diaphanous*" should be understood.

Using the GFA, we can say that in the context of developmental biology, gene kinds are grounded in genome kinds: gene kinds are kinds of loci that are individuated on the genomes of particular species. The relation of grounding here is hierarchical and goes in a downward direction (if we take the smaller entities as constituting the lower level, as typically done in "levels thinking"). But the relation is not one of downward causation, and hence understanding it in terms of metaphysical dependence is appropriate—it is a particular sort of part–whole relation. Genomes have epistemological priority over genes, in the sense that genes are individuated as parts of a developing organism's genome. That is, an organism's genome can be *decomposed* into genes, and as such, genes are parts of genomes. Ontologically, genomes have priority in the sense that genomes are not *constituted* by genes in the way that, for example, cars are assembled from parts that have existed prior to the whole car. That is, genes exist only as parts of genomes and do not have an independent existence of their own.[16]

Even though my focal question was in what sense we can say that functional kinds and functionally defined levels of organization are nonarbitrary, I have so far not said much about nonarbitrariness. On the GFA, the nonarbitrariness (or naturalness) of the gene kinds that feature in developmental biological research is indicated by their successful use for realizing the aims of the classificatory program in which they feature. For example, that the *D. melanogaster windbeutel* gene is a nonarbitrary grouping of entities can be seen from the fact that developmental biologists successfully refer to the *D. melanogaster windbeutel* gene in the explanation of certain developmental phenomena (e.g., dorsoventral specification in fruit fly embryos; Konsolaki & Schüpbach, 1998). But this is not an *explanation* of the kind's nonarbitrariness. The kind's nonarbitrariness is explained by its being grounded in aspects of the world. As we have seen, this grounding occurs by way of a hierarchical relation between the grounded kinds (gene kinds, such as the *D. melanogaster windbeutel* gene) and the grounding kinds (genome kinds, such as the *D. melanogaster* genome) that are located at neighboring levels of organization. To be a little more concrete about this grounding relation, the higher-level entity (a particular organism's genome) *constrains* which entities (genes) can be individuated at the lower level, as the lower-level entities are individuated as loci on the higher-level entity. While the levels are related by part–whole relations between the entities existing at the two levels, the metaphysical dependence of the lower on the higher level in this case thus consists of a constraining relation.[17] While lower-level entities are directly constrained by the higher-level entities of which they are parts, the relation of constraining between higher-level and lower-level kinds is derivative from the relation between entities. In this derivative sense, we can say that gene kinds are constrained by (i.e., metaphysically depend on) genome kinds. For such an explanation to work, whatever does the grounding must be assumed to have a similar naturalness or nonarbitrariness as that what is grounded. In addition, the way in which these two elements are related (in this case, by means of a downward relation between two levels of organization where the higher-level kinds ground the lower-level kinds) must be assumed to be nonarbitrary, too. The nonarbitrariness of gene kinds thus indicates that the genome kinds and the two associated levels of organization that feature in this particular research context are nonarbitrary, too.

To deepen this example, let me briefly turn to a different context of investigation, namely, classical transmission genetics. As on the GFA kinds are relative to specific classificatory programs, examining different contexts is relevant. Although genes are individuated differently in classical genetics from how they are individuated in developmental biology, important similarities obtain. For one, in classical genetics, genes are individuated by reasoning backward, too. When the notion of the gene was first introduced by the Danish biologist Wilhelm Johannsen in the early twentieth century, the concept was intended as a fully instrumental unit. As Johannsen put it, the term "gene" should express "the simple picture … that a property of the developing organism is or can be (co-)determined by 'something' in the gametes. No hypothesis about the nature of this 'something' should be formulated or supported" (Johannsen, 1909, p. 124, my translation).[18] Johannsen emphasized that the newly introduced term should be hypothesis free in that it should not be taken as having any implications regarding the nature of genes in general, leaving open the possibility that different genes have completely different natures. A gene, in Johannsen's view, could be *any* factor that codetermined an organism's phenotype and could be passed on from generation to generation.

In classical genetics, Johannsen's gene concept continued to serve as the core gene concept, although it underwent modifications as the field developed further. It soon became clear that genes do not map in a one-to-one manner on organismal traits but that the relation between genes and traits was a many-to-many relation: one trait is often codetermined by many genes and one gene often codetermines many traits. The view of genes involved here is functional (genes are for traits), but it is not a view on which genes straightforwardly cause traits. Rather, genes featured indirectly in the explanation of organismal traits in the way that *differences* between two organisms' phenotypes were explained by *differences* in their genotype. As Waters (1994) writes, "The basic dogma of classical genetics was that gene differences cause phenotypic differences.... What were studied were character differences, not characters, and what explained them were differences in genes, not the genes themselves" (p. 172). Genes, then, are functional units, the function of which is identified by means of differences in their effects (Keller, 2013; Waters, 1994, 2007, 2017). Indeed, this interpretation fits well with Johannsen's view, in which the notion of codetermination played an important role.

As in developmental genetics, genes here are individuated by reverse reasoning, albeit in this case not from functional molecular products but from traits or, more precisely, trait differences.[19] This allows for a similar interpretation of the nonarbitrariness of gene kinds in this particular research context, even though the grounding relation obtains between kinds at different levels of organization (in this case, the organism level, at which traits are located, and the gene level). The *D. melanogaster white* gene, for example, is grounded in the eye color differences between fruit flies, which are explained by allele differences. Even though the individuation of traits is not a trivial matter and the breakdown of organisms into separate traits depends on how researchers analyze organisms, the trait difference between two fruit flies (where one has red eyes and the other white eyes) is sufficiently nonarbitrary to serve as a ground for gene kinds.

Note that the account presented here is quite far from a complete metaphysical explanation of the nonarbitrariness (or naturalness) of gene kinds. The nonarbitrariness of gene kinds is *indicated* by their successful use in particular contexts of investigation, which in turn is explained metaphysically by invoking a grounding relation. As kinds are relative to research contexts, so too are the corresponding levels (see also Love, 2012) as well as the relevant grounding relations. The grounding relations presuppose the nonarbitrariness of the factors that do the grounding, which in turn must be grounded in further aspects of the world, and so on. So, this is a partial explanation that asks for further iterations, not a full-blown account of how gene kinds are grounded. But at the very least, this account provides a little more clarity on how to think about the nonarbitrariness of gene kinds, of genome kinds and of the levels of organization at which these kinds are located.[20]

Note, too, that the relations between the levels of organization discussed at the beginning of this section are not all of the same sort. Developing multicellular organisms are systems that start with a single cell that, through a sequence of cell divisions, results in a fully developed organism. In this sense, the part–whole relationship between an organism's cells and the organism itself is decompositional (we can analyze the organism as a system of cells) as well as constitutive (the organism results from cells being put together into a system, albeit not in the same way as a car results from the parts being assembled). The relation between genes and genomes is decompositional but *not* constitutive. With respect to "levels thinking," thus, two things follow from the example discussed above. First, even when only examining one particular area of investigation in terms of levels of organization and on the assumption of a fairly simply levels structure (in this case: genes as parts of genomes, as parts of cells, as parts of organisms), it still may be necessary to recognize different kinds of part–whole relations holding between the various levels in the domain under consideration. Second, one way in which part–whole relations may vary is that some part–whole relations are both decompositional and constitutive, whereas others are only decompositional. For metaphysics more generally, this entails pluralism about part–whole relations.[21]

13.6   Conclusion

In this chapter, I have explored the metaphysical connections between practices of grouping entities into kinds and locating those kinds at particular levels of organization. I have suggested that these practices are interconnected, as kinds are located at levels of organization and levels in turn are characterized by kinds. Applying a new account of kinds to the example of gene kinds, I have tried to show how the nonarbitrariness of kinds and levels can be understood in terms of grounding, how kinds at different levels of organization can be metaphysically connected, and how such a connection between kinds can result in derivative metaphysical connections between levels (where kinds at one level are grounded in kinds at another level, and the former level is derivatively grounded in the latter level). This is not to say that kinds and levels are objective, real elements of the structure of the world. Rather, they are the results of human ordering practices aimed at particular epistemic targets that have *some* objective aspect that explains their functionality and can be understood in terms of grounding.

The exploration presented in this chapter is far from complete: I have considered only one example and did not discuss every metaphysical detail of that example that may have been important. But I hope to have shown how "kinds thinking" and "levels thinking" are connected at a metaphysical level and how using a thoroughly naturalistic account of kinds—the GFA—can help to achieve some clarity about the metaphysical basis not only of kinds but also of levels. Note that I do not want to suggest that the grounding relations between kinds, and derivatively between levels, are always of the same sort. Any thoroughly naturalistic account of kinds and levels should resist focusing on one sort of grounding relation to the exclusion of others, and more clarity on the various ways in which kinds and levels in the various sciences are grounded must come from examinations of a multitude of concrete cases. The considerations presented here could serve as a unified perspective from which such cases can be analyzed.

Acknowledgments

I am indebted to James DiFrisco, Marc Ereshefsky, Markus Eronen, Jon Umerez, and the participants of the 36th Altenberg Workshop in Theoretical Biology for comments and criticism.

Notes

1. Ereshefsky (2001) introduced the notion of "classificatory programs" to refer to such sets of principles underlying scientific classifications. See below for further discussion (see also Ereshefsky & Reydon, 2015, in press).

2. For overviews, see, for example, Wimsatt (1994), Kim (2002, 2003), Eronen (2015), and Eronen and Brooks (2018) on levels and Koslicki (2008), Hawley and Bird (2011), Magnus (2012), and Bird and Tobin (2018) on kinds.

3. See Brooks (2017), DiFrisco (2017), and Potochnik (2017, chap. 6; this volume) for discussions of skepticism about levels and defenses of the concept.

4. But see Batterman (this volume) on levels defined by order parameters for interesting examples of such "kind-less" levels.

5. This argument resembles the infamous "miracle argument" for scientific realism (Putnam, 1975, p. 73), but my concern will not be with realism about kinds and levels. I will address only one aspect of the larger realism question, namely, the question of nonarbitrariness.

6. This means that functional kinds are natural kinds or, more precisely, that the category of functional kinds is a subcategory of the category of natural kinds. I will not explicitly argue for this claim in the present chapter, but it should be clear that it follows from my account of functional kinds and levels.

7. But the traditional interpretation is controversial. Philosophers of biology have long debated the nature of taxa in the Linnaean system, and the majority view is that taxa aren't kinds at all but individuals, such that the traditional interpretation fails. This debate is important, but I will ignore it here as I only use the Linnaean system as a quick example to illustrate "kinds thinking" and "levels thinking" in science.

8. I am using "things" in a broad sense, including material objects, living systems, events, phenomena, and so on, covering whatever exists. I do not want to defend a particular preferred ontology but hold an agnostic position regarding what the furniture of the world consists of.

9. For a recent, brief overview of the connection between realism and mind independence regarding kinds, see Khalidi (2016, section 1). Note, though, that Khalidi opposes the view that natural kinds must be mind independent.

10. My discussion of this example draws on discussions that I presented elsewhere (Reydon, 2009b, 2016).

11. For reasons of space, I can give only a brief sketch of the GFA here. For details, see Ereshefsky and Reydon (in press).

12. See Waters (2017) for a recent defense of a similar (possibly stronger) view.

13. Admittedly, this is open-ended. Open-endedness is a virtue here, though, as it makes the GFA thoroughly naturalistic by allowing it to cover a multitude of ways in which kinds can depend on the world. Committing to one particular way of metaphysical dependence would make the GFA insufficiently naturalistic in the same way as other accounts of natural kinds are insufficiently naturalistic.

14. "Genes that aid spinal cord healing in lamprey also present in humans, MBL team discovers" (posted on February 15, 2018): https://www.eurekalert.org/pub_releases/2018-01/mbl-gta011218.php (accessed February 22, 2018). The research article itself is more subtle and does not speak of humans and lampreys having the same genes (Herman et al., 2018).

15. This example is also discussed in Reydon (2009b), on which the following passage draws.

16. A car's parts are constitutive of the car, as cars are built by putting the parts together into a functioning whole. This is not how natural genomes and genes relate to each other: genes are parts of genomes, but not in a constitutive way, only in a decompositional way. This point is similar to Mayr's criticism of early genetics as "beanbag genetics" (Mayr, 1963, p. 263; see Dronamraju, 2011, for a detailed discussion of Mayr's criticism). Note that for artificial genomes, the situation is different: in synthetic biology, genomes are built by assembling genetic parts into functioning wholes.

17. Several authors have advanced the idea that higher levels impose constraints on lower levels but typically do not think of this in terms of constraints on the individuation of lower-level entities. See Eronen and Brooks (2018, section 3.3).

18. "Bloß die einfache Vorstellung soll Ausdruck finden, daß durch 'etwas' in den Gameten eine Eigenschaft des sich entwickelnden Organismus bedingt oder mitbestimmt wird oder werden kann. Keine Hypothese über das Wesen dieses 'etwas' sollte dabei aufgestellt oder gestützt werden" (Johannsen, 1909, p. 124).

19. The distinction parallels Moss's (2003) distinction between the "gene-D" and "gene-P" concepts.

20. A question that I have to set aside for reasons of space is how gene and genome kinds, and the corresponding levels, are related between research contexts. The GFA conceives of kinds as local (i.e., relative to specific classificatory programs), but there should be more to say, as scientists in practice often transgress the boundaries between research programs and develop overarching taxonomies.

21. For reasons of space, I cannot elaborate this issue here. Suffice it to note that pluralism about part–whole relations is widely endorsed. Various taxonomies of part–whole relations are available, often constructed in the context of applied/formal ontology, data handling in engineering, and cognitive science (e.g., Motschnig-Pitrik & Kaasbøll, 1999; Winston et al., 1987; for cases from biology, see Winther, 2011). Part–whole relations are generally discussed in terms of (de)composition and constitution (e.g., Motschnig-Pitrik & Kaasbøll, 1999).

References

Adams, M. D., Celniker, S. E., Holt, R. A., et al. (2000). The genome sequence of *Drosophila melanogaster*. *Science, 287*, 2185–2195.

Bernstein, S. (2016). Grounding is not causation. *Philosophical Perspectives, 30*, 21–38.

Bird, A., & Tobin, E. (2018). Natural kinds. In E. N. Zalta (Ed.), *The Stanford encyclopedia of philosophy*. Retrieved from https://plato.stanford.edu/archives/spr2018/entries/natural-kinds/

Bliss, R., & Trogdon, K. (2016). Metaphysical grounding. In E. N. Zalta (Ed.), *The Stanford encyclopedia of philosophy*. Retrieved from https://plato.stanford.edu/archives/win2016/entries/grounding/

Brooks, D. S. (2017). In defense of levels: Layer cakes and guilt by association. *Biological Theory, 12*, 142–156.

Correia, F., & Schnieder, B. (Eds.). (2012). *Metaphysical grounding: Understanding the structure of reality*. Cambridge, UK: Cambridge University Press.

DiFrisco, J. (2017). Time scales and levels of organization. *Erkenntnis, 82*, 795–818.

Dronamraju, K. (2011). *Haldane, Mayr, and beanbag genetics*. New York, NY: Oxford University Press.

Ereshefsky, M. (2001). *The poverty of the Linnaean hierarchy: A philosophical study of biological taxonomy*. Cambridge, UK: Cambridge University Press.

Ereshefsky, M., & Reydon, T. A. C. (2015). Scientific kinds. *Philosophical Studies, 172*, 969–986.

Ereshefsky, M., Reydon, T. A. C. (in press). The grounded functionality account of natural kinds. In W. Bausman, J. Baxter, O. Lean, A. Love, & C. K. Waters (Eds.), *From biological practice to scientific metaphysics*. Minneapolis: University of Minnesota Press.

Eronen, M. I. (2015). Levels of organization: A deflationary account. *Biology and Philosophy, 30*, 39–58.

Eronen, M. I., & Brooks, D. S. (2018). Levels of organization in biology. In E. N. Zalta (Ed.), *The Stanford encyclopedia of philosophy*. Retrieved from https://plato.stanford.edu/archives/spr2018/entries/levels-org-biology/

Fine, K. (2012). Guide to ground. In F. Correia & B. Schnieder (Eds.), *Metaphysical grounding: Understanding the structure of reality* (pp. 37–80). Cambridge, UK: Cambridge University Press.

Gerstein, M. B., Bruce, C., Rozowsky, J. S., Zheng, D., Du, J., Korbel, J. O., Emanuelsson, O., Zhang, Z. D., Weissman, S., & Snyder, M. (2007). What is a gene, post-ENCODE? History and updated definition. *Genome Research, 17*, 669–681.

Griffiths, P. E., & Stotz, K. (2006). Genes in the postgenomic era. *Theoretical Medicine and Bioethics, 27*, 499–521.

Griffiths, P. E., & Stotz, K. (2007). Gene. In D. L. Hull & M. Ruse (Eds.), *The Cambridge companion to the philosophy of biology* (pp. 85–102). Cambridge, UK: Cambridge University Press.

Griffiths, P. E., & Stotz, K. (2013). *Genetics and philosophy: An introduction*. Cambridge, UK: Cambridge University Press.

Hacking, I. (1990). Natural kinds. In R. B. Barrett & R. F. Gibson (Eds.), *Perspectives on Quine* (pp. 129–141). Oxford, UK: Basil Blackwell.

Hacking, I. (2007). Natural kinds: Rosy dawn, scholastic twilight. In A. O'Hear (Ed.), *Philosophy of science* (pp. 203–239). Cambridge, UK: Cambridge University Press.

Hall, B., Limaye, A., & Kulkarni, A. B. (2009). Overview: Generation of gene knockout mice. *Current Protocols in Cell Biology*, unit 19.12.

Harrison, P. M., Kumar, A., Lang, N., Snyder, M., & Gerstein, M. (2002). A question of size: The eukaryotic proteome and the problems in defining it. *Nucleic Acids Research, 30*, 1083–1090.

Hawley, K., & Bird, A. (2011). What are natural kinds? *Philosophical Perspectives, 25*, 205–221.

Herman, P. E., Papatheodorou, A., Bryant, S. A., Waterbury, C. K. M., Herdy, J. R., Arcese, A. A., Buxbaum, J. D., Smith, J. J., Morgan, J. R., & Bloom, O. (2018). Highly conserved molecular pathways, including Wnt signaling, promote functional recovery from spinal cord injury in lampreys. *Scientific Reports, 8*, 742.

International Human Genome Sequencing Consortium. (2001). Initial sequencing and analysis of the human genome. *Nature, 409*, 860–921.

Johannsen, W. (1909). *Elemente der exakten Erblichkeitslehre (deutsche wesentlich erweiterte Ausgabe in fünfundzwanzig Vorlesungen)* [Elements of the exact theory of heredity (German edition, considerably extended, in 25 lectures)]. Jena, Germany: Verlag von Gustav Fischer.

Keller, E. F. (2013). Genes as difference makers. In S. Krimsky & J. Gruber (Eds.), *Genetic explanations: Sense and nonsense* (pp. 34–42). Cambridge, MA: Harvard University Press.

Khalidi, M. A. (2016). Mind-dependent kinds. *Journal of Social Ontology, 2*, 223–246.

Kim, J. (2002). The layered model: Metaphysical considerations. *Philosophical Explorations, 5*, 2–20.

Kim, J. (2003). Supervenience, emergence, realization, reduction. In M. J. Loux & D. W. Zimmerman (Eds.), *The Oxford handbook of metaphysics* (pp. 556–584). New York, NY: Oxford University Press.

Konsolaki, M., & Schüpbach, T. (1998). *windbeutel*, a gene required for dorsoventral patterning in *Drosophila*, encodes a protein that has homologies to vertebrate proteins of the endoplasmic reticulum. *Genes & Development, 12*, 120–131.

Koslicki, K. (2008). Natural kinds and natural kind terms. *Philosophy Compass, 3/4*, 789–802.

Love, A. C. (2012). Hierarchy, causation and explanation: Ubiquity, locality and pluralism. *Interface Focus, 2*, 115–125.

Ludwig, D. (2018). Letting go of 'natural kind': Toward a multidimensional framework of non-arbitrary classification. *Philosophy of Science, 85*, 31–52.

Lynch, E. D., Lee, M. K., Morrow, J. E., Welsh, P. E., Léon, P. E., & King, M.-C. (1997). Nonsyndromic deafness DFNA1 associated with mutation of a human homolog of the *Drosophila* gene *diaphanous*. *Science, 278*, 1315–1318.

MacLeod, M., & Reydon, T. A. C. (2013). Natural kinds in the life sciences: Scholastic twilight or new dawn? *Biological Theory, 7*, 89–99.

Magnus, P. D. (2012). *Scientific enquiry and natural kinds: From planets to mallards*. Houndmills, UK: Palgrave Macmillan.

Mayr, E. (1963). *Animal species and evolution*. Cambridge, MA: Harvard University Press.

Moss, L. (2003). *What genes can't do*. Cambridge, MA: MIT Press.

Motschnig-Pitrik, R., & Kaasbøll, J. (1999). Part-whole relationship categories and their application in object-oriented analysis. *IEEE Transactions on Knowledge and Data Engineering, 11*, 779–797.

Pei, B., Sisu, C., Frankish, A., Howald, C., Habegger, L., Mu, X. J., Harte, R., Balasubramanian, S., Tanzer, A., Diekhans, M., Reymond, A., Hubbard, T. J., Harrow, J., & Gerstein, M. B.. (2012). The GENCODE pseudogene resource. *Genome Biology, 13*, R51.

Portin, P. (1993). The concept of the gene: Short history and present status. *Quarterly Review of Biology, 68*, 173–223.

Potochnik, A. (2017). *Idealization and the aims of science*. Chicago, IL: University of Chicago Press.

Putnam, H. (1975). *Mathematics, matter and method: Philosophical papers* (Vol. 1). Cambridge, UK: Cambridge University Press.

Racioppi, C., Valoroso, M. C., Coppola, U., Lowe, E. K., Brown, C. T., Swalla, B. J., Christiaen, L., Stolfi, A., & Ristoratore, F. (2017). Evolutionary loss of melanogenesis in the tunicate *Molgula occulta*. *EvoDevo, 8*, 11.

Reydon, T. A. C. (2009a). How to fix kind membership: A problem for HPC-theory and a solution. *Philosophy of Science, 76*, 724–736.

Reydon, T. A. C. (2009b). Gene names as proper names of individuals: An assessment. *British Journal for the Philosophy of Science, 60*, 409–432.

Reydon, T. A. C. (2010). How special are the life sciences? A view from the natural kinds debate. In F. Stadler (Ed.), *The present situation in the philosophy of science* (pp. 173–188). Dordrecht, Netherlands: Springer.

Reydon, T. A. C. (2016). From a zooming-in model to a co-creation model: Towards a more dynamic account of classification and kinds. In C. E. Kendig (Ed.), *Natural kinds and classification in scientific practice* (pp. 59–73). London, UK: Routledge.

Trogdon, K. (2013). An introduction to grounding. In M. Hoeltje, B. Schnieder, & A. Steinberg (Eds.), *Varieties of dependence: Ontological dependence, grounding, supervenience, response-dependence* (pp. 97–122). München, Germany: Philosophia.

Venter, J. C., Adams, M. D., Myers, E. W., et al. (2001). The sequence of the human genome. *Science, 291*, 1304–1351.

Waters, C. K. (1994). Genes made molecular. *Philosophy of Science, 61*, 163–185.

Waters, C. K. (2007). Causes that make a difference. *Journal of Philosophy, 103*, 551–579.

Waters, C. K. (2017). No general structure. In M. H. Slater & Z. Yudell (Eds.), *Metaphysics and the philosophy of science: New essays* (pp. 81–107). New York, NY: Oxford University Press.

Wieben, E. D. (2003). Primer on medical genomics, Part VII: The evolving concept of the gene. *Mayo Clinic Proceedings, 78*, 580–587.

Wimsatt, W. C. (1994). The ontology of complex systems: Levels of organization, perspectives, and causal thickets. *Canadian Journal of Philosophy, 20*, 207–274.

Winston, M. E., Chaffin, R., & Herrmann, D. (1987). A taxonomy of part-whole relations. *Cognitive Science, 11*, 417–444.

Winther, R. G. (2011). Part-whole science. *Synthese, 178*, 397–427.

14 Control Hierarchies: Pattee's Approach to Function and Control as Time-Dependent Constraints

Jon Umerez

Overview

This work presents an approach to understand the origin and nature of level formation and control processes in the biological realm and to account for the specific interlevel relations they give rise to, which adopts the form of biological *organization*. It is argued that the idea of constraint-mediated control hierarchy developed after Pattee is an appropriate way to address both complementary demands posed, on the one hand, by the challenge of finding a plausible replacement for the familiar but controversial "universal hierarchical ordering" and, on the other, by the emerging program of reconstructing the concept of level of organization. This proposal is consistent with new revisions both of the neo-mechanist view and of the skeptical and deflationary reactions that initially triggered the concern motivating the approach here defended. The conclusion supports the need to incorporate into the current debates the idea of hierarchical control based on very specific constraints in order to clarify interlevel relations when dealing with living systems.

Hierarchical control is the most universal organizational principle of living matter.
—Pattee (1971a, p. 47)

14.1 Introduction: Some Motivations

As many of the contributions to this volume acknowledge, we have recently seen a surge in interest regarding issues of levels of organization and hierarchy in philosophy of science. This renewed interest is due in great part to the neo-mechanist approach (without forgetting the continuous work in other areas such as, for instance, evolutionary hierarchical theory; see Tëmkin, this volume). However, this same interest has also provoked a skeptical reaction highlighting presumed problems and inadequacies of the very concepts of *level* or *level of organization* and, subsequently, of *hierarchy* itself (i.e., Eronen, 2013, 2015; Eronen & Brooks, 2018; Potochnik & McGill, 2012; Thalos 2013).

The analysis provided in this chapter responds firstly to a longstanding motivation regarding the related problems, on the one hand, about how to understand the origin and nature of control processes and of level formation in the biological realm and, on the other hand, about how to account for the specific interlevel relations they give rise to, in the form of biological *organization* (Umerez, 1994). More recently, this interest has become coupled with concerns regarding the generalized depiction of interlevel relation in terms of plain composition and the resulting deflationary or skeptical reaction (Umerez, 2016).

Simultaneously, this chapter also seeks to respond to a double challenge. On the one hand, I consider how to tackle Potochnik and McGill's (2012, p. 133, n. 5) challenge of finding a plausible replacement for "universal hierarchical ordering" and, on the other, how to contribute to Brooks's (2017, p. 153) program of reconstructing the concept of level of organization by presenting a usage with its explicit aim. I will argue that the idea of *constraint-mediated control hierarchy* developed after Pattee is one way to address both complementary demands. First, this is a good (if not the only) instance of a sound concept of hierarchical relation among levels that is not ambiguous or vague in its characterization and that does not entail any given universal ordering. Second, it has the very specific and explicit aim, triggered by the question regarding the origin of life, to help understand the nature of living phenomenology stemming from the nonliving physical realm. But, at the same time—and due precisely to the restricted and specific character of both the original account and its purpose—it might be of more general application to other cases, and above all, it may constitute the bottom and common ground to characterize the specificity of living organization as such.

I am confident that this proposal is consistent with new revisions both of the neomechanist view and of the skeptical and deflationary reactions that are central to my concerns. We have simultaneously seen the incorporation by Bechtel of the idea of control (Winning & Bechtel, 2018) into his previous views, the defense (or at least admission) by Eronen (Brooks & Eronen, 2018) of the scientific and heuristic value of the concepts of levels and hierarchies, and the proposal by other authors (Brooks, 2017; DiFrisco, 2017; Green & Batterman, 2017) of specific uses or meanings of those concepts, for instance, with the help of the notion of scale, but without attempting to abandon or replace them.

In order to achieve these goals, I will proceed in this chapter as follows. I will briefly recap my proposal to reconstruct the meaning of level of organization by means of an articulation of several concepts defining kinds of relations, among which *organization* is both the most inclusive (encompassing all) and the most specific (fulfilling all the conditions), including the relation of control. Next, in order to complement my understanding of the concept of "level of organization" with one analogous to the concept of "hierarchy," I will introduce in some detail the starting points of Pattee's contribution to hierarchy theory, placed in the wider context of his early theoretical development and centered on his elaboration of the concept of constraint. I will then propose an articulation of the concepts of hierarchy, distinguishing among different kinds of interlevel relations that are defined based on the type of constraint present. In the last section, I will reiterate the need to incorporate the idea of hierarchical control into the current debate in order to clarify interlevel relations when dealing with living systems. Finally, I will extract some general consequences and conclusions derived from this approach.

14.2   Levels, Composition, and Organization

I have recently argued, first, that some of those negative or reticent reactions (skeptical and deflationary) are derived from a problematic view that privileges particular conceptions of the ideas of level and hierarchy, which I fail to see as representative, and from a serious oversight of other, more constructive approaches (although these approaches are often admittedly outside standard philosophical literature) (Umerez, 2016, pp. 67–68). Referring to the "widespread suspicion in the philosophical literature regarding the legitimacy of considering inter-level relations," especially from top down, I have maintained that "an oversight of the very relevant scientific approaches and a restricted rendition of the concept of levels of organization are two motives" for such rejection of levels and hierarchy (Umerez, 2016, pp. 65–66).

Next, I have also argued that in some current renditions of the idea of (biological) level, the issue of "organization" is taken for granted. In particular, the basic notion of "composition" employed turns out to be too general and weak to account for "organization" (as in "levels of") (Umerez, 2016, pp. 80–81). In addition to more specific concerns about the concept of composition (such as Eronen's 2013 and, more detailed, 2015 criticism of the notion of "component" and "same-level criterion"), I have addressed, in an opposite direction, a more basic concern implying that the mechanist approach of, for instance, Craver and Bechtel (2007) is not actually a purely compositional one, but involves other traits or characteristics that make it a very special kind of composition (i.e., *organizational*). This reading, on the one hand, might constitute a response to certain criticisms against levels (such as Eronen's) but, on the other hand, it could also compromise the exclusion of interlevel relations from causal consideration (upward and downward) by virtue of being allegedly just constitutive in a plain sense. A similar suspicion and possible way out also affects some aspects of the proposal of organizational closure as "an emergent causal regime" using the notion of "configuration" by Mossio et al. (2013) and Moreno and Mossio (2015, chap. 2).[1]

In both cases, the central concepts involved are always introduced with explicit qualifications: *mechanisms* are *organized* collections of activities and entities, and *configurations* are whole sets of inherent AND *relational* properties of the constituents.[2] The difficult issue is where to ground the organizational or configurational nature of those groups of components such that they are different from "mere aggregations" BUT are accounted for just in terms of composition (this being the premise used to dismiss relations of downward causation—general or reflexive—reinterpreted as merely constitutive).

In Umerez (2016) I made an attempt to clarify the notion of level, that is, *level of organization*, in biology (recapped from Umerez, 1994). In this work I provided a reconstruction of the concept, articulated around five features that indicate kinds of relations between elements giving rise to some specific form of level arrangement. Those features include *composition, integration, emergence, control,* and, encompassing them all, *organization.* The reconstruction is articulated because it is built around a common but abstract criterion—kind of relation—and goes from the more general to the more specific, providing a progressive specification of the concept of level in which each additional form of relation includes the previous one, but not vice versa, in a manner that is reminiscent of Salthe's (1993) representation of a specification hierarchy: {composition {integration {emergence {regulation {organization}}}}}.

I supplemented this with a detailed examination of what is involved in the relation of composition considering four (nonexhaustive) aspects: (1) nestedness, (2) partial ordering, (3) homogeneity and heterogeneity of component parts, and (4) the discreteness or continuity of the arrangement. According to this analysis of composition, I claim that it does not permit, just by itself, distinguishing organized sets of components from mere aggregates (too unspecific a relation) and that, therefore, something else is involved, since neither "organized collections" nor "configurations" can be taken as primitives or as given. This "something else involved" amounts precisely to nontrivial causal relations at different levels: specific and precise interlevel causal relations that bring about *organization*:

> Therefore, when considering organization, we are referring to complex systems implying some form of interrelation among elements that goes well beyond mere composition: such systems manifest integrated global emergent properties, capable of regulating the behavior (dynamics) of their constituents. (Umerez, 2016, p. 75)

This brings us to the next issue with respect to current accounts and debates about levels and hierarchies: that of how to interpret the relation among levels. But first, let us introduce Pattee's approach with some detail, in order to examine how the concepts of constraint and control hierarchy were proposed and used with the purpose of understanding and explaining biological organization and complexity. While he was not, of course, the only one to address the issue in these terms, Howard Pattee is commonly acknowledged to have opened this path and to have provided a well-grounded and comprehensive account of these notions and related concepts. For instance, both groups of authors mentioned in this section refer to the work of Pattee and, in particular, to his elucidation of hierarchical control based on the concept of constraint. In the case of the organizational approach of Moreno and colleagues, the influence of Pattee has always been acknowledged as one of the central cornerstones of the very approach in the collective work of the IAS Research group, particularly in their application of the concept of constraint and other related ideas (like that of dynamical decoupling). In the mechanist camp, explicit recognition by Bechtel of this inspiration is found in virtually all of his recent publications incorporating control.[3]

Other authors are also involved in this recovery, for instance, Hooker or Korn, to mention just two cases that single out Pattee's contribution and, specifically, his analysis and elaboration of the concept of constraint. Dealing with various inadequate hierarchical descriptions, Korn (2002) states that "the one exception to this confusion is the work of Pattee (1969) in which a hierarchy is described as a descending arrangement of constraints" (p. 200), reaffirming this credit in Korn (2005, p. 138). More influential in the philosophical discussion is the assessment by Hooker (2013), who, in a paper that highlights the role played by constraints to understand complexity, said that "it was Pattee who emphasised the importance of constraints, especially of such coordinated constraints, to biological organisation and evolution" (p. 761).

14.3   Pattee and the Wider Context of the Debate

As I anticipated in the introduction, this analysis is concerned with the conceptual role of the ideas of control and constraint, in the specific rendition of them formulated by Howard Pattee, because I think it may help to illuminate two relevant issues. On the one hand, it illustrates the possibility of a hierarchical approach with levels of organization free of some of the problems of definition and practical use that have been attributed to both terms (level of organization, hierarchy). And, on the other, it simultaneously suggests an alternative and precise way to interpret the notion of interlevel relation and the derived issue of causation.

Let us recall that the deflationary distrust and disaffection with the concepts of level of organization and hierarchy are justified with the indictment of their alleged ambiguity, plurality, and potential confusion of meanings. I have previously questioned, as have others (Brooks, 2017), whether this reaction is justified or instead is derived from focusing on particular accounts of those concepts while overlooking others. Besides, this very suspicion is not in itself novel and has its own antecedents. I have previously noted (Umerez, 1994), among many others, for instance, Bunge's stern analysis of both terms "level" and "hierarchy" in the 1960s/1970s, a time that witnessed a great interest on issues related to hierarchy and levels of organization.

But here I would rather recall Marjorie Grene's participation in the 1968 *Symposium on Hierarchical Structures* (Whyte et al., 1969). Grene had at that time already addressed the issue of levels in biology (Grene, 1967), dealing with ontological as well as epistemological problems regarding "levels of reality." In her short contribution, she starts by sharing a very general impression of most of the participants, stating: "To a philosophical observer, the first lesson of this conference was that the use of the word 'hierarchy' is strikingly equivocal. In diverse disciplines, it signifies diverse concepts" (Grene, 1969, p. 56). As the symposium distributed its content into three parts—inorganic, organic, and artifact hierarchies (following an initial one dealing with the concept itself)—Grene continues her analysis by comparing biologists' to astronomers' usage of the term. She notes,

Biologists on the other hand, when they worry about hierarchical organization in living things, are concerned about (at least) doubly determinate systems: systems such that an arrangement of the elements comprising the systems constraints the behavior of the elements themselves, the controlling order thus constituting an upper level (not necessarily larger, however) in relation to the elements so ordered, which, in turn constitute a lower level of the system. Dr. Pattee's definition characterizes this kind of situation quite clearly. (Grene, 1969, p. 56)

She ends the observation by adding that the interest of people working on artifacts is closer to the biological sense.

There is another key insight by Grene I would like to recall here. When dealing with Bunge's strict analysis and its eliminativist corollary (Bunge, 1969), Grene (1969) points out that "by this excision [of looser meanings of hierarchy] he would, it seems, eliminate *all* the uses actually made of the term by all the scientists concerned.... Such a drastic revision of scientific–and philosophical–vocabularies, however, appears ill-advised" (p. 57). The reasons she gives for not following Bunge's exhortation are, I think, as accurate now as they were then: on the one hand, she thinks that philosophers should not (she says "cannot") try to be prescriptive regarding the use of scientific terms, and on the other, she argues that a clear disclosure and explanation of the different meanings in different contexts (or usages) should solve the problems involved in the term's meaning. She also adds a more substantive reason, which might perhaps be susceptible to stronger contemporary criticism, yet is worth considering: she thinks that the several meanings have something in common: "In every case there is *ranking* of some kind; in most cases, moreover, it is a ranking of *real* entities or of levels of organization" (Grene, 1969, p. 57).

Regarding this last appraisal, I would like to explore the issue, not of what hierarchies in general have in common, but of what the hierarchical arrangement in living systems has at its core in the terms of the epigraph of this chapter: *control as the basis of organization.*

14.4   Pattee's Early Introduction of Control and Constraint

Let us then examine some of the contributions made by Pattee, one of the main theorists in the 1960s/1970s, a time when the issues of hierarchy and levels of organization, if we may say so, reached a peak. Pattee has not defended or advocated anything like a traditional layer-cake view (an object of much of the current criticism) or a merely compositional approach (focus of most of the deflationary views). Pattee, together with other contemporaries (for instance, Robert Rosen), set out a different research program establishing that the central problem in the study of hierarchies and levels of organization was not as much the structure, specifying which levels are in a fixed ordering, but the relation between levels, understanding how they originate and interact (*hierarchical interfaces*).

In fact, Pattee has not endorsed any version of those textbook modern "scala naturae" pictures depicting the series from atoms to biosphere. The few times in which he presents what most closely resembles a "list" of hierarchical levels of control (levels of organization), for instance, in Pattee (1971b), where, under the heading of "biological examples of the problem," he includes the following: genetic code, developmental controls, integrated cognitive systems, and language structures (pp. 257–260). Even here (as in general) he is more concerned with the "hierarchical interface" than with defining the levels themselves. More recently Pattee (2009), when recalling the role of complementarity in his work on hierarchies, comments on the need to specify the domain being analyzed:

The concept of complementary hierarchical models is essential for modeling complex systems (e.g., Pattee 1973a). Every level of biological organization requires a different operational definition of information and interpretation. It should be clear that a model of the cell's interpretation of molecular information in the gene must be different from a model of the brain's interpretation of the information in this sentence. It is a waste of time arguing over concepts like information, interpretation, and function without specifying the domain of the model. (p. 296)

The same final warning also applies to the very concepts of level or hierarchy.[4] In this sense, Pattee was very clear about his specific aim quite early on. As he has stated more than once, his interest in hierarchies stemmed from the necessity to address conceptual difficulties found in his research on the *origins of life*. As a physicist dealing with the "highly unlikely and somehow arbitrary constraints which harness these laws [of Physics] to perform specific and reliable functions" (Pattee, 1970, p. 117), he focused on studying the origin and nature of hierarchical controls in biological systems (and limited accordingly his use and analysis of the idea of hierarchy). He explicitly introduces this specific aim at the opening of his first published work on hierarchies:

The origin of life problem is the context in which I began thinking about hierarchies. The origin of life is perhaps the most mysterious hierarchical interface of all, but at the same time I believe it may present one of the most instructive approaches to general hierarchical control problems. This is because the lower level pre-life processes are ordinary physics and presumably subject only to precise laws which do not include extra hierarchical rules or constraints. However, to be recognized as "alive" a collection of matter must exhibit some additional integrative function by exerting a collective control over the individual molecules. This integrative function is what characterizes *hierarchical control*. (Pattee 1969a, p. 161)

Let us then address the question of control in his work, an issue Pattee has dealt with since his first papers. Indeed, Pattee (1961) deals already with genetic controls in quite a standard way. But in his contribution to a volume on *Advances in Enzymology and Related*

Subjects of Biochemistry, he points already to the concept of *control* in order to understand the origin of life:

Consequently we may expect that the origin of life problem will shift away from the evolution of the building blocks and the elementary operations of joining them together, to the more difficult problem of the *evolution of control* in complex organizations. This problem is more difficult because the idea of "control" is not defined in the same sense as we can define biochemicals. ... From this point of view, the question of the origin of life becomes the problem of understanding elementary molecular control processes, and of formulating a theory of the evolution of molecular control. (Pattee, 1965b, pp. 405–406)

At the other end of his publishing trajectory, Pattee eventually introduced the three key concepts in the title of the book that compiles some of his papers—law, life, and language—in terms of control. In particular, the first two appear as opposites, law as absence of control, and life as characteristic of "*individual* organisms with *variable heritable controls*" (Pattee, 2012a, p. 3). In between, we find a corpus of work devoted to understanding life by clarifying the kind and nature of those processes of control[5]—and to elucidating the means, processes, or devices by which this relation of control among levels is exerted and materialized physically in biological systems in the form of *constraints*.

The first use of the concept of constraint, already suggestive of what became his characteristic treatment, is found in the chapter on "Physics, Automata, and the Origin of Life" contributed to the Symposium on Fundamental Biological Models, entitled *Natural Automata and Useful Simulations*, the proceedings of which he edited (together with Edgar A. Edelsack, Louis Fein, & Arthur B. Callahan) in a book with the same title of the symposium (Pattee et al., 1966). This paper is an attempt to study hereditary processes from a physical standpoint and describes the molecular models of replication based on tactic copolymerization that he was developing during those years.[6] He introduces the concept of constraint to help define hereditary models, distinguishing already between holonomic and nonholonomic constraints (see next section), applied to simple mechanical models (from a inclined plane with one mass rolling down, to another with several masses rolling down, and to a more complex one that includes an escapement delaying the movement of the masses) or other automata models as growing helical chains, which he had worked on previously as *molecular sequence computers* (Pattee, 1961).

Next, he introduces the idea of a *classification process* as necessary for understanding hereditary processes (Pattee, 1967). This new concept is defined in terms of the operations of nonholonomic constraints, and he already singles out the enzyme as the main materialization of this kind of constraint in natural (not artificial) domains.

In terms of molecular reactions this non-holonomic condition implies that the number of energetically possible states is larger than the number of reactions actually available to the system. Now the chemical reactions which are available as distinct from those which are energetically possible can differ only in the activation energy and entropy, so that we are led to associate classification or hereditary propagation with the control of rates of specific types of reactions. Of course in cells the epitome of such specificity and catalytic control of reactions is the enzyme. (Pattee, 1967, p. 416)

Then, as he has recalled several times, Grene introduced him to Polanyi's work (1968), and the ideas he had been developing in previous years regarding the specificity of living systems in terms of the basic difference between law-based and constraint-based behavior or phenomena began to adopt the form of an explicit hierarchical theory:

What I call symbolic constraints on a lawful dynamics Polanyi (1968) calls special boundary conditions that "harness" dynamical laws. My hierarchy theory was an elaboration of Polanyi's concept of the functionally irreducible hierarchical levels of boundary conditions. (Pattee, 2012a, p. 24)

From this moment on, the next decade would see the elaboration of his ideas about hierarchies and levels of organization in an attempt to rigorously characterize the nature and origin of interlevel relations adopting the form of control (functional hierarchy). He would use all these concepts derived from mechanics in order to accomplish this goal.

14.5   Interlude: Basics from Mechanics

As has been repeatedly noted, Pattee grounds his approach on very basic concepts pertaining to the physical sciences, particularly mechanics. In order to account for the dynamics (behavior) of any physical system (biological systems being a special case of physical systems), he reminds us that, in principle, we simply need to know the initial conditions and the laws of motion. Leaving aside for a moment the issue of measurement implicit in the procurement of those initial conditions (subsequently expressed as records and therefore of a symbolic nature), we should, at least, address the question of an *alternative description*, which is at the core of Pattee's hierarchical perspective and is derived from the use of the concept of constraint, as is done in physics.

Physical laws and initial conditions: It is well known that, in physics, initial conditions and laws of motion provide an exhaustive description of the possible behavior—future and past—of a mechanical system. In principle, there is no need to add any other condition, as this would be redundant. The choice of the coordinates of the system that specify its configuration or state space defines all the possible degrees of freedom of the system or all the possible trajectories of its elements. These coordinates establish the variables of the motion equations of the system. Laws of physics are usually expressed in the form of differential equations with respect to time based on energy principles or in the form of other equations derivable from these (except, e.g., the mathematical relation between classic and quantum laws). From the principles of conservation or invariance, we derive the laws of movement, which are independent from any particular configuration of initial conditions. Laws of motion, then, tell us that the state of the system will change in time in a certain way. In this sense, laws of nature are, in principle, inexorable and incorporeal.

Constraints: Any other description of the behavior of a physical system that is not a detailed microscopic description requires additional, or auxiliary, conditions. In principle, this situation is solved in classical mechanics by introducing additional equations called constraints. Such constraints are the result of relatively permanent forms or structures that limit the degrees of freedom of the system. The purpose of this addition is to simplify the study, description, or calculation of the behavior of a given system at macroscopic scales (while quantum mechanics can resort to the notion of steady state, classical mechanics need them to explain solid bodies and to explain other kind of constraints such as boundary conditions imposed onto matter in certain experimental settings and the fabrication of machines and artifacts). Constraints, unlike laws, must be the consequence of what we call some form of material structure such as molecules, membranes, or any kind of surface. These structures may be static or time dependent, but in any case, it is important to realize that they are made of matter that obeys fundamental laws of nature, in addition to behaving as constraints. What does this mean? If the laws of movement are complete and inexorable, what else can be said? Why is an equation of constraint not redundant or inconsistent? Here we have to realize that constraints cannot be treated as fundamental properties of matter. Unlike laws, that, as noted, are inexorable and incorporeal, constraints might be accidental or arbitrary and must have some distinctive physical materialization in the form of a structure of sorts. This is why we say that constraints are always *alternative descriptions* of part of the system (the same as measurements, incidentally). That is, constraints cannot be expressed in the same language as the microscopic description of matter. In fact, a constraint selectively ignores microscopic degrees of freedom in order to provide a simplified prediction or explanation of motion. In other words, the concept of constraint must represent a selective loss of detail, and therefore, in the realm of physics, the forces of constraint are inevitably linked to a new hierarchical level of description.

Thus, one of Pattee's main contributions is to distinguish and characterize two kinds of hierarchical relations—structural and functional—based respectively on two kinds of physical constraints: *holonomic* constraints are auxiliary conditions that limit permanently the degrees of freedom of a system and are, therefore, the basis for structural hierarchies, while *nonholonomic* constraints are variable auxiliary conditions that limit in time the number of degrees of freedom of the system, being the basis for the functional hierarchies typical of living systems. The latter are dynamical structures that establish time-dependent relations among degrees of freedom but introduce a different temporal scale (see Umerez, 1994; Umerez & Mossio, 2013, p. 494).

14.6 Pattee's Hierarchical Approach

With these conceptual elements imported from physics, Pattee launched himself into developing a hierarchy theory fitting his interests in theoretical biology. This is in part why I claim that in Pattee's work on hierarchy, we can find the foundation and potential development of a naturalist(ic) approach to biological hierarchical organization (Umerez, 2001, pp. 165–166). Accordingly, following the discussion above, he primarily developed the foundations of such a theory in a compact series of publications at the end of the 1960s and the beginning of the 1970s, culminating in an edited volume on the subject (Pattee, 1973a). It goes without saying that after this, he continued expanding and refining his account.

In his initial contribution to the previously mentioned 1968 *Symposium on Hierarchical Structures* he stated that the main basic characteristics of hierarchical controls are that they are autonomous in the sense that they produce their own rules, which are not externally imposed and are a fully-fledged part of the physical world; furthermore, they have a specific effect on individual elements of the collection (out of which they have arisen) and produce some integrated function of the collection as such (Pattee, 1969a, pp. 162–163). In a later paper, he lists more specifically "Some Properties of Control Hierarchies" (see Pattee, 1972, pp. 11–18).

With this original motivation in mind, Pattee develops an increasingly complex approach grounded on those basic concepts and distinctions taken from the language of physics (law/ rule, initial conditions/boundary conditions/constraints, dynamics/record [measurement], rate dependent/rate independent, etc.) and extends it to more general epistemological issues (matter/symbol, observer, *epistemic cut*, etc.) in his attempt to give an account of the specificity of fully natural but nontrivial interlevel relations in biological systems. For this, he takes the cell as the basic instance and the enzymatic reaction as the paradigmatic process.

In a very illuminating conversation with Robert Rosen (and Raymond Somorjai) on theoretical biology, conducted by physicists Paul Buckley and David Peat (who in their Preface justify including a round-table discussion on theoretical biology in a book "otherwise concerned with the problems of physics" because the subject "has all the intellectual challenge and excitement associated with physics in the twenties"), Pattee makes clear the precise scope and aim of his view:

My concept of hierarchy is very much more *limited*; it has to do with the alternation between descriptions and constructions, the idea being that at one level we have dynamical systems in the very general sense, that is, state descriptions and rate-dependent transformations between states which we then describe (or define, if you like) by the proper choice of observables. At another level we have rate-independent global or asymptotic structure, or, if you like, singularities or instabilities. One can label these things differently.... The two modes of description, which I call dynamic and syntactic, are complementary in the logical sense.... This is what I mean by hierarchy: it's an alternation of levels of description upon systems. Another way to say it is an alternation of continuity and discreteness. One can think of the external system as being continuous, obeying dynamics and having instabilities in the sense of continuous systems where infinitely small causes have large observable effects. Now, the *description*, on the other hand, is always a *course grained view* of this, a discrete view. (Pattee, 1979a, p. 118)

14.7   Articulation of Concepts of Hierarchy

Based on the distinctions introduced by Pattee, we can conduct an articulated analysis of the concept of hierarchy, similar to the one previously applied to the concept of level of organization, by using the kind of interlevel relation and, in particular, the type of constraint as a criterion of clarification of the intended meanings (Umerez, 1994).

Hierarchy, as *generic interlevel relation*, may be defined as *a transitive and asymmetric relation of partial order among the levels (more than one) that constitute a system*. This means that the elements entering the mutual relation are subsystems (intralevel relation) that are related among themselves as levels (interlevel relation), in a successive order within which such levels are lower with respect to the immediate higher level ("lower" and "higher" meaning just a relative position of succession in a partial ordering). Without this (mutually) relative order, there is no hierarchical disposition in any operative, defined, or nontrivial sense of the concept. I emphasize that these are merely the minimal conditions that any hierarchy must meet, since it implies only a generic relation of composition in which lower levels are related to higher ones insofar as the former constitute (make up) the latter. This relation then runs in one direction, that is, is unidirectional from lower to higher, without allowing us to talk properly of interaction (interrelation) in its fullest sense. The idea of hierarchy as a generic relation among levels is usually applied in the formulation of models of complex systems. In this sense, it is used to describe such systems out of subsystems (or elements) that are related among them by virtue of, for instance, the size or the number or elements, the intensity of connecting forces, processes or behaviors, temporal scales, or any other trait that might define the levels. Next, we take into consideration further conditions for adequately obtaining interactive interlevel relations, which are no longer so common and affect a more restricted range of hierarchical systems. The idea of interaction among levels as a more specific relation stems from the necessity to account for the reverse effect on higher levels on lower ones.

14.7.1   Hierarchy as Interaction: Global Constraint

Hierarchy, as *interactive* interlevel relation, is defined as *a relation of constraint in which higher levels limit globally and permanently (some) degrees of freedom of lower levels*. According to its characteristic, we will call this constraint and the kind of hierarchy it produces *structural*. This definition must be understood as complementary to the previous one, since it derives from it and therefore presupposes and specifies it. This means that not all hierarchies should meet this definition, but those that do must simultaneously meet the former one. This definition, on the one hand, *presupposes* the previous one because it is in the former that the sense of the lower/higher relation is determined and, on the other hand, specifies the former because it limits the generic relation to one of constraint. Constraint implies a composition relation among the components that produce it and adds a relation of limit onto those very components. Therefore, this relation, in contrast to the one previously described, is bidirectional. But it must be noted that it continues to be asymmetrical and transitive. It is just that now there will be a double asymmetry: one in the sense of composition and another one in the sense of the limitation. There are, then, two disjoint complementary relations that provide a fuller sense of interlevel relation.

The reference to this double transitivity introduces a further issue regarding the understanding of these kinds of constraints. Recalling the mechanical functioning of constraints, we stated that, in theory, laws and initial conditions are not only necessary but also sufficient elements to obtain a total description of the trajectories of a system. The introduction of constraints as auxiliary conditions is grounded on the possibility of making an alternative description of some of the variables of a given system. It is an alternative description that implies a simplification of the system under observation, because part of its dynamical details is ignored. The next case involves the other kind of constraint that represents the stricter notion of interlevel relation.

14.7.2   Hierarchy as Control: Specific Constraint

Let us now consider the concept of hierarchy derived from an interlevel relation in the form of an individualized constraint on specific elements. This is based on the sense of nonholonomic constraint. This type of specific constraint implies a variable action that imposes temporal limitations on certain degrees of freedom. Consequently, hierarchy taken as control relation can be defined as *a relation of constraint in which higher levels limit individually and in a variable way the degrees of freedom of specific components of lower levels*. According to its character, we will call this constraint and the kind of hierarchy it produces *functional*. Its variable and specific nature means that it operates a reduction of degrees of freedom that is selective and, therefore, should be understood in functional terms. In this regard, it is important to note that constraints, particularly functional ones, impose a reduction in the degrees of freedom of the elements of the system that revert to a simultaneous increase in the total degrees of freedom (of the system itself). Pattee often cites an accurate formulation by Stravinsky in his *Poetics of Music* that grasps the wider sense of this apparently paradoxical principle: "The more constraints one imposes, the more one frees one's self of the chains that shackle the spirit... and the arbitrariness of the constraint serves only to obtain precision of execution" (quoted in Pattee, 1971b, p. 256; 1973b, p. 74; 1981, p. 127).

In less poetic terms, the source of this increase in degrees of freedom stems from the classificatory action of variable constraints. By acting selectively and temporarily, they introduce new distinctions that, although reducing the original variables of the system, extend the state space with respect to other additional parameters. That constraining action adopts the form of selection and classification in a way that it creates new distinctions where there were none previously. When the degrees of freedom of individual elements are limited, particularly when it is done specifically through control constraints, the variety of differentiated behaviors increases along with, in general, the variety of options of the global system. Therefore, the general significance of constraint, as an alternative description, acquires in this case an even greater relevance, since, in contrast to structural constraints that are integrable in the general coordinates of the system, functional constraints are not and instead entail a redefinition of the system. In sum, in this case, constraints produce not only a reduction in degrees of freedom in the system but also a classification among its elements.

14.8   Hierarchy, Interlevel Relation, and Control

Having revised the conceptual constituents of constraint-mediated control hierarchy, we can come back to the initial issue and discussion opened at the beginning of the chapter. I argue that depriving the hierarchical arrangement or the relations among levels (of organization) of any causal relevance has been another factor promoting deflationary accounts that quite naturally follow from that renunciation. However, as advanced in the first sections, some of the authors who provided an alibi to this reaction have begun to revise or widen their views by explicitly incorporating the issue of control into their approaches. I will concentrate here, again, on Bechtel, as well as on Moreno and coauthors.

Whether dealing with the compositional (Bechtel) or configurational (Moreno et al.) aspects of organization, each asserted that interlevel causation is not necessary because it collapses into constitutive relations. Instead, causation was reserved for intralevel relations within a system or to generic relations between different systems, which then might be at different levels.

As is well known, Craver and Bechtel (2007) clearly asserted that "[on] our view, the interlevel relationship is only constitutive" (p. 562), and interpreted interlevel causes (both bottom-up and top-down) as "mechanistically mediated effects." Mechanistically mediated effects are "hybrids of causal and constitutive relations," in which causal relations are exclusively intralevel, and interlevel relations are only constitutive. In response to understandable presuppositions, Bechtel disclosed his former reluctance:

Mechanisms, after all, behave as they do because of the operations of their parts, and is natural to construe this relation as causal. Certainly, there are relations between levels within a mechanism... . But there are good reasons not to characterize the relation between a mechanism and its parts causally. The notion of causation commonly entails a number of associations that do not comport well with interlevel relations. (Bechtel, 2008, p. 153)

In turn, Mossio et al. (2013) have interpreted part–whole relations as a kind of "relational mereological supervenience," involving in a complex way an intermediate stage of *configurations*. In this view, organizational closure is defined as a network of constraints where emergent causal powers are derived from constraints but without proper interlevel causation, which is taken to be a constitutive relation. They do this because of their constitutive interpretation of relational supervenience and a configurational interpretation of emergence (where, as we have seen before, configurations include both the inherent and relational properties of the constituents in certain contexts).

These consequences regarding the nature of interlevel relations are consistent with their understanding of composition seen above. Nevertheless, when considering issues of control (or regulation), Moreno (et al.) are willing to assume that causal relations may be identified between the controller and the controlled, situated at different levels within the system. This was already present in Moreno and coauthors' previous work (summarized in Moreno & Mossio, 2015, pp. 28–37) and has also recently been incorporated by Bechtel into his extended "account of causality in biological mechanisms" (Winning & Bechtel, 2018, p. 290).

Bich et al. (2016), in turn, offer a detailed analysis of biological regulation and argue that we need to distinguish between *dynamical stability* and *regulation* as two different forms of control mechanisms that help to maintain biological organization in the face of perturbations.[7] The second regulation implies a hierarchical distinction between the controller and the controlled, which are located at different levels and are of a different nature:

> Our thesis is that the robust and adaptive behaviour of living systems requires more complex mechanisms than those ensuring basic network stability; mechanisms that specifically rely on the asymmetry or hierarchical distinction between controlled and controlling subsystems. We propose to use the term "regulation." (Bich et al., 2016, p. 239)

Both kinds of control (in their terms) are exerted through different kinds of constraints, respectively constitutive or regulatory, implementing first-order control in the case of dynamical stability or second-order control in the case of regulation (also their terms). But again, the main point is that the latter requires a distinct module or subsystem that is *dynamically decoupled* from the regulated one, acting at different rates and thus independently (Bich et al., 2016, pp. 239, 253–258).

I cannot discuss here in detail whether this view of strict regulation (control in my terms) is consistent with an in-principle denial of the existence of intrasystem interlevel causation (what Mossio & Moreno dubbed *nested causation*) or to which extent this view of regulation through hierarchical constraints (since it is not intralevel) should be interpreted as a constitutive relation (unless is taken as external to the system). The discussion should, rather, center on whether and in which cases such control (regulatory) mechanisms (i.e., the constraints) are formed out of those very components and activities of the lower-level dynamics that they contribute to control (i.e., regulate). Moreno and Mossio (2015) say that, in the absence "of any compelling argument in favour" of "the possibility that biological organisation might involve nested causation," they defend the constitutive interpretation (pp. 58–61). I wish here instead to recognize the importance of including the notion of control as a fundamental hierarchical kind of relation among different levels of organization.

The same applies to Bechtel's incorporation of hierarchical control into his account. In a recent paper, he identifies as a limitation of the neo-mechanist approach the fact that it has not recognized the importance of control (Winning & Bechtel, 2018, p. 307) and proposes to include it as a third type of relation:

Mechanist philosophers have construed this relationship in terms of either composition or causation (i.e., causal production). But there is a third type of relationship that often exists between processes in mechanisms—control—which has important implications for understanding the nature of mechanisms and mechanistic explanation that have not yet been fully appreciated. (Winning & Bechtel, 2018, p. 288)

In an accompanying footnote, the authors remark that classifying control as a third type of relationship does not imply a suggestion "that control is non-causal." Adopting this third kind of relationship calls for a review of interlevel relations as well:

The notion of top-down causation has been fraught with controversy. Much of this turns on the notion of levels employed. What is it for one entity or causal process to be located at a higher level than another? In the context of biology and neuroscience, an important sense of level arises in the context of control–a controller is at a higher level than the system it controls. (Bechtel, 2017b, p. 203)

In this accompanying footnote, besides crediting Pattee, Bechtel states that hierarchical control is a very important notion regarding top-down causation (Bechtel, 2017b, p. 221). The inclusion of control relations in the mechanist account and its consideration as a form of top-down causation by Bechtel does not necessarily imply by itself a change in his previous position about interlevel causes (within a system). It may, however, at least make room for further conceptual schemes beyond composition and intralevel causality.

Moreover, we see in both the mechanist and the organizational approaches that the relation of control through material constraints plays a key role in identifying and accounting for levels in a very specific and operative way, as controller and controlled. This notion of levels is neither ambiguous and limited to controversial compositional interpretations nor related to layer-cake and universal presuppositions (see, e.g., Brooks, 2017, for the latter). Thus, regardless of the constitutive interpretation of some interlevel relations, the main point here is the relevance of constraints and of control hierarchies in dealing with biological organization. In fact, understanding how constraints and control hierarchies work might help us to better elucidate the nature of interlevel relation with respect to causation, which is the subject of much confusion and controversy.

14.9   Conclusion

In this chapter, I have introduced Pattee's approach, discussed its relevance for current debates, and used some of his insights to extract consequences regarding the nature and interpretation of hierarchical organization and relations among levels of organization in biological systems. One of these consequences is to place the relation of control, originated as an autonomous nonholonomic constraint, at the very center and basis of the idea of organization, as used in "biological organization" or "levels of organization." Another is that it may help to assess whether and in which sense we may characterize this relation as causal.

In a sense, this approach is one response (among potentially others) to the demand for a description of specific and precise usages of hierarchy and levels of organization. Moreover, even if it is somehow restrictive in its definition and range of application (biological systems), it aspires to have a certain degree of more general applicability. This potential generality derives from the fact that, as far as we limit ourselves to levels of organization in biological systems, we can postulate that in all cases we have to refer, at least implicitly, to levels in the terms stated in this chapter: as having at its core a relation of control as an organizational principle. One final consequence, then, would be that by incorporating the idea of control through material constraints, we can make sense of the relation between levels in a system in both directions, upward and downward, each being different but complementary. In both cases, we are now able to characterize specific relations of constitution and specific relations of control in ways that are mutually dependent, bidirectional, and asymmetric because when the direction of the specific relation is reversed in our analysis, it is due to a different criterion or principle.

Acknowledgments

I thank James DiFrisco, Dan Brooks, and Bill Wimsatt for their invitation to participate in the workshop and Dan Brooks and Ilya Tëmkin for their thorough and extremely helpful review of the manuscript. Funding for this work has been provided by grant GV/EJ IT228-19 from the Basque government (Eusko Jaurlaritza), research project FFI2014-52173-P from the Ministerio de Economía y Competitividad, and research project PID2019-104576GB-I00 from the Ministerio de Ciencia e Innovación.

Notes

1. This and other differences of interpretation are part of the ongoing debate among members of my own research group (IAS-Research), in the development of the specific *organizational approach* to biological systems that we share.

2. See Love (this volume) for a different account of specific stable configuration states as levels of organization. This account relates the recognition of such configurations specifically to transitions and the relationship between cell and tissue levels, exposed and defined through manipulation practices.

3. Bechtel (2017a, p. 256), Bechtel (2017b, p. 221, n.1), Bechtel (2018, p. 575), and Winning and Bechtel (2018, p. 290).

4. I have previously adopted and emphasized this aspect: "The biological scope of our approach explains the restricted use of these terms to the cases of hierarchies in which levels interact not in a trivial or just lawful way, but in a specific way as autonomously originated rules exerting an arbitrary control over the lower level dynamics" (Umerez & Moreno 1995, p. 142).

5. And other more complex epistemic and semiotic aspects that turn out to be intricately connected, but which I will not be able to elaborate here. As I have said elsewhere, there is no proper substitute to reading Pattee's papers (Umerez, 2001, p. 160). For further analysis of some of those aspects, see also Umerez (1995, 1998, 2009).

6. He has developed some of those automata models in other papers, such as Pattee, 1961, 1965a, 1965b, and he describes them again in Pattee (1968a) and in his contributions to the first and second meetings of the symposia *Towards a Theoretical Biology*, organized by Waddington (Pattee, 1968b, 1969b).

7. I will not dispute here the choice of words, making "control" the generic term encompassing both kinds of processes and redefining "regulation" in a stricter sense. It seems that we could easily take the reverse option by using regulation as the general and reserving control for the more specific and demanding, as we more frequently see in the literature.

References

Bechtel, W. (2008). *Mental mechanisms*. London, UK: Routledge.

Bechtel, W. (2017a). Explicating top-down causation using networks and dynamics. *Philosophy of Science, 84*, 253–274.

Bechtel, W. (2017b). Top-down causation in biology and neuroscience: Control hierarchies. In M. Paolini & F. Orilia (Eds.), *Philosophical and scientific perspectives on downward causation* (pp. 203–224). London, UK: Routledge.

Bechtel, W. (2018). The importance of constraints and control in biological mechanisms: Insights from cancer research. *Philosophy of Science, 85*(4), 573–593.

Bich, L., Mossio, M., Ruiz-Mirazo, K., & Moreno, Á. (2016). Biological regulation: controlling the system from within. *Biology and Philosophy, 31,* 237–265.

Brooks, D. S. (2017). In defense of levels: Layer cakes and guilt by association. *Biological Theory, 12,* 142–156.

Brooks, D. S., & Eronen, M. I. (2018). The significance of levels of organization for scientific research: A heuristic approach. *Studies in History and Philosophy of Biological & Biomedical Sciences, 68/69,* 34–41.

Bunge, M. (1969). The metaphysics, epistemology and methodology of levels. In L. L. Whyte, A. G. Wilson, & D. Wilson (Eds.), *Hierarchical structures* (pp. 17–28). New York, NY: American Elsevier.

Craver, C. F., & Bechtel, W. (2007). Top-down causation without top-down causes. *Biology and Philosophy, 22,* 547–563.

DiFrisco, J. (2017). Time scales and levels of organization. *Erkenntnis, 82*(4), 795–818.

Eronen, M. I. (2013). No levels, no problems: Downward causation in neuroscience. *Philosophy of Science, 80*(5), 1042–1052.

Eronen, M. I. (2015). Levels of organization: A deflationary account. *Biology and Philosophy, 30*(1), 39–58.

Eronen, M. I., & Brooks, D. S. (2018). Levels of organization in biology. In E. N. Zalta (Ed.), *The Stanford encyclopedia of philosophy*. Retrieved from https://plato.stanford.edu/archives/spr2018/entries/levels-org -biology/

Green, S., & Batterman, R. (2017). Biology meets physics: Reductionism and multi-scale modeling of morphogenesis. *Studies in History and Philosophy of Biological and Biomedical Sciences, 61,* 20–34.

Grene, M. (1967). Biology and the problem of levels of reality. *The New Scholasticism, 41,* 427–449.

Grene, M. (1969). Hierarchy: One word, how many concepts? In L. L. Whyte, A. G. Wilson, & D. Wilson (Eds.), *Hierarchical structures* (pp. 56–58). New York, NY: American Elsevier.

Hooker, C. (2013). On the import of constraints in complex dynamical systems. *Foundations of Science, 18,* 757–780.

Korn, R. W. (2002). Biological hierarchies, their birth, death and evolution by natural selection. *Biology and Philosophy, 17,* 199–221.

Korn, R. W. (2005). The emergence principle in biological hierarchies. *Biology and Philosophy, 20,* 137–151.

Moreno, Á., & Mossio, M. (2015). *Biological autonomy: A philosophical and theoretical enquiry*. Dordrecht, Netherlands: Springer.

Mossio, M., Bich, L., & Moreno, Á. (2013). Emergence, closure and inter-level causation in biological systems. *Erkenntnis, 78,* 153–178.

Pattee, H. H. (1961). On the origin of macromolecular sequences. *Biophysical Journal, 1,* 683–710.

Pattee, H. H. (1965a). The recognition of hereditary order in primitive chemical systems. In S. Fox (Ed.), *The origins of prebiological systems* (pp. 385–405). New York, NY: Academic Press.

Pattee, H. H. (1965b). Experimental approaches to the origin of life problem. *Advances in Enzymology, 27,* 381–415.

Pattee, H. H. (1966). Physical theories, automata and the origin of life. In H. Pattee, E. Edelsack, L. Fein, & A. Callahan (Eds.), *Natural automata and useful simulations* (pp. 73–104). Washington, DC: Spartan Books.

Pattee, H. H. (1967). Quantum mechanics, heredity and the origin of life. *Journal of Theoretical Biology, 17,* 410–420.

Pattee, H. H. (1968a). Automata theories of hereditary tactic copolymerization. In A. D. Ketley (Ed.), *The stereochemistry of macromolecules* (Vol. 3, pp. 305–331). New York, NY: Marcel Dekker.

Pattee, H. H. (1968b). The physical basis of coding and reliability in biological evolution. In C. H. Waddington (Ed.), *Towards a theoretical biology 1: Prolegomena* (pp. 67–93). Edinburgh, UK: Edinburgh University Press.

Pattee, H. H. (1969a). Physical conditions for primitive functional hierarchies. In L. L. Whyte, A. G. Wilson, & D. Wilson (Eds.), *Hierarchical structures* (pp. 161–177). New York, NY: American Elsevier.

Pattee, H. H. (1969b). Physical problems of heredity and evolution. In C. H. Waddington (Ed.), *Towards a theoretical biology 2: Sketches* (pp. 268–284). Edinburgh, UK: Edinburgh University Press.

Pattee, H. H. (1970). The problem of biological hierarchy. In C. H. Waddington (Ed.), *Towards a theoretical biology 3: Drafts* (pp. 117–136). Edinburgh, UK: Edinburgh University Press.

Pattee, H. H. (1971a). The recognition of description and function in chemical reaction networks. In R. Buvet & C. Ponnamperuma (Eds.), *Chemical evolution and the origin of life* (pp. 42–50). New York, NY: North Holland.

Pattee, H. H. (1971b). Physical theories of biological co-ordination. *Quarterly Reviews of Biophysics, 4*(2/3), 255–276.

Pattee, H. H. (1972). The nature of hierarchical controls in living matter. In R. Rosen (Ed.), *Foundations of mathematical biology* (Vol. 1, pp. 1–22). New York, NY: Academic Press.

Pattee, H. H. (1973a). Physical problems of the origin of natural controls. In A. Locker (Ed.), *Biogenesis, evolution, and homeostasis* (pp. 41–49). Heidelberg, Germany: Springer.

Pattee, H. H. (1973b). The physical basis and origin of hierarchical control. In H. H. Pattee (Ed.), *Hierarchy theory* (pp. 73–108). New York, NY: Georges Braziller.

Pattee, H. H. (1973c). Postscript: Unsolved problems and potential applications of hierarchy theories. In H. H. Pattee (Ed.), *Hierarchy theory* (pp. 129–156). New York, NY: Georges Braziller.

Pattee, H. H. (1979a). Discussion with R. Rosen & R. Somorjai. In P. Buckley & D. Peat (Eds.), *A question of physics—Conversations in physics and biology* (pp. 84–123). Toronto, Canada: University of Toronto Press.

Pattee, H. H. (1981). Symbol-structure complementarity in biological evolution. In E. Jantsch (Ed.), *The evolutionary vision* (pp. 117–128). Boulder, CO: Westview.

Pattee, H. H. (2009). Response by H. H. Pattee to Jon Umerez's paper: Where does Pattee's "How does a molecule become a message?" belong in the history of biosemiotics? *Biosemiotics, 2*, 291–302.

Pattee, H. H. (2012a). Introduction. What these papers are about. In H. H. Pattee & J. Raczaszek-Leonardi (Eds.), *Laws, language and life* (pp. 3–30). Dordrecht, Netherlands: Springer.

Pattee, H. H. (2012b). *Laws, language and life: Howard Pattee's classic papers on the physics of symbols with contemporary commentary* (J. Raczaszek-Leonardi, Ed.). Dordrecht, Netherlands: Springer.

Pattee, H. H., Edelsack, E., Fein, L., & Callahan, A. (Eds.). (1966). *Natural automata and useful simulations*. Washington, DC: Spartan Books.

Polanyi, M. (1968). Life's irreducible structure. *Science, 160*, 1308–1312.

Potochnik, A., & McGill, B. (2012). The limitations of hierarchical organization. *Philosophy of Science, 79*(1), 120–140.

Salthe, S. N. (1993). *Development and evolution: Complexity and change in biology*. Cambridge, MA: MIT Press.

Thalos, M. (2013). *Without hierarchy: The scale freedom of the universe*. Oxford, UK: Oxford University Press.

Umerez, J. (1994). *Autonomous hierarchies: A study on the origin and nature of the processes of control and level formation in complex natural systems* [in Spanish]. Unpublished doctoral dissertation, University of the Basque Country (UPV/EHU).

Umerez, J. (1995). Semantic closure: A guiding notion to ground artificial life. In F. Morán, A. Moreno, J. J. Merelo, & P. Chacón (Eds.), *Advances in artificial life* (pp. 77–94). Berlin, Germany: Springer Verlag.

Umerez, J. (1998). The evolution of the symbolic domain in living systems & artificial life. In G. van der Vijver, S. Salthe, & M. Delpos (Eds.), *Evolutionary systems: Biological and epistemological perspectives on selection and self-organization* (pp. 377–396). Dordrecht, Netherlands: Kluwer.

Umerez, J. (2001). H. Pattee's theoretical biology: A radical epistemological stance to approach life, evolution and complexity. *BioSystems, 60*(1/3), 159–177.

Umerez, J. (2009). Where does Pattee's 'How does a molecule become a message?' belong in the history of biosemiotics? *Biosemiotics, 2*(3), 269–290.

Umerez, J. (2016). Biological organization from a hierarchical perspective: Articulation of concepts and inter-level relation. In N. Eldredge, T. Pievani, E. Serrelli, & I. Tëmkin (Eds.), *Evolutionary theory: A hierarchical perspective* (pp. 63–85). Chicago, IL: University of Chicago Press.

Umerez, J., & Moreno, Á. (1995). Origin of life as the first MST: Control hierarchies and interlevel relation. *World Futures, 45*(2), 139–154.

Umerez, J., & Mossio, M. (2013). Constraint. In W. Dubitzky, O. Wolkenhauer, H. Yokota, & K.-H. Cho (Eds.), *Encyclopedia of systems biology* (pp. 490–494). New York, NY: Springer.Whyte, L. L., Wilson, A. G., & Wilson, D. (Eds.). (1969). *Hierarchical structures*. New York, NY: Elsevier.

Winning, J., & Bechtel, W. (2018). Rethinking causality in biological and neural mechanisms: Constraints and control. *Minds & Machines, 28*(2), 287–310.

15 Phenomenological Levels in Biological and Cultural Evolution

Ilya Tëmkin

Overview

Human history is undeniably shaped by culture, a unique attribute of *Homo sapiens*, and, possibly, of other species of the hominid lineage. A product of cognitive evolution, culture is a form of information that is symbolically encoded, transmitted irrespective of genealogical relatedness, and amenable to accumulation and progressive development. Incorporating cultural transmission into an explanatory evolutionary framework necessitates an expansion of the nested compositional hierarchical model of the biological realm to include cognition-based entities that emerge at the organismal level and comprise a hierarchy of sociocultural assemblages. Formalizing the structure and properties of sociocultural entities, and elucidating their relationships with genealogical lineages and economic systems (the two kinds of biological individuals) across levels of organization set the ground for teasing apart causality and providing explanations for the patterns and trends—both biological and cultural—in the evolution of our species.

15.1 Introduction

Contemporary evolutionary theory is committed to the existence of nested compositional hierarchies in the biological realm and attempts to explain natural phenomena as products of the dynamics of energy and matter exchange, as well as information flow within and across hierarchical levels. While incipient at the dawn of evolutionary biology (Eldredge, 2016), level-based thinking did not take shape until the second half of the twentieth century, when the theoretical studies of biological complexity (e.g., Pattee, 1973; Simon, 1962) and new ideas in philosophy of biology (e.g., Ghiselin, 1974; Hull, 1976, 1978) fueled the revision of the explanation for large-scale historical patterns in the fossil record (Eldredge, 1982; Salthe, 1975; Vrba & Eldredge, 1984). These developments were unified and formalized in a largely theoretical body of work by the mid-1980s (Eldredge, 1985, 1986; Eldredge & Salthe, 1984; Salthe, 1985). The beginning of the new millennium has seen a renaissance of the hierarchy theory of evolution with the emergence of a new synthesis, integrating advances in genomics, complex systems science, and recent philosophical developments, that not only did provide a more complete theoretical framework for explaining complex patterns and processes of biological evolution but also formed an operational framework for a growing number of empirical studies from genomics to paleo-biology (Eldredge et al., 2016; Tëmkin & Eldredge, 2015).

One of the fundamental assumptions of the hierarchy theory of evolution is that living systems engage in two principal classes of processes: economic interactions through matter and energy exchange, and propagation accompanied by information transfer from ances-tors to descendants. As the ability and opportunity to reproduce depend on the economic, or ecological, success of a living system, the interplay of the dynamic interactional and informational processes is at the heart of evolutionary processes.

It has long been acknowledged that cumulative culture provides an additional pathway of information transmission that is not channeled through genealogical descent, so that human evolution is uniquely a consequence of change in the intertwined cultural and biological modes of information transfer. Attempts to account for dual inheritance by anthropologists, archaeologists, and sociologists were largely built on modifying quantita-tive population genetics models to capture culture-specific modes of transmission among individuals within populations (reviewed by Serrelli, 2016) and explaining large-scale, historical cultural patterns in terms of specific underlying "microevolutionary" processes (Blute, 2010; Mesoudi, 2011). Such approaches typically ignore the hierarchical structure of living systems, including that of human sociocultural groups and, consequently, tend to reduce causality to lower-level phenomena and conflate attributes and processes char-acteristic of entities at incommensurate scales. For instance, various forms of cultural expression, such as speech, artifacts, and beliefs, are frequently regarded as cultural equiva-lents of the phenotype, or "extended phenotype" (e.g., Mesoudi, 2011). Such a view is logically inconsistent in light of a hierarchical perspective because the property of the phenotype in biology is restricted to organismal and suborganismal levels of organization, whereas neither artifacts nor beliefs pertain exclusively to individual organisms and can be associated with population-level entities. Instead, they have individual existence and can be replicated forming lineages incongruent with those formed by organismal traits. Such use of imprecise or inconsistent biological evolutionary metaphors in the domain of cultural evolution stems from the failure to recognize the hierarchical structure of and interconnectedness among the sociocultural and biological entities (Tëmkin, 2016).

Despite efforts to integrate a hierarchical perspective in the field of cultural evolution (Eldredge, 2009; Eldredge & Grene, 1992; Hurt et al., 2001), the project of formalizing an explicit hierarchical representation of sociocultural organization is still in its infancy. This contribution attempts to clarify the underlying ontology of sociocultural systems, their rela-tionships to the entities of the economic and genealogical hierarchies of the biological realm, and implications for elucidating evolutionary phenomenology—a step toward creating a sound framework for an overarching causal theory of sociocultural evolution.

15.2   The Architecture of Biological Complexity

The following discussion of the principles of biological hierarchical systems and the summary of the hierarchy theory of evolution are largely based on Tëmkin and Eldredge (2015) and Tëmkin and Serrelli (2016), so the reader is referred to these publications for comprehensive account and references.

15.2.1 The Nested Compositional Hierarchy of Biological Systems

A *biological system* is a complex network of entities directly or indirectly interacting via energy and matter transfer, and comprising an integrated whole with systemwide properties. Such systems are individuals capable of development, growth, senescence, and demise that comprise a *nested compositional hierarchy* in the context of spatial scale. In this type of ordered organization, *levels* are classes of parts and wholes, with their rank assigned according to clustering of their constituting entities in the multidimensional space of the members' properties. Most generally, levels represent local maxima of predictability and regularity in such a phase space (Wimsatt, 1976). In the following discussion, the term "biological hierarchy" is used in the narrow sense of a nested compositional hierarchy of biological systems. The term "focal level" designates a level at which a particular phenomenon is observed, whereas the notions of *higher* (or *upper*) and *lower levels* refer to the more inclusive and less inclusive levels relative to the focal level, respectively.

15.2.2 Hierarchical Dynamics

In a hierarchy, two principal types of interactions arise from scalar differences in process rates, resulting in the difference between strong interactions with high-frequency dynamics within levels and weak interactions with low-frequency dynamics among levels. Entities at a given level interact directly with each other in the same dynamic process, whereas entities at different levels interact only in an aggregative fashion. Such nontransitivity of direct effects across levels establishes the levels as quasi-independent systems allowing for investigating dynamics of individual levels in their own right and ignoring within-level interactions when considering interlevel dynamics (see Woodward, this volume).

The *intralevel* interactions are governed by common topological features of complex networks that are isomorphic across levels. Biological systems exhibit a hierarchy of control, a high degree of modularity, a high clustering coefficient, a heavy tail in the degree distribution, and a short mean path length—properties responsible for the emergent property of *robustness*, an exceptionally high degree of tolerance against random failures and external perturbations.

The *interlevel* interactions are governed by a system of dual control, where the upper and the lower levels mutually affect each other indirectly through downward and upward causation, respectively. In *upward causation*, interactions of entities at a lower level establish *initiating conditions* that manifest in either aggregate or emergent fashion. In *downward causation*, interactions of entities at a higher level exert constraints or determine boundary conditions, thereby establishing the direction of control, affecting simultaneously all the component subsystems.

Taken together, the processes occurring at a focal level can simultaneously be initiating conditions for upper and boundary conditions for lower levels and, in turn, be affected by boundary and initiating conditions established by these levels, respectively. Noncontiguous levels may affect the dynamics at the focal level indirectly through cascading upward and downward effects across levels. The complexity of biological dynamics stems from the synergetic effect of idiosyncratic processes at different organizational levels and the dynamics of interlevel interactions.

15.2.3   Information and Energy Flow in Biological Systems

Biological systems are characterized by the propensity to endure and expand (grow and reproduce) that manifests in two distinct kinds of processes: living entities *interact* with their environment through regulated dynamic energy and matter exchange with their environment, and they *propagate* by transmitting heritable information pertaining to their structure and function through successive generations. The former amounts to the maintenance of structural integrity and complexity of a system, whereas the latter enables living systems to expand far beyond their physical size constraints and to endure long past a limited temporal span and eventual demise.

In biological systems, the informational and dynamic processes are typically decoupled, resulting in differentiated and specialized sets of entities at each hierarchical level: *replicators* and *interactors*, respectively. Interactors are causally related to replicators in such a way that the survival of the former is causally responsible for the differential propagation of the latter. Replication results in the formation of *lineages*, a type of biological individual that is incorporated into a nested hierarchy based on genealogical descent. Unlike systems that are composed of interacting parts and capable of development, lineages comprise variant sublineages amenable to variational evolution (Caponi, 2016).

An overarching conceptual framework of biology that integrates the dynamics of energy/ matter exchange and transmission of heritable information recognizes two interconnected nested hierarchies: (1) the economic, or ecological, hierarchy of biological systems-interactors and (2) the genealogical, or reproductive, hierarchy of replicating lineages (figure 15.1). The levels and entities of the two hierarchies lack an exact one-to-one correspondence, despite the fact that some levels are congruent and contain the same or overlapping classes of individuals. This dual hierarchical model captures the spatial and temporal dimensions of the organic realm, underscoring a fundamental dissimilarity between dynamic systems and informational lineages: economic individuals form a hierarchy of control, allowing for perpetual directional interactions and feedback loops, whereas the time-irreversible flow of information through genealogical lineages reflects the historical nature of biological systems and yields a static hierarchy of classification.

15.2.4   The Hierarchy Theory of Evolution

The complexity of evolutionary phenomena is a synergetic outcome of complex network responses to temporally decoupled perturbations at different levels of the economic hierarchy and their cascading effects on differential survival of lineages in the genealogical hierarchy. The prevalent pattern of long-term stability of biological systems is a consequence of the universal systemic robustness, typical of biological complex networks, which results in homeostatic maintenance of a steady state. Hierarchical nesting of biological systems amplifies the ability of such systems to buffer considerable extrinsic stress. Perturbations that are strong enough to compromise the system's buffering mechanisms disrupt the dynamics of energy and matter flow, resulting in cascading effects across the levels of the economic hierarchy. As the interaction of individuals in the economic hierarchy with their environment is causally related to differential propagation of genealogical individuals, the flow of information through lineages might elicit an evolutionary response. A change in the transmitted information at a given level has an effect on the diversity

ECONOMIC HIERARCHY   GENEALOGICAL HIERARCHY

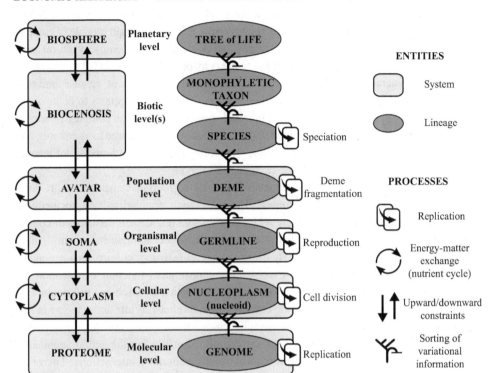

Figure 15.1
The dual hierarchical model of biological systems. The economic (ecological or interactional) hierarchy of biological systems represents dynamics of matter and energy exchange and, generally, corresponds to the spatial dimension of life. The genealogical (informational or evolutionary) hierarchy of propagating lineages represents transmission of heritable information, corresponding to the temporal dimension of life.

pattern of genealogical individuals: variation can either increase through the origin of new genealogical variants or decrease through sorting, where only a subset of variants persists. Because the fidelity of replication by genealogical individuals ensures the long-term persistence of a given system through descendants, a change in information flow creates novel initiating conditions in terms of the (re)assembly of a biological system following a perturbation event. Ultimately, evolution is the process of regaining equilibrium by a biological system in a new state in response to extrinsic perturbations that occurs at the nexus between the economic and genealogical hierarchies, where the dynamic interactions in the former are translated into a historical pattern of the latter.

15.3   The Human Dimension

15.3.1   The Concept of Culture

Culture is the symbolically encoded information transmitted through social learning and amenable to accumulation and progressive improvement. It manifests primarily in stereotypical behaviors (social norms), shared values, ideas, concepts, and the technosphere (material culture), generally corresponding to the transmissible or replicable elements (TREs) of Wimsatt (2018). Cultural information is distributed among individual members of a population-level cultural group, and stored and encoded at the organismal level by the brain capable of complex cognitive processes. Culture defines the integrity and boundary of a given sociocultural group and identity (whether conscious or not) of its members by prescribing a set of behavioral norms that guide the interactions of its members with the physical environment, representatives of other cultural groups, and among themselves. Thus defined, culture appears to be unique to *Homo sapiens* and, possibly, some extinct hominid species. Even though complex patterns of social learning and behavior in other animal species are sometimes referred to as cultures, they lack the defining cumulative aspect of culture—an ability to expand and modify information through successive generations (Boesch & Tomasello, 1998).

As a body of knowledge and social norms shared by a group of conspecific individuals, culture is a population-level phenomenon. However, such population-level sociocultural groups cannot be identified exclusively with either avatars (ecological networks of conspecific individuals) or demes (groups of interbreeding conspecific organisms comprising a segment of a reproductive lineage). By means of social learning, culture, as a product of cognitive processes, simultaneously executes the genealogical function of retention and transmission of information, as well as the ecological function by imposing behavioral constraints on economic interactions among constituting individuals. While not stripping human groups of their biological attributes, an overlying culture fuses avatar- and deme-like properties of interaction and replication, at the same time reconfiguring the actual agencies of biological reproductive dynamics and mainly trophic economic interactions. Therefore, sociocultural groups are uniquely hominid population-level units that integrate attributes of both demes and avatars (Eldredge, 2009; Eldredge & Grene, 1992). The difference in emphasis on either the economic or genealogical aspects of human groups is reflected in the historical distinction between sociology and anthropology, respectively (Blute, 2010).

15.3.2 Cultural Transmission

Cultural transmission is a set of social learning processes producing a cumulative body of knowledge, a product of successive elaborations by many individuals through multiple generations that is distributed among members of a sociocultural group and typically recorded in technosphere. Functionally, cultural transmission is particularly effective for adapting to temporally and spatially variable and unpredictable environments (Richerson & Boyd, 2000), because not only does it provide means for rapid dissemination and elaborations of newly acquired adaptive behaviors, but it also retains the memory of past adaptations that can be recalled rather than reinvented when necessary conditions arise. Cultural transmission is, therefore, an adaptive, population-level aggregate trait resulting from cumulative affect of a variety of social learning mechanisms that operate at the level of individual organisms.

Knowledge originates at the organismal level in the process of *individual learning* through one's reflection about external stimuli and sources, as well as through reelaboration on socially acquired knowledge in light of personal experience. Such information can be sorted, rapidly disseminated, and amplified through *social learning*, where knowledge is acquired (whether consciously or not) by the learner from conspecifics, resulting in an acquisition of a novel behavioral response by the former. Social learning occurs sequentially throughout an organism's life cycle and facilitated by other individuals and social institutions (Wimsatt, 2018; Wimsatt & Griesemer, 2007), so that the costs of time, energy, and potential risks associated with trial and error in individual learning can be minimized, requiring little additional, specialized, neural machinery and cognitive skills (Richerson & Boyd, 2000). For instance, in most social species, younger and more naive individuals typically learn to navigate their environment by following more experienced members of a social group that expose them to a highly selected set of stimuli that then lead to acquisition of specific behaviors by reinforcement.

While social learning is not unique to hominids, the frequency of specific social learning mechanisms requiring different cognitive skills on the part of individual learners differentiates human from nonhuman social learning. It has been argued that the cumulative aspect of cultural transmission in humans and presumably their hominid ancestors involves primarily imitative learning, possibly facilitated by active instruction, in which the learner not only comprehends both the demonstrator's goal and the strategy used in pursuing that goal but also is capable of aligning this goal and strategy with his or her own (Tomasello, 2000; Tomasello et al., 1993). This innate cognitive ability of humans to perceive intentionality in conspecific individuals and to attribute mental states (such as beliefs, emotions, and knowledge) to others constitutes *theory of mind*, a defining cognitive trait that evolved in the hominid lineage.

15.3.3   Language

One of the principal forms of cultural transmission in humans is linguistic instruction. In the context of a social network at the population level, the ability to share mental images with others using language leverages the value of acquiring novel knowledge and skills, as well as facilitates formation of social alliances (Heinrich, 2000; Pinker, 2010). Evolutionarily, transfer of information by language (either orally, or in the form of a writing system and/or other aspects of the technosphere), pertaining to a complex useful skill provides means of multiplying the benefits of applying the skill with little or no cost and fosters the development of cooperation among participating parties. This occurs because sharing knowledge through language is a low-cost process energy-wise at the organismal level: it neither deprives the speaker of his or her knowledge or possessions (unlike the sharing of commodities) and is associated with little energy expenditure on speaking, whereas the listeners economize on the time and energy spent on developing the skill on their own through trial and error. At the population level, the sharing individuals benefit from an enhanced well-being of the group and favorable reputation in the eyes of their peers. As both the sharing and receiving parties profit from their exchange at a low cost, communication by language provides the incentive and the means to cooperate and promotes social integration and the development of reciprocal altruism over the long run, facilitating, in turn, formation of novel levels of the sociocultural hierarchy (see below).

15.3.4   The Technosphere

The *technosphere* is a physical form of culturally transmitted information, represented by material culture, and domesticated and cultivated organisms. Elements of the technosphere comprise hierarchically nested entities that characterize human population-level entities at different scales. As representations of distributed cognitive processes of their makers, although they are static and incapable of replicating on their own, artifacts are both components and mediators of thought processes (Theiner et al., 2010; Wilson & Clark, 2009; Wolf, 2007).

The historical pattern of origin, differential persistence, and demise of elements of the technosphere is rooted in the behavior and, ultimately, the cognitive processes of individuals who design, create, and share the underlying concepts and ideas with other organisms at the population level through social learning. The fabrication and operation of material cultural entities typically require sophisticated sequences of behavior and depend on both individual and social learning. The purposeful design of tools necessitates a rationale, typically supported by intuitive theories that capture folk understandings of physics (objects, motion, and the forces), geometry (places and directions), and biology (generative forces that govern life cycle and physiological processes) (Pinker, 2010). Elements of material culture that represent linguistically encoded information (such as artifacts containing written or audiovisual records) are particularly significant as they greatly facilitate the dissemination of cultural information across vast geography and spanning multiple generations.

Arguably, the most critical attribute of a sociocultural group, the technosphere, is a human-mediated vehicle of niche construction, allowing for the development of progressively more effective ways to exploit the environment for the economic benefit of the group and, increasingly, reproduction. A highly developed technosphere has the capacity to dramatically modify the environment at unprecedented spatial scales and within an extremely short period of time relative to biological evolutionary time scales, as attested by present-day ecological crisis, accelerated rate of extinction, and alteration of global climate.

The history of technosphere development reflects a general trend toward a greater independence of the economy of human groups from biocenotic control, commencing with the advent of agriculture about 10,000 years ago (Eldredge, 1995). The corollary of this trend is a progressive replacement of biological ecosystems with anthropogenic ones, often characterized by fundamentally altered physical landscape, biological diversity, and climate, where humans are chiefly responsible for regulating the flow of energy and matter.

Artifacts archive technological know-how and record changes of conceptual innovations in the elements of their design, reflecting the cumulative aspect of cultural transmission and displaying variation over time. Even though it does appear meaningful to talk about evolution of manmade objects, artifacts evolve only in the metaphorical sense, being merely physical manifestations of the origin and sorting of mental representations across the levels of the cognitive hierarchy (see below).

15.4   The Sociocultural/Cognitive Hierarchy

15.4.1   The Sociocultural Hierarchy

Culture defines human social groups as both economic systems and replicating lineages, although not being coincident with either avatars or demes—their biological population-level counterparts. In terms of economic interactions, culture modulates interactions among individual organisms within and across cultural groups, as well as between individuals and their environment via downward causation through social norms and traditions. As a genealogical entity, culture accumulates, stores, and transmits information that is encoded symbolically, so that information can be shared within a generation and across generations of biologically unrelated individual organisms. In contrast, heritable information in biological lineages is typically encoded and transmitted from ancestors to descendants (either genetically or epigenetically) or between nonrelatives (as in horizontal gene transfer in prokaryotes) in the form of physical data storage mediums, such as nucleic acid sequences and other molecular determinants. Moreover, culture may channel transmission of genealogical information within and among sociocultural groups at the organismal level by imposing specific behavioral norms on reproductive activities, such as endogamy and exogamy. The complex relationship between genealogical (biological) and cultural modes of information transmission is often described by a *dual inheritance* model (Cavalli-Sforza & Feldman, 1981) that acknowledges the differences between the two mechanisms and postulates that human evolution is shaped by the interplay of cultural and biological modes of information transfer.

The uniqueness of sociocultural systems requires an expansion of the economic-genealogical dual hierarchical model to include cognition-based entities that emerge at the organismal level and comprise a hierarchy of population-level assemblages (figure 15.2). Hopefully, formalizing the hierarchical structure of sociocultural entities and their relationship to the genealogical and economic hierarchies of the biological realm will provide an important conceptual tool for elucidating causality in biological and cultural evolution of our species. The conviction

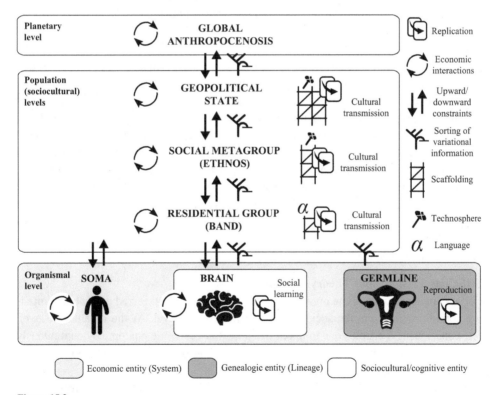

Figure 15.2
The triadic sociocultural hierarchical model of cognition-based entities. The sociocultural/cognitive entities engage in both energy/matter exchange and information transmission. See text for details.

that a hierarchical perspective holds the promise of providing greater explanatory power stems from the fact that the emergence of cognition and, consequently, of complex sociocultural groupings is ultimately a product of biological evolution, the most comprehensive account of which is based on an inherently hierarchical perspective.

15.4.2   The Organismal Level

The Human Brain and Cognition

Cultural information is stored and encoded at the organismal level by the brain capable of complex cognitive processes. At the organismal level, large brains require a trade-off between the benefits of increased intelligence, on one hand, and somatic and reproductive (i.e., the economic and genealogical, respectively) detriments on the other, with the upper size limit imposed by constructional and/or physical constraints. Economically, the main-tenance of large brains demands greater energy expenditure through metabolism and places the greater burden on the female anatomy due to risks associated with bearing of and giving birth to offspring with a large head. With respect to reproductive activity, large brains are associated with longer life span and delayed sexual maturity, the latter increasing the chances of accidental death before reaching reproductive age. Therefore, in the hominid ancestors, the investment in intelligence must have amply compensated for the increased demands for somatic maintenance and late onset of reproductive age. However, such a trade-off cannot be accounted for by phenomena solely at the organismal level.

The extended childhood has an upward effect at the population level, affecting both its demic and avatar aspects. In terms of economic interactions, prolonged childhood favors social learning by necessitating a close bond between juveniles and adults (across multiple generations) through extensive nurturing, time-consuming acquisition of skills through apprenticeship, and the mastery of the social environment to ensure the economic success of the individual (Pinker, 2010; Richerson & Boyd, 2000). With respect to reproductive activity, extended childhood is hypothesized to have contributed to biparental care, long-term serial monogamy, and decoupling sexuality from reproduction (making the former a subject of social negotiation; Pinker, 2010). Taken together, these factors contribute to greater complexity of population social structure. The increasing complexity of social life, in turn, exerted downward constraints through sorting of individuals in favor of greater intelligence as an organism-level adaptation to navigating entangled social relationships and ensuring economic success. This positive interlevel feedback loop might have ulti-mately resulted in the emergence of a cognitive evolutionary novelty—self-consciousness—in hominids, the lineage of unprecedented social and behavioral complexity.

Self-Consciousness and Theory of Mind

Certain cognitive traits at the organismal level have been hypothesized to establish initiat-ing conditions for complex social life at the population level. At the organismal level, self-consciousness enables one to maximize potential economic outcome by continuously modulating one's behavior based on the understanding of events and behavior of others in terms of mediating intentional and causal forces.

The capacity for grammatical language, one of the most powerful mechanisms of cul-tural transmission, is believed to be causally related to self-consciousness, linking intel-ligence at the individual organism level with cognitively mediated auditory communication mechanism at the population level (Howe & Courage, 1997; Macphail, 2000). Thus, as a communication device, grammatical language is an emergent population-level property of the evolution of self-consciousness at the level of individual organisms. Enabled by the emergence of the capacity for causal thinking, language learning gives positive cognitive feedback that greatly improves causal understanding, enhancing social cooperation through mutual behavior management via the use of symbols, and increasing the capacity to respond flexibly and persistently to environmental events (Rumbaugh et al., 2000).

Another cognitive key innovation that evolved in the primate lineage and associated with self-consciousness is the ability to understand social relationships and their categories, especially the third-party social relationships (Dasser, 1988; Tomasello, 2000; Tomasello & Call, 1994). Possibly, it has evolved through generalization of mental processes responsible for establishing relational categories in the physical (spatial) domain that manifest, for example, in discrimination learning problems involving oddity, transitivity, and relation matching (Thomas, 1980). In effect, the evolution of the cognitive ability to navigate a complex physical environment became a preadaptation for maneuvering a complex social landscape.

Despite the fact that nonhuman primates are sufficiently intelligent to modulate their behavior in response to behavior of individuals in their social group by detecting and tracking their mental states, they can neither represent mental states of others nor perceive or understand underlying causes of the dynamic relations among conspecifics as intentional agents pursuing specific goals (Sterelny, 2000; Tomasello, 2000). The social behavior of nonhuman primates is based on understanding conspecifics as autonomous animate beings and can be explained by associative learning derived from past experience. In humans, the evolution of self-consciousness transformed complex social life into cultural life: the *theory of mind*, or cognitive ability of humans to perceive intentionality in conspecifics, and to represent and attribute mental states to other humans is a defining social cognitive trait crucial for the emergence, maintenance, and development of culture.

The increased complexity of social life at the population level exerted a positive feedback by downward constraint, shaping further development of cognition at the individual level. Comprehending causal relationships and intentionality makes social learning by imitation, particularly when facilitated by linguistic instruction, considerably more efficient because the learner not only understands the purpose and the strategy of achieving the goal but can also modify the strategy or devise one of his or her own to achieve the desired outcome in response to the population-level social environment. It has been hypothesized that theory of mind facilitated social learning through stimulus enhancement, by signaling what is important in the environment as inferred from heightened attention to others' interests (Sterelny, 2000). Moreover, this cognitive facility provides further economy in reducing problem-solving time through mental rehearsal that requires understanding and anticipation of the range of outcomes of actions yet to be made (Dunbar, 2000). As the survival and success of an individual in the social context and in a given ecological environment depend on specific knowledge and behavioral skills, the cognitive facilities and mechanisms of information transfer that facilitate the acquisition of such knowledge and skills will be selected. The transmission, storage, and accumulation of such information by individual humans, incentivized by mutual benefits of social integration and communication via language, mark the emergence of elementary culture. Once emerged, the progressive cumulative aspect of culture drives its further development, calling for more sophisticated means of preserving and transmitting a growing body of cultural information at higher levels.

15.4.3 Population Level (Sociocultural Levels)

The term "population-level entity" is adopted here to designate any supraorganismal sociocultural group below the level of species. It is possible to distinguish a continuum of intergrading and overlapping levels for such human groupings depending on their size scale, degree of sociality, and kin relations (such as individual families, patri- or matrilineages, clans, phratries, and others) that represent alternative or qualitatively different, culture-specific sociocultural architectures. The boundaries of some such groupings are hard to define, especially given the transient nature of some of them. The present analysis focuses on the compositional hierarchy of, arguably, least ambiguous and universal sociocultural systems that arise from the integration of lower-level systems and are characterized by novel, level-specific properties and relations. Future analyses might extend, alter, or redefine the proposed scheme, as the more rigorous criteria for entification of sociocultural systems get developed.

Residential Group (Band)

A *band* is the smallest-scale human social group identified by residential proximity. Bands are dynamic interaction networks of approximately thirty individuals united by cooperative tasks, such as sharing or trading of food, favors, and skills; childrearing; gathering, hunting, and defense; transportation of possessions; and construction and maintenance of living spaces. The hominid bands typically comprise primary kin (typically comprising less than 10 percent of the band members), relations by marriage, and unrelated individuals, as well as by novel coresidence patterns, emphasizing bilateral kin associations, brother–sister affiliation, affinal alliances, and coresidence with nonrelatives characterized by frequent cooperative interband relations and migration, facilitated by monogamous pair bonding and paternal recognition within cooperatively breeding social units. Unlike other primates, the hominid bands have a much more diverse and complex social structure and composition of coresidents, resulting in a characteristically low intraband genetic relatedness (Hill et al., 2011). Not only did integration of nonrelatives into a band result in an alteration in an ancestral residential structure, but it also stimulated the development of a diversity of cooperative interactions among individuals, including mutualistic sharing, reciprocal altruism, and deferring to dominant individuals. As these modes of cooperation provide alternative, incompatible pathways for distribution of resources, the social network of a band represents a dynamic equilibrium of partnerships, alliances, and coalitions that emerge, dissolve, or shift in response to the economic (resource availability) and genealogical (kinship) context (Fiske, 1991). These aspects of human band social structure are displayed by modern hunter–gatherer societies that likely closely parallel that inferred for early humans who lived as foragers for 95 percent of our species' history, until the beginning of the Holocene.

As the principal systemic factor that integrates and stabilizes a small-scale human social system, social cooperation networks provide a downward constraint on the individual residents' behaviors that are rare or absent in other primates: shared intentionality, disposition to teaching and learning, widespread preference for equitable distribution, and penalizing those that show deviant behavior (reviewed by Hill et al., 2011). The display of such "prosocial" and "other-regarding" behaviors on the part of individual band members creates a positive feedback loop that fosters greater integration of the social system through novel interactions and interdependence of the band members in the transfer of goods and know-how. In turn, population-level constraints result in an increased proportion of individuals who are cognitively better equipped to solve problems of complex social life, thus reconfiguring initiating conditions at the organismal level and enabling the development of ever more complex and highly attuned interactions. The context of a highly intricate and subtle fabric of social interactions in small-scale residential bands was likely the condition that channeled cognitive evolution at the organismal level toward the capacity for understanding and anticipating the intentional behavior of others in both cooperative and competitive interactions by representing the mental states that generate those actions (i.e., the theory of mind).

The evolution of complex and flexible cooperative interactions, requiring constant negotiations and shifting alliances, would not be possible without a developed theory of mind at the organismal level, allowing for classifying and evaluating the actions of others (to ascertain their generosity, competence, and other traits), displaying a variety of moral emotions (such as sympathy, anger, and guilt) that impact cooperation, and establishing one's own reputation (reviewed by Pinker, 2010). Stable behavioral patterns that emerge in the process of recurrent interactions become formalized in a variety of elemental social norms (e.g., politeness, ritual, and taboo) guiding the population-level dynamics.

Despite the fact that the evolution of the theory of mind and complex social behavior at the individual level enable successful navigation of complex social interactions and facilitate social learning of successful innovations, the small population size of residential bands makes the spread and fixation of cultural innovations unlikely, as shown by the loss of innovations in modern isolated populations due to infrequent interaction between potential models and imitators and to stochastic events (Hill et al., 2011). Mathematical models that explore various parameters affecting cultural transmission suggested that an increase of interaction network size beyond constraints of single band size is necessary for the development of culture (Boyd & Richerson, 1996; Henrich, 2004), a hypothesis corroborated by the expansion of a social network size due to interband interactions, as evidenced by the emergence of long-distance transfer of tools and raw materials in the middle Pleistocene (Hill et al., 2011).

Whereas residential bands were likely the necessary historical precursors of larger units, or sociocultural metagroups, once the latter emerged, the former were eventually dissolved, and the ensuing sociocultural dynamics became transferred to higher levels. Despite the fact that the hunter–gatherer bands do not comprise modern "ethne" (see below), recognizable small-scale residential groups (broadly understood) most certainly do, such as a small group of fellow villagers, members of a local congregation, or residents of an old neighborhood. While such groupings still might be quite common, they become obliterated by progressive urbanization and an ever-increasing rate of migration.

Sociocultural Metagroup (Ethnos)
Ethnos (*ethne*, pl.) is a time-extended, dynamic social network of interacting residential groups (bands), the integrity and stability of which is maintained by shared culture and characterized by collective group identity. The cognitively based and typically unconscious (self-evident or "natural" to its members) ethnic self-identity (i.e., the concept of "we" or "us") manifests in behavioral norms, marking the boundary of the ethnos by dictating the rules of acceptance or rejection of individuals as part of the group. An important characteristic of ethne

resulting from the cumulative nature of cultural transmission is the collective self-awareness of historicity, typically reflected in the origination myths and calendar systems. Ethnic self-identity can be viewed as a population-level parallel of self-consciousness at the level of individual organism, capable of autobiographic memory. Whereas human residential bands—despite apparent differences—are comparable to primate social groupings of comparable size, the metagroup social structure of ethne is uniquely human.

Enabled by cognitive evolution of self-consciousness at the organismal level and first expressed through complex social interactions in primitive residential bands, culture gained its key property—the capacity for accumulation—at the level of ethnic sociocultural metagroups. In small-scale residential social groups, individual acts of creative intelligence are likely to pass unobserved or do not get a sufficient number of imitators to faithfully preserve the innovation due to a smaller network size of interacting individuals. The increased size and a greater density of ethnic communities reconfigured social interaction networks into more complex, modular systems characterized by specialization of trade, division of labor, and more pronounced social stratification based on the hierarchy of control. These novel sociocultural metagroup-level architectures—in the form of organizations, institutions, and manufacturing infrastructures—function as social interaction scaffolds (Wimsatt, 2013, 2018), necessary to ensure fixation and faithful transmission of cultural variants and canalize their further enhancements and innovations. The cultural variants—ideas, actions, and inventions—arise from independent individual learning or modification of previously learned and socially transmitted knowledge, and they are selected based on their utility or some other factor by other members of the group. Because of continuous modification, the degree to which cultural innovations are individual or social/cooperative in origin varies widely.

As an expanding body of information that preserves gained valuable knowledge—the recipes of economic success and persistence of the ethnos—culture becomes the ethnos' most valuable population-level attribute. Due to the fact that cultural information is distributed among individual members of an ethnos at the organismal level, it is hypothesized that cultural norms, complex social interactions, and diversity of languages evolved as a means to control the flow of cultural information in such a way that allows for an exchange ideas, knowledge, and technology, while minimizing chances of such information being exploited (primarily by visual theft) by members of competitor ethne (Pagel, 2012).

Culturally transmitted behavioral norms are typically codified into *traditions*, which govern reproductive (i.e., endogamy and exogamy) and economic (e.g., transfer of goods and favors) aspects of individuals comprising an ethnos. These stereotypical behaviors are inherently complex, because they depend on the specific context of the global social network of the ethnos, including relations among individuals within subethnic residential groups, relations among individuals of different subethnic groups, and relations of individuals of different ethne.

Sociocultural metagroups might have originated through an integration of smaller-scale residential bands in response to alterations of the environment, the exploitation of which demanded cooperation and larger population sizes, such as big game hunting or massive plant gathering (as during the Neolithic agricultural revolution), so that the formation and historical fate of ethne are shaped by the interactions with specific landscapes and biocenoses, or ethnic "homelands" (Gumilev, 1990). In certain circumstances, such as migration following an environmental deterioration or enforced resettlement, an ethnos can transition to a different biocenosis, however, typically preserving an imprint of the homeland in its culture.

As historical individual entities, ethne arise from integration of preexisting ethnic groups or fragmenting into constituting subethnic groupings (remaining in diaspora), and they become extinct through extermination or cultural assimilation when one ethnos becomes absorbed by another, a process frequently accompanied by a progressive loss of the former's culture. The integration can take the form of mixing, in which constituting ethne merge into relatively unstable and short-lived larger groupings characterized by a mosaic of diverse cultural elements, or merging, leading to the formation of a new ethnos, whose newly developed culture replaces those of the component ethne (Gumilev, 1990).

The development and economic success of an ethnic group are largely affected by the state of its technosphere: some ethne remain in relative stasis, typical of groups with limited capacity of exploiting their natural environment (such as contemporary hunter–gatherer societies), whereas others actively modify their environment (sometimes to the point of nonreversible deterioration) through extensive use of technology, allowing for greater carrying capacity and growth. The reticulate dynamics of interethnic interactions (analogous to that of demes) may obscure the discreteness of individual ethnic groups, particularly in the interethnic contact zones.

Geopolitical State (Nation)

The *geopolitical state* is a mosaic of different ethne maintained by an integrated economy and a centralized political hierarchy of control. The concept roughly corresponds to the notions of "nations," "cultures," or "civilizations" in sociology, archeology, and anthropology. The unified culture of a geopolitical state either emanates from that of a dominating ethnos (enforced by a direct rule or implicitly imposed by a prevailing number of individuals) or represents an amalgam of cultures of multiple ethne. Whereas historically the central control typically belonged to an oligarchy (a small group of people distinguished by wealth, kinship, nobility, military power, or religious control) or a dynastical aristocracy (hereditary elite with the actual or nominal power attributed to a sovereign ruler, or monarch), many other forms of government with complex differentiated power structure developed over a course of human history. In highly developed geopolitical states, the hierarchy of control is further differentiated and modularized, adopting a great variety of forms that vary widely in scale and social attributes (e.g., tribal unions, social classes, political parties, philosophical schools, guilds, and ecclesiastical structures). The highly reconfigurable mosaic of nationwide institutions, organizations, and manufacturing infrastructures, often merged in various combinations, generates scaffolding mechanisms for preserving and transmitting cultural information pertaining to the power structure of the state and comprising generalized social policies and laws that may supersede or reflect certain cultural norms and traditions of constituting ethne.

Geopolitical states are characterized by spatial heterogeneity of the social network distribution and centers of technosphere development, primarily dictated by the landscape and other features of the environment. Institutions of learning and other physical repositories of information (such as libraries and archives) serve as primary expandable external information stores and usually are concentrated in most technosphere-laden regions, the urban areas. As focal points of trade and immigration, cities exert a downward effect on both the genealogical dynamics at the organismal level, accelerating genetic mixing of constituting ethne, and the economic interactions, resulting in disproportionate accumulation of wealth due to concentration of intellectual and technological innovation, education, and jobs (McKinney, 2016; Moretti, 2012).

A critical feature of geopolitical entities is the capacity for unrestricted expansion. Lower-level sociocultural groups display a restricted or regulated growth: residential units are constrained by the availability of resources provided by local ecosystems, whereas ethne are prevented from unchecked growth by the necessity to maintain their cultural integrity that can be threatened by extensive interethnic interactions, such as cultural assimilation. In contrast, geopolitical systems provide new mechanisms for expansion and exercising authority across vast geography by means of expandable bureaucracy and delegation of sanction (Volk, 2017). As the boundaries of geopolitical states are not defined by specific ethnic composition, the states can and do expand by incorporating lower-level sociocultural entities into its network, irrespective of ethnic identity, via assimilation or warfare. With great size comes great complexity: territorial expansion and population size increase drive the development of the hierarchy of social institutions of governance and ever increasing specialization—and, hence, increased efficiency—of the technosphere. This, in turn, provides a positive feedback on further political integration, which can only be limited by the presence of competing geopolitical states. Despite the rise and fall of geopolitical entities, there is a general trend toward integration into larger units of greater power: where the dynamic equilibrium among them is maintained by international military pacts, trade treaties, and cultural alliances.

15.4.4   Species Level: The Global Anthropocenosis

The *global anthropocenosis* is the planet-wide cultural interaction network of nation-states and the totality of their technosphere attributes. (A related concept of the anthroposphere, or global human-mediated ecosystem, cannot be consistently used in the context of a hierarchy composed of entirely biotic entities, as it incorporates the abiotic component of the human-mediated environment.) The global anthropocenosis can also be expressed in terms of lower-level constituting ethne, or the *ethnosphere*, or cultural mosaic of all interacting sociocultural metagroups (Gumilev, 1990). Genealogically, it represents the entire human species; economically, it is a globally distributed metapopulation, engaged in the energy and matter transfer with the Earth's economic system at large. Unchecked by biocenotic control of smaller-scale ecological systems, human activity—unlike that of any other species—has become a force of planetary magnitude, affecting Earth's geochemical, hydrological, and climatological dynamics; reshaping landscapes; and altering the composition and distribution of the biota. The rise of prominence of *Homo sapiens* as the biological system of unprecedented scale and agency demarcates the beginning of the new geological age, the Anthropocene.

Defining the anthropocenosis as a distinct sociocultural entity of global scale is warranted because its cohesiveness is amply expressed by technosphere-enabled globalization of human activities with regard to information transmission, as well as energy and matter exchange mechanisms, including communication technology (e.g., the Internet, telecommunication, air travel network, ground transport infrastructure), the global market, and coordinated exploitation of shared oceanic resources, just to name a few. International institutions and organizations—such as the United Nations, the World Health Organization, the World Bank, or Interpol—function as scaffolding enterprises by establishing international policies necessary for mediating the peaceful relationships among geopolitical entities through the fostering of cooperation in trade, security, cultural exchange, and for facing common health and economic challenges affecting the well-being of species across the globe. The downward economic effects of political processes that take place at the international level propagate downward to the level of individual organisms, taking the form of economic crises, stagnation, or growth of wealth affecting individuals' economic well-being both directly and indirectly. An example of the downward cultural effects of progressive international globalization at the organismal level includes biocultural homogenization, including the loss of linguistic diversity (Rozzi, 2012, 2013).

15.5 Conclusion

The main objective of this work was to incorporate cultural transmission into evolutionary theory by establishing the ontology of cognition-based entities—cognitive brain, residential group (band), metagroup (ethnos), geopolitical state (nation), and the anthropocenosis—and integrating them into the nested compositional hierarchical model of the organic world (figure 15.2). The proposed perspective connects human cultural phenomena to the two primary processes at work in biological evolution—replication and self-maintenance—and the associated genealogical and economic hierarchies of the organic realm. Not exhaustive by any means, the analysis presents a conceptual framework for causal analysis by teasing apart intra- and interlevel processes and examining their synergetic effects in the hopes of providing explanations for the patterns and trends in the evolution of our species and making informed predictions regarding its future trajectories. For example, the dual hierarchical model of evolution has been recently used to link the emergence of symbolic intelligence and the burst of cultural innovations to the last dispersal of *Homo sapiens* out of Africa (Parravicini & Pievani, 2016). Whereas this "final wave" model makes a convincing argument for a causal connection between climatic and biogeographic factors, on one hand, and early human biological population-level dynamics, on the other, the lack of cognition-based sociocultural entities precluded establishing a causal nexus between these processes and the symbolic behavior beyond mere association. The analysis of the interlevel dynamics of cognition-based entities in the context of the genealogical and economic hierarchies can potentially provide deeper insights into the interpretation of hominid evolution, shed light upon the emergence of art, religion, and science, as well as the origin of specific cultural attributes of different ethne; and inform methodologies for reconstructing cultural changes.

Acknowledgments

I thank Niles Eldredge, Emanuele Serrelli, Jon Umerez, and James DiFrisco for their invaluable suggestions for improving the manuscript. I am most grateful to Kholidakhon Shermukhamedova for encouragement and support, as well as the inspiration for this project and beyond.

References

Blute, M. (2010). *Darwinian sociocultural evolution: Solutions to dilemmas in cultural and social theory.* Cambridge, UK: Cambridge University Press.

Boesch, C., & Tomasello, M. (1998). Chimpanzee and human culture. *Current Anthropology, 39,* 591–604.

Boyd, R., & Richerson, P. J. (1996). Why culture is common, but cultural evolution is rare. *Proceedings of the British Academy, 88,* 77–93.

Caponi, G. (2016). Lineages and systems: A conceptual discontinuity in biological hierarchies. In N. Eldredge, T. Pievani, E. Serrelli, & I. Tëmkin (Eds.), *Evolutionary theory: A hierarchical perspective* (pp. 47–62). Chicago: University of Chicago Press.

Cavalli-Sforza, L. L., & Feldman, M. W. (1981). *Cultural transmission and evolution: a quantitative approach.* Princeton, NJ: Princeton University Press.

Dasser, V. (1988). A social concept in Java monkeys. *Animal Behaviour, 36*(1), 225–230.

Dunbar, R. I. M. (2000). Causal reasoning, mental rehearsal, and the evolution of primate cognition. In C. Heyes & L. Huber (Eds.), *The evolution of cognition* (pp. 205–219). Cambridge, MA: MIT Press.

Eldredge, N. (1982). Phenomenological levels and evolutionary rates. *Systematic Zoology, 31,* 338–347.

Eldredge, N. (1985). *Unfinished synthesis: Biological hierarchies and modern evolutionary thought.* New York, NY: Oxford University Press.

Eldredge, N. (1986). Information, economics, and evolution. *Annual Review of Ecology and Systematics, 17,* 351–369.

Eldredge, N. (1995). *Dominion.* New York, NY: H. Holt.

Eldredge, N. (2009). Material cultural macroevolution. In A. M. Prentiss I. Kuijt, & J. C. Chatters (Eds.), *Macroevolution in human prehistory: Evolutionary theory and processual archaeology* (pp. 297–316). New York, NY: Springer.

Eldredge, N. (2016). The checkered career of hierarchical thinking in evolutionary biology. In N. Eldredge, T. Pievani, E. Serrelli, & I. Tëmkin (Eds.), *Evolutionary theory: A hierarchical perspective* (pp. 1–16). Chicago: University of Chicago Press.

Eldredge, N., & Grene, M. (1992). *Interactions: The biological context of social systems.* New York, NY: Columbia University Press.

Eldredge, N., Pievani, T., Serrelli, E., & Tëmkin, I. (Eds.). (2016). *Evolutionary theory: A hierarchical perspective.* Chicago, IL: University of Chicago Press.

Eldredge, N., & Salthe, S. N. (1984). Hierarchy and evolution. *Oxford Surveys in Evolutionary Biology, 1,* 184–208.

Fiske, A. P. (1991). *Structures of social life: The four elementary forms of human relations: communal sharing, authority ranking, equality matching, market pricing.* New York, NY: Free Press.

Ghiselin, M. T. (1974). A radical solution to the species problem. *Systematic Zoology, 23,* 536–544.

Gumilev, L. N. (1990). *Etnogenez i biosfera Zemli* (Ethnogenesis and the biosphere of Earth). Moscow, Russia: Gidrometeoizdat.

Heinrich, B. (2000). Testing insight in ravens. In C. Heyes & L. Huber (Eds.), *The evolution of cognition* (pp. 289–305). Cambridge, MA: MIT Press.

Henrich, J. (2004). Demography and cultural evolution: How adaptive cultural processes can produce maladaptive losses: The Tasmanian case. *American Antiquity, 69*(2), 197–214.

Hill, K. R., Walker, R. S., Božičević, M., Eder, J., Headland, T., Hewlett, B., Magdalena Hurtado, A., Marlowe, F., Wiessner, P., & Wood, B. (2011). Co-residence patterns in hunter-gatherer societies show unique human social structure. *Science, 331*(6022), 1286–1289.

Howe, M. L., & Courage, M. L. (1997). The emergence and early development of autobiographical memory. *Psychological Review, 104*(3), 499–523.

Hull, D. L. (1976). Are species really individuals? *Systematic Zoology, 25,* 174–191.

Hull, D. L. (1978). A matter of individuality. *Philosophy of Science, 45*(3), 335–360.

Hurt, T. D., VanPool, T. L., Rakita, G. F. M., & Leonard, R. D. (2001). Explaining the co-occurrence of traits in the archaeological record: A further consideration of replicative success. In T. D. Hurt & G. F. M. Rakita (Eds.), *Style and function: Conceptual issues in evolutionary archaeology* (pp. 51–67). Westport, CT: Bergin and Garvey.

Macphail, E. M. (2000). The search for a mental Rubicon. In C. Heyes & L. Huber (Eds.), *The evolution of cognition* (pp. 253–271). Cambridge, MA: MIT Press.

McKinney, M. L. (2016). Hierarchy theory in the Anthropocene: Biocultural homogenization, urban ecosystems, and other emerging dynamics. In N. Eldredge, T. Pievani, E. Serrelli, & I. Tëmkin (Eds.), *Evolutionary theory: A hierarchical perspective* (pp. 334–350). Chicago, IL: University of Chicago Press.

Mesoudi, A. (2011). *Cultural evolution: How Darwinian theory can explain human culture and synthesize the social sciences.* Chicago, IL: University of Chicago Press.

Moretti, E. (2012). *The new geography of jobs.* Boston, MA: Houghton Mifflin Harcourt.

Pagel, M. (2012). Adapted to culture. *Nature, 482,* 297–299.

Parravicini, A., & Pievani, T. (2016). Multi-level human evolution: Ecological patterns in hominin phylogeny. *Journal of Anthropological Sciences, 94,* 167–182.

Pattee, H. H. (Ed.). (1973). *Hierarchy theory: The challenge of complex systems.* New York, NY: G. Braziller.

Pinker, S. (2010). The cognitive niche: Coevolution of intelligence, sociality, and language. *Proceedings of the National Academy of Sciences, 107*(Suppl. 2), 8993–8999.

Richerson, P. J., & Boyd, R. (2000). Climate, culture, and the evolution of cognition. In C. Heyes & L. Huber (Eds.), *The evolution of cognition* (pp. 329–346). Cambridge, MA: MIT Press.

Rozzi, R. (2012). Biocultural ethics: Recovering the vital links between the inhabitants, their habits, and regional habitats. *Environmental Ethics, 34,* 27–50.

Rozzi, R. (2013). Biocultural ethics: From biocultural homogenization toward biocultural conservation. In R. Rozzi, S. Pickett, C. Palmer, J. J. Armesto, & J. B. Callicott (Eds.), *Linking ecology and ethics for a changing world: Values, philosophy, and action* (pp. 2–32). Berlin, Germany: Springer.

Rumbaugh, D. M., Beran, M. J., & Hillix, W. A. (2000). Cause-effect reasoning in humans and animals. In C. Heyes & L. Huber (Eds.), *The evolution of cognition* (pp. 165–183). Cambridge, MA: MIT Press.

Salthe, S. N. (1975). Problems of macroevolution (molecular evolution, phenotype definition, and canalization) as seen from a hierarchical viewpoint. *American Zoologist, 15,* 295–314.

Salthe, S. N. (1985). *Evolving hierarchical systems: Their structure and representation.* New York, NY: Columbia University Press.

Serrelli, E. (2016). Evolutionary genetics and cultural traits in a "body of theory" perspective. In F. Panebianco & E. Serrelli (Eds.), *Understanding cultural traits: A multidisciplinary perspective on cultural diversity* (pp. 179–199). New York, NY: Springer.

Simon, H. A. (1962). The architecture of complexity. *Proceedings of the American Philosophical Society, 106,* 467–482.

Sterelny, K. (2000). Primate worlds. In C. Heyes & L. Huber (Eds.), *The evolution of cognition* (pp. 143–162). Cambridge, MA: MIT Press.

Těmkin, I. (2016). Homology and phylogenetic inference in biological and material cultural evolution. In F. Panebianco & E. Serrelli (Eds.), *Understanding cultural traits: A multidisciplinary perspective on cultural diversity* (pp. 287–313). New York, NY: Springer.

Těmkin, I., & Eldredge, N. (2015). Networks and hierarchies: Approaching complexity in evolutionary theory. In E. Serrelli & N. Gontier (Eds.), *Macroevolution: Explanation, interpretation and evidence* (pp. 183–226). New York, NY: Springer.

Těmkin, I., & Serrelli, E. (2016). General principles of biological hierarchical systems. In N. Eldredge, T. Pievani, E. Serrelli, & I. Těmkin (Eds.), *Evolutionary theory: A hierarchical perspective* (pp. 19–25). Chicago, IL: University of Chicago Press.

Theiner, G., Allen, C., & Goldstone, R. L. (2010). Recognizing group cognition. *Cognitive Systems Research, 11*(4), 378–395.

Thomas, R. K. (1980). Evolution of intelligence: An approach to its assessment. *Brain, Behavior and Evolution, 17*(6), 454–472.

Tomasello, M. (2000). Two hypotheses about primate cognition. In C. Heyes & L. Huber (Eds.), *The evolution of cognition* (pp. 165–183). Cambridge, MA: MIT Press.

Tomasello, M., & Call, J. (1994). Social cognition of monkeys and apes. *American Journal of Physical Anthropology, 37*(Suppl. 19), 273–305.

Tomasello, M., Kruger, A. C., & Ratner, H. H. (1993). Cultural learning. *Behavioral and Brain Sciences, 16*(3), 495–511.

Volk, T. (2017). *Quarks to culture: How we came to be.* New York, NY: Columbia University Press.

Vrba, E. S., & Eldredge, N. (1984). Individuals, hierarchies and processes: Toward a more complete evolutionary theory. *Paleobiology, 10,* 146–171.

Wilson, R. A., & Clark, A. (2009). How to situate cognition: Letting nature take its course. In M. Aydede & P. Robbins (Eds.), *The Cambridge handbook of situated cognition* (pp. 55–77). Cambridge, UK: Cambridge University Press.

Wimsatt, W. C. (1976). Reductionism, levels of organization, and the mind-body problem. In G. G. Globus, G. Maxwell, & I. Savodnik (Eds.), *Consciousness and the brain: A scientific and philosophical inquiry* (pp. 205–267). New York, NY: Plenum Press.

Wimsatt, W. C. (2013). Entrenchment and scaffolding: An architecture for a theory of cultural change. In L. R. Caporael, J. R. Griesemer, & W. C. Wimsatt (Eds.), *Developing scaffolds in evolution, culture, and cognition* (pp. 77–105). Cambridge, MA: MIT Press.

Wimsatt, W. C. (2018). Articulating Babel: The geography of cultural evolution. In A. C. Love & W. C. Wimsatt (Eds.), *Beyond the meme: Development and structure in cultural evolution (pp. 1–41)*. Minneapolis: University of Minnesota Press.

Wimsatt, W., & Griesemer, J. (2007). Reproducing entrenchments to scaffold culture: The central role of development in cultural evolution. In R. Sansom & R. Brandon (Eds.), *Integrating evolution and development: From theory to practice* (pp. 227–323). Cambridge, MA: MIT Press.

Wolf, M. (2007). *Proust and the squid: The story and science of the reading brain.* New York, NY: HarperCollins.

Contributors

Jan Baedke
Ruhr University Bochum

Robert W. Batterman
University of Pittsburgh

Daniel S. Brooks
University of Cincinnati

James DiFrisco
KU Leuven

Markus I. Eronen
University of Groningen

Carl Gillett
Northern Illinois University

Sara Green
University of Copenhagen

James Griesemer
University of California, Davis

Alan C. Love
University of Minnesota

Angela Potochnik
University of Cincinnati

Thomas A. C. Reydon
Leibniz Universität Hannover

Ilya Tëmkin
Smithsonian Institution

Jon Umerez
University of the Basque Country

William C. Wimsatt
University of Chicago

James Woodward
University of Pittsburgh

Index